The Climate Demon

Climate predictions – and the computer models behind them – play a key role in shaping public opinion and our response to the climate crisis. Some people interpret these predictions as "prophecies of doom," and some others dismiss them as mere speculation, but the vast majority are only vaguely aware of the science behind them. This book provides an overview of the strengths and limitations of climate modeling. It covers historical developments, current challenges, and future trends in the field. The accessible discussion of climate modeling requires only a basic knowledge of science. Uncertainties in climate predictions and their implications for assessing climate risk are analyzed, as are the computational challenges faced by future models. The book concludes by highlighting the dangers of climate "doomism," while also making clear the value of predictive models, and the severe and very real risks posed by anthropogenic climate change.

R. SARAVANAN is Head of the Department of Atmospheric Sciences at Texas A&M University. He is a climate scientist with a background in physics and fluid dynamics and has been a lead researcher using computer models of the climate for more than thirty years. He built an open-source simplified climate model from scratch and has worked with complex models that use the world's most powerful supercomputers. He received his Ph.D. in Atmospheric and Oceanic Sciences from Princeton University and his M.Sc. in Physics from the Indian Institute of Technology, Kanpur. He carried out postdoctoral research at the University of Cambridge and subsequently worked at the National Center for Atmospheric Research in Boulder, Colorado. Saravanan has served on national and international committees on climate science, including the National Research Council (NRC) Committee on the Assessment of Intraseasonal to Interannual Climate Prediction and Predictability, and the Science Steering Committee of the Prediction and Research Moored Array in the Atlantic (PIRATA). He recently helped create the TED-Ed animated short "Is the Weather Actually Becoming More Extreme?"

"If you wish to correctly interpret climate modeling results, read *The Climate Demon*. Saravanan's brilliant and humorous book helps both scientists and the general public objectively understand strengths and limitations of climate predictions."

Samuel Shen, Distinguished Professor, San Diego State University

"A wide-ranging guided tour of the modern science of climate prediction, told by a leading expert without jargon or mathematics, and illuminated by history, philosophy, technology and even literature."

Richard C.J. Somerville, Scripps Institution of Oceanography, University of California, San Diego

"Output from climate models completely underpins our policies on cutting emissions and making society more resilient to future climate – issues that will be affecting everyone on the planet in years to come. I thoroughly recommend this book if you want to understand the science behind these all-important [climate] models."

Tim Palmer, University of Oxford

"R. Saravanan is probably the most knowledgeable person in the world to write this book. This is a book that is accessible to the average college graduate and even many with less formal education. The pace of the reading is smooth, with many metaphors and analogies taken from a wide variety of sources … all whimsical but hitting his points perfectly. Readers will enjoy these features. This is a first rate, well-researched summary and analysis of how predictions in climate science work hand-in-hand with high-tech empirical data and detailed climate simulation models along with their enabler: the modern supercomputer. All this without a single equation and only a handful of figures used for clarification or sometimes just to draw a smile. This is a must read for natural and social scientists from all walks of life, as well as policymakers and managers in all sectors who have any interest in climate science and its inferences. The book makes the point that comprehensive climate models are our only means of making serious forecasts into the future. It is stressed that the most likely method of mitigating serious climate change is to curb our production of carbon dioxide. The book considers these necessities while at the same time giving a sober analysis of the limitations of climate model forecasts. The book is not a 'Doomsday' book, nor does it advocate doing nothing, but rather following a rational course, using our best tools, given to us by ever-improving computers and climate models."

Gerald R. North, Texas A&M University

The Climate Demon

Past, Present, and Future of Climate Prediction

R. SARAVANAN
Texas A&M University

CAMBRIDGE
UNIVERSITY PRESS

University Printing House, Cambridge CB2 8BS, United Kingdom

One Liberty Plaza, 20th Floor, New York, NY 10006, USA

477 Williamstown Road, Port Melbourne, VIC 3207, Australia

314–321, 3rd Floor, Plot 3, Splendor Forum, Jasola District Centre, New Delhi – 110025, India

103 Penang Road, #05–06/07, Visioncrest Commercial, Singapore 238467

Cambridge University Press is part of the University of Cambridge. It furthers the University's mission by disseminating knowledge in the pursuit of education, learning, and research at the highest international levels of excellence.

www.cambridge.org
Information on this title: www.cambridge.org/9781316510766
DOI: 10.1017/9781009039604

© R. Saravanan 2022

This publication is in copyright. Subject to statutory exception and to the provisions of relevant collective licensing agreements, no reproduction of any part may take place without the written permission of Cambridge University Press.

First published 2022

A catalogue record for this publication is available from the British Library.

Library of Congress Cataloging-in-Publication Data
Names: Saravanan, R. (Ramalingam), author.
Title: The climate demon : past, present, and future of climate prediction / R. Saravanan, Texas A&M University.
Description: Cambridge, UK ; New York, NY : Cambridge University Press, 2022. | Includes bibliographical references and index
Identifiers: LCCN 2021019837 (print) | LCCN 2021019838 (ebook) | ISBN 9781316510766 (hardback) | ISBN 9781009018043 (paperback) | ISBN 9781009039604 (epub)
Subjects: LCSH: Weather forecasting–Research. | Weather forecasting–Mathematical models. | Weather forecasting–Computer simulation. | Climatology–Research. |
BISAC: SCIENCE / Earth Sciences / Meteorology & Climatology
Classification: LCC QC995.46 .S27 2022 (print) | LCC QC995.46 (ebook) | DDC 551.63–dc23
LC record available at https://lccn.loc.gov/2021019837
LC ebook record available at https://lccn.loc.gov/2021019838

ISBN 978-1-316-51076-6 Hardback
ISBN 978-1-009-01804-3 Paperback

Additional resources for this publication at https://r.saravanan.us/climate-demon

Cambridge University Press has no responsibility for the persistence or accuracy of URLs for external or third-party internet websites referred to in this publication and does not guarantee that any content on such websites is, or will remain, accurate or appropriate.

To my parents,
Ramalingam and Kamala

Contents

Figures

Preface

I was originally trained as a physicist. Thirty years ago, for my Ph.D., I switched fields to climate science, which is a natural application of physical principles. I remain conflicted between the very different philosophies of physics and climate science, especially with regard to the use of models for prediction. F. Scott Fitzgerald observed that intelligence is perhaps the "ability to hold two opposed ideas in mind at the same time and still retain the ability to function." This book aspires to intelligence in that sense: It celebrates climate modeling even as it critiques it.

To determine the value of a fundamental physics constant, physicists do not form a committee to aggregate differing estimates of the constant and come up with an officially sanctioned value. Rather, if one group of physicists measures a fundamental constant and comes up with a number, it is expected that all other groups of physicists measuring the constant will come up with almost the same number, once the field has matured. The replication of measurements is a cornerstone of the scientific method; it ensures objectivity, and it is what distinguishes science from more subjective disciplines of study.

But in the world of climate science, it is not uncommon to survey models (or even experts), statistically aggregate the numbers they provide (which may vary by a factor of two or more), and only then estimate a climate parameter and its uncertainty. A different analysis technique or model may yield a different estimate for the parameter. While most scientists (and models) are in general agreement on the causes of and the future of climate change, it is hard for us to obtain very precise numbers for measures of climate change.

Why can't climate science be more precisely replicable? The short answer is that climate scientists are trying to solve a complex and urgent problem, to which it is impossible to apply the standard techniques of basic physics, such as experimenting repeatedly with the planet's climate (this is also true of branches of physics like cosmology[1]). Climate change is a messy problem, but we can't just ignore it and move on to a cleaner problem, unless we are prepared to move to a different planet (which, at the time of writing, is not a viable option). Exact replication of predictions is hard in climate science, and, in some ways – as I explain in this book – exact replication is not even desirable in the short term.

We can also ask a related question: How accurate does our knowledge of climate parameters have to be before we can take action to mitigate climate change? We need to know the gravitational constant very accurately to plan interplanetary space travel. But it may be sufficient to know a climate parameter (for example, how much the Earth is likely to have warmed by the year 2100) within a factor of two. If we know that the climate impacts of increased carbon dioxide are going to be seriously bad, we must act quickly, even if we are not sure whether the impacts will be terrible or two times as terrible.

It is worth explaining the intricacies of climate prediction to the general public. There are some who appear to be making life-changing decisions based on the belief that climate predictions are prophecies of doom. But the making of climate predictions is not unlike the making of sausage – the product may be nicely packaged, but the process underlying the product is quite messy. I hope that acquainting readers with the many uncertainties inherent in climate predictions will lead them to treat these predictions less as a cause for despair and more as a call for stronger action to tackle global warming.

As a climate scientist, I have built simple models to understand climate phenomena, and I have worked with complex models to make climate predictions. In November 2016, I was attending a workshop at Princeton University on the topic of modeling hierarchies when I happened to purchase a copy of the *Old Farmer's Almanac* at Labyrinth Books, across from the campus. The folksy predictions in the $7.95 almanac were a stark contrast to the multimillion-dollar predictions discussed at the workshop. I started to think about the philosophical basis for model predictions; two years later, in October 2018, I gave a talk titled "On the Complications of Simplified Models" at the retirement symposium, also at Princeton University, for my Ph.D. adviser, Isaac Held. Several chapters in this book grew out of that talk. The discussion of model tuning (Section 10.1) is directly motivated by a question posed during the symposium – a younger member of the audience asked something to the effect of, "Why should tuning be considered a bad thing in climate models, since it is definitely a good thing in music?" This question made me realize that a philosophical analysis of climate modeling would be useful to scientists as well as the general public.

This book is about places I have been in and people I have met. At the time, I had little awareness of the historical significance of these places or the scientific achievements of these people; only many years later did I begin to appreciate their significance. For example, I took classes for two years in Princeton University's Jones Hall, with little knowledge of its past role in the history of computing! I have been fortunate to interact with some of the pioneers of climate modeling mentioned in the book. Suki Manabe, Kirk Bryan, Akira Kasahara, and Warren Washington were teachers and colleagues. I have also been in the physical presence of several others mentioned in the book, such as Freeman Dyson, Joe Farman, Ed Lorenz, Joe Smagorinsky, and Phil Thompson. But, as a junior scientist, I never really spoke to them beyond the occasional exchange of pleasantries.

I apologize in advance for this book being US-centric and, in particular, Princeton-centric, for narrative reasons and also due to my own familiarity with people and places. Many important people and scientific developments in the history of climate prediction, especially outside the United States, are not covered. This book is not meant to be a comprehensive or authoritative history of either climate change or climate prediction. There are other books that serve that purpose.

I apologize also to philosophers of science. I am not a philosopher – just a scientist who thinks about how science is done. I aim to apply some elementary philosophical concepts to climate science. I have restricted myself to limited philosophical terminology whose interpretation is generally self-evident from the normal English meaning of the words. For readability, I have, in most cases, dropped the technical qualifiers used in philosophy; I use "deductivism" instead of hypothetico-deductivism, and "determinism" instead of causal determinism.

This book is not about the narrow topics of my own scientific research. It is about my broad philosophical understanding of climate modeling, shaped by more than three decades of working with climate models, both simple and complex. While this book is about both philosophy and climate prediction, it is neither an academic treatise on philosophy nor a scientific textbook on climate prediction. Also, it is not a comprehensive book about climate change, although many aspects of climate change are discussed in the context of climate prediction.

Acknowledgments

I would like to acknowledge the influence of all of my faculty colleagues and collaborators at Texas A&M University, especially Jerry North and Andy Dessler.

I also owe much to the many climate-related discussions that took place with and among my former colleagues and mentors at the National Center for Atmospheric Research in Boulder, Colorado; at the UK Universities Global Atmospheric Modelling Programme; at the Department of Applied Mathematics and Theoretical Physics at the University of Cambridge; and at the Geophysical Fluid Dynamics Laboratory in Princeton, New Jersey. Finally, I would like to thank my physics colleagues at the Indian Institute of Technology, Kanpur, especially Kalyan Banerjee and J. K. Bhattacharjee, as well as Mohini Mullick and P. R. K. Rao, who introduced me to the philosophy of science.

The following people graciously took the time to read portions of the book draft: my colleagues Clara Deser, Andy Dessler, Isaac Held, Suki Manabe, Jim McWilliams, Onuttom Narayan, Jerry North, Susan Solomon, Mark Taylor, and Debbie Thomas, as well as my two children Shilpa and Shiva Saravanan. Comments from these readers have greatly improved the content of the book. In particular, detailed suggestions from Shilpa helped improve the readability of the text.

There are several people whose expression of specific ideas (in parentheses) over the years influenced this book, although they are not explicitly mentioned or cited in the text: Jeff Anderson (compensating errors), Joe Barsugli (modeling philosophy), Byron Boville (constraining power), Ken Bowman (Manabe koan), Kevin Hamilton (sunspots), Boris Hanin (machine learning), Rol Madden (p-hacking), Michael McIntyre (conservation principles), Cécile Penland (statistical modeling), Prashant Sardeshmukh (model errors), and Mark Taylor (supercomputing).

The writing of this book has benefited greatly from transcripts of oral interviews recorded and made publicly available by Spencer R. Weart, author of *The Discovery of Global Warming*, and Paul N. Edwards, author of *A Vast Machine: Computer Models, Climate Data, and the Politics of Global Warming*. Both are excellent books that cover the topics of global warming and computer modeling much more comprehensively than this one.

The Oral History Project of the American Meteorological Society, archived at the National Center for Atmospheric Research, was the source of transcribed interviews with Susan Solomon and Warren Washington. The Oral History Project of the American Institute of Physics, archived at the National Center for Atmospheric Research, was the source of transcribed interviews with Akira Kasahara and Syukuro Manabe.

I would like to acknowledge the editorial and production team members who helped bring this book to fruition: Matt Lloyd, Sarah Lambert, Elle Ferns, Shaheer Husanne, and Cara Chamberlain.

Finally, I would like to thank my wife, Manjula Hosalli, for being a constant source of inspiration and support.

Introduction

On the night of September 7, 1900, the residents of Galveston, Texas, then a coastal city of 37,000 located near Houston, went to sleep thinking it was a night like any other. The next morning, they woke up to the deadliest natural disaster in the history of the United States: the Great Galveston Hurricane.[1] Galveston residents received no warning of the impending calamity. A 15-foot-high storm surge, driven by winds exceeding 130 miles per hour, inundated the city. More than 8,000 lives were lost.

The US Weather Bureau had been trying to track the hurricane. On September 4, the Bureau was informed, via telegraph, that a hurricane had passed over Cuba. Based on previous experience, meteorologists believed that hurricanes moving away from Cuba, toward the Florida Straits, would curve back toward Florida and track northeastward. But, unbeknownst to meteorologists, this hurricane veered westward toward Texas. This was before radio communication and satellites, so even a ship that observed the hurricane on its route couldn't send out a warning. The people of Galveston never saw it coming.

More than a century later, on August 25, 2017, another powerful storm, Hurricane Harvey, made landfall on the Texas coast, about 180 miles south of Houston. The Houston metropolitan area experienced unprecedented amounts of rainfall: A year's worth of rain fell in just a few days, totaling to more than 60 inches (150 cm). More than 100 people lost their lives.[2] Due to the incessant rain and the resulting flooding, Harvey became the costliest natural disaster in US history. But it was far from the deadliest, despite the large affected population. Satellites and computer models helped mitigate the disaster.

In 1900, forecasters had to rely only on their experience of past hurricanes to make predictions. The Great Galveston Hurricane had been sighted in Cuba, but it followed an unexpected path afterward, which the forecasters failed to predict. In 2017, Hurricane Harvey's track was forecast days ahead using computer models of the atmosphere, fed by information from satellite observations in space. Even though the amount of rainfall was unprecedented, computer models were able to predict it well in advance.[3] The people of Houston saw it coming. They had time to prepare.

There are a few lessons to be learned from these two Texas hurricanes. (1) Unprecedented events will occur. (2) Experience alone cannot help us prepare for them. (3) Models can predict unprecedented events. (4) Good observations are needed to feed information to the models.

Hurricane Harvey is an example of an unprecedented event that was accurately predicted. However, hurricane forecasts are not perfect. In September 2005, Houston underwent the largest evacuation in US history because it was expected that Hurricane Rita would make landfall close by. Rita changed course slightly at the last minute, missing the city; far more Houstonian lives were lost due to the evacuation process, fraught by gridlock as it was, than due to the hurricane itself.

The accurate forecast of Hurricane Harvey made it possible to prepare for the hurricane's landfall, but it is still debated whether the city of Houston was as prepared as it could have been. No evacuation order was issued, due to the memory of the failed evacuation during Hurricane Rita. And the prodigious amount of rain, as well as the resulting flooding, surprised residents and emergency responders. Even when science accurately predicts an impending disaster, policy decisions on how to deal with the disaster are not easy to make. They require a great deal more than scientific insight. These policy decisions contain implicit or explicit moral choices that depend on human psychology and the socioeconomic status of the affected communities. Poorer communities may lack the transportation and alternative accommodation options that allow wealthier communities to evacuate.

Global warming is an unprecedented event in human history. It is a slow-motion disaster; we can see it coming, but not as clearly as we can see a hurricane on a weather map. Surface temperatures around the globe have been increasing since the late nineteenth century. This warming signal was initially weak, hidden amidst the background noise of the natural climate variations that have occurred for eons. But in recent decades, as the pace of warming has accelerated, the signal has emerged from the noise. Scientific studies, using a combination of data and models, attribute this warming signal to human activities. Since the Industrial Revolution, the burning of fossil fuels like coal, oil, and natural gas has led to the emission of gases such as carbon dioxide that warm the globe through a phenomenon known as the *greenhouse effect*.

The evidence for global warming comes from multiple lines of scientific inquiry – fossil records of climate variations that occurred in the distant past, statistical analyses of recent temperature measurements, and model predictions of future climate, among others. There are different uncertainties and errors associated with each line of inquiry, but, together, they lead to some robust conclusions. We predict that the Earth will continue to warm as we continue to use fossil fuels and that this warming will negatively impact society and the environment. Some regions of the world are already feeling the harmful effects of this climate crisis; if the warming continues, many more will follow suit. Warmer temperatures likely contributed to the unprecedented rainfall amounts associated with Hurricane Harvey. We can stop global warming by switching from fossil fuels to other sources of energy, ones that do not emit the gases responsible for the greenhouse effect. This switch needs to be made as quickly as possible. The longer we wait, the harder it becomes to reverse the harmful impacts of the accumulated greenhouse effect.

There is no simple solution for the problem of global warming. It will be hard for society to wean itself off fossil fuels. The need for energy pervades nearly every sphere of human activity, from home heating and cooling, to automobiles and aviation, to industrial manufacturing. Efforts to mitigate global warming, or adapt to it, will impact different countries and different strata of society in very different ways. To deal with the climate crisis in an equitable fashion, we must address complex geopolitical challenges and moral quandaries. The wealthy can escape from a regional disaster like a hurricane by evacuating to a safer location, leaving the poor and the vulnerable behind. There is no such escape from global warming, which will ultimately affect us all.

Purpose of the Book

Climate prediction is one of the most ambitious prediction efforts ever undertaken in human history. Computer models of climate, like the weather models used to forecast Hurricane Harvey, use the world's most powerful supercomputers to generate predictions that are widely disseminated and analyzed. For instance, the research arm of the global consulting giant McKinsey released a report in 2020 assessing the impacts of global warming. The report included dire predictions, such as the one that, under a high carbon-dioxide-emission scenario,

> urban areas in parts of India and Pakistan could be the first places in the world to experience heat waves that exceed the survivability threshold for a healthy human being, with small regions projected to experience a more than 60 percent annual chance of such a heat wave by 2050.[4]

Risk assessments like the one made by McKinsey are increasingly used by governments and businesses to make decisions that will transform the global economy. But these assessments present highly processed information, which involves multiple stages of knowledge transfer: Economists and social scientists create scenarios that climate scientists use to make predictions, which are then fed back to economists and social scientists so that they can assess impacts and risk. Due to limited data and imperfect models, there are uncertainties involved at each stage. Not all of this may be made apparent when the final risk assessments are presented.[5]

The vast majority of people around the world trust the climate predictions that form the basis of risk assessments like the one from McKinsey, and they accept the need for urgent actions to mitigate carbon emissions. But there is a minority that holds contrarian views. Some contrarians question the reality of global warming itself, arguing that the climate changes we see are just noise and that there is no long-term warming signal at all. Others accept that there is some warming, but question climate predictions, arguing that models overestimate the warming and that "[f]uture warming is likely to be at the lowest bound of these model results."[6] In the contrarian view, there is no need to take urgent action to mitigate carbon emissions.

At the other end of the ideological spectrum, we have the emergence of a fringe phenomenon known as "climate doomism," where people mistakenly believe that climate models predict an inevitable climate apocalypse that will soon make our planet uninhabitable. This belief can inspire a range of negative emotions, ranging from debilitating anxiety to paralyzing despair, and can lead to people making life-changing decisions based on their beliefs. Since climate predictions are conditional on future human actions, such a fatalist belief could become a self-fulfilling prophecy if it becomes widespread and causes us to fail to act out of despair.

Climate predictions are accorded gravitas due to their scientific foundation and the sophisticated infrastructure that produces them. These predictions do not possess the certainty that doomers ascribe to them. They are, however, much more certain than the contrarians make them out to be. An important goal of this book is to inform those outside the field of climate science about the strengths and limitations of climate prediction by explaining the scientific knowledge that lies between the ideological extremes while motivating the need for urgent action.

We ask the following questions about the models used for climate prediction:

How were these models built?
What is the science behind them?
What is their philosophical basis?
Why should we trust them?

This book attempts to answer these questions by tracing the history of climate prediction and extrapolating it into the future. Along the way, it addresses some fundamental philosophical questions relating to the enterprise of climate modeling.

Different types of models are used to make predictions that directly affect society. Weather models make very accurate forecasts of air temperature and hurricane tracks for the next several days. Economic models predict, not quite so accurately, how the global economy will evolve over the next several years. These two types of models are fundamentally very different. Weather models solve the well-known equations of physics that govern the motion of air. But we don't have accurate equations that describe the behavior of people and corporations. So economic models are often based on statistical approximations, derived from data, on how people and corporations behave.[7] This means long-term economic predictions are far less reliable than short-term weather predictions.

On the modeling spectrum, climate models fall somewhere in between weather models and economic models. At their core, climate models are very similar to weather models in that they solve well-known equations governing the motion of air and water, derived from physics. But they also include approximate formulas to describe more complex processes, such as the formation of clouds and the impact of industrial haze. In order to make long-term predictions, climate models require additional assumptions about societal and economic behavior, which have nothing to do with the laws of physics. All this renders climate predictions more uncertain than weather predictions.

Despite the uncertainties, we have confidence in the basic aspects of global warming predicted by models – and these alone justify taking swift and effective action to mitigate carbon emissions. This action needs to be far stronger than the modest steps being taken now. The incomplete nature of models only means that there is still deep uncertainty, which is not widely appreciated, regarding predictions of extreme climate change scenarios. This uncertainty is by no means a reason for inaction, even if contrarians try to present it as such. If anything, the uncertainty adds urgency to the need for mitigation. Awareness of this uncertainty may also change the minds of those who believe that mitigation is futile because climate predictions foretell certain doom.

Structure of the Book

This book does not discuss all aspects of the global warming problem. It focuses on a specific topic: the role of climate prediction. It presents a frank assessment of the strengths and weaknesses of climate models used to make predictions. Although the shortcomings of climate models are enumerated, the book argues that these models are the best tools we have to apportion blame for the climate change that has already occurred and to rationally plan for climate change in the future.

The book is divided into three parts. It begins in the 1940s with Part I, which outlines the early history of computer models of weather and climate. Computers themselves were invented in part to make complex tasks such as weather and climate prediction possible. The first digital weather forecast was made in 1950 using a simple weather model. Soon after, complex computer models proved to be of great practical value in improving the quality of weather forecasts. Climate prediction, on the other hand, remained confined to curiosity-driven academic research until the signal of global warming was recognized in the late 1980s. At this time, climate models gained new prominence as practical tools for studying and predicting global warming.

All weather predictions are now made using complex computer models. But scientists make climate predictions using a variety of models ranging from the simple to the highly complex. Predictions from the simpler models have mostly qualitative value because they do not consider all important inputs. The more complex models are more comprehensive, and their predictions have more quantitative value. However, media coverage of climate predictions does not always distinguish between different types of model predictions, nor does it mention the different types of uncertainties associated with them. Simplicity is not always a virtue when it comes to climate modeling. Climate is the result of a complex interaction between the atmosphere, the land, the ocean, and the biosphere; complexity is inherent to the system. But how much complexity is essential for a model to make useful climate predictions?

These issues are addressed in Part II, which discusses current challenges in climate modeling using numerous analogies, including an 800-year-old problem faced by the city of Pisa that can still teach us about scientific complexity. The arc of climate model improvement, the progression from "unknown unknowns" to "known unknowns" to "known knowns" is described. Also introduced is the notion of "unknown knowns," which corresponds to misinformation or disinformation associated with the spread of contrarian views on climate. This is followed by a discussion of the communication of climate change to the public and the associated difficulties in translating the details of climate predictions from scientific to natural language.

Part III describes trends in climate modeling driven by new science and technology, including geoengineering, machine learning, and the demise of Moore's Law of ever-faster computer chips. The book ends with a discussion of the philosophical concept of Pascal's Wager, the tradeoff between making modest sacrifices in the present in return for a large, promised reward in the future. This leads to a discussion of the impacts and risks of climate change. To confront the climate crisis, we must take steps now that will affect future generations, for better or for worse. Science can enumerate the value judgments and moral choices we need to make, but it cannot make those choices for us.

Multiple philosophical themes run through the book. One is the distinction between data-driven (or inductivist) science and hypothesis-driven (or deductivist) science. Another is the interplay between reductionist and emergent views of the climate system. This manifests as the tension between the simplicity and complexity of models, which this book ties back to the concept of Occam's Razor. A third theme is the conflict between the predictability of determinism and the unpredictability of chaos. The concept of determinism is embodied in the notion of Laplace's Demon, a hypothetical intellect that can compute the trajectory of every atom in the universe and which therefore has complete knowledge of the past and the future.

The title of the book refers to another recurring philosophical theme in the book, the notion of a Climate Demon. It is the climate analog of Laplace's Demon, and serves as a metaphor for a model that accurately calculates the trajectory of future climate.

PART I

The Past

1

Deducing Weather

The Dawn of Computing

Our climate prediction story begins in the historic town of Princeton, New Jersey. The town was the site of a battle during the Revolutionary War; after the war, in 1783, Princeton served for four months as the provisional capital of the United States. Today it is best known as the home of Princeton University. Through the center of town runs Nassau Street, which began as part of a Native American trail that later became a stagecoach route between New York and Philadelphia. This street marks the divide between "town" and "gown": To the north lives the general population of Princeton, and to the south lies the picturesque campus of Princeton University with its neo-Gothic architecture.

On Nassau Street's north side, across from the university, there is a wonderful independent bookstore called Labyrinth Books. Here, for the affordable price of $7.95, you can buy yourself a whole year's worth of weather and climate forecasts. These forecasts are found in the annual edition of a little book called the *Old Farmer's Almanac*, which has proudly provided this service every year for more than two centuries.[1] For each of eighteen different regions of the United States, the almanac lists quantitative predictions of the average temperature and precipitation for each month of the year, as well as qualitative weather forecasts for individual periods of these months. The almanac even features a thoughtfully placed hole in its top left corner so that it can be hung from a nail in the barn or the outhouse, enabling convenient perusal of its folksy weather-related articles and tables.

Historically, the almanac used a secret formula for weather forecasts, devised in 1792 by its founder, Robert B. Thomas. This formula is based on the premise that "nothing in the universe happens haphazardly, that there is a cause-and-effect pattern to all phenomena."[2] Thomas believed that the sun had an effect on the Earth's weather, and he credited Italian astronomer Galileo Galilei's seventeenth-century study of sunspots as a key part of his secret formula. Farmers – the almanac's original target customers – needed to know, for instance, when the first snow of the year would fall in order to plan their harvests. Even townsfolk could benefit from advance knowledge of sunny days on which to plan a wedding or a picnic. The longevity of the almanac demonstrates its success in catering to the practical needs of farmers and townsfolk over the years. But the venerable almanac does have competition in the folksy prediction market: Since 1887, a groundhog named Phil has been prophesying the start of spring in the town of Punxsutawney, Pennsylvania.[3] We shall make further acquaintance of this furry forecaster later on.

For decades, while the *Old Farmer's Almanac* and, later, Punxsutawney Phil made their prognostications, the "gown" part of Princeton – the scientists in the ivory towers of the university – remained largely silent on the subjects of weather and climate prediction. That began to change in the late 1940s, with the invention of the digital computer.

***

Let us walk south from Labyrinth Books through the university campus, past Nassau Hall (which served as the capitol building while Princeton was briefly the nation's capital). We eventually arrive at Jones Hall. This well-appointed building with carved oak paneling modeled after colleges at Oxford University currently houses the departments of East Asian Studies and Near Eastern Studies. Let us go back in time to 1931, when this building was built. It was then the new home of the mathematics department. It also had a different name, Fine Hall, after the beloved and well-respected math professor Henry B. Fine.[4] (The building's name was changed to Jones Hall in 1969, when the mathematics department moved to a new building. That building inherited the name Fine Hall.)

One of the occupants of Fine Hall in 1931 was the man who helped design it, a mathematician named Oswald Veblen.[5] Mathematics faculty were generally burdened with heavy teaching duties. Therefore, Veblen nursed a grand vision of a new science institute, a true ivory tower, in which faculty would focus only on research.[6] He managed to persuade an educator named Abraham Flexner, who was already in the process of setting up an institute of this nature, to locate it in Princeton. Known as the Institute for Advanced Study (IAS), it would bestow renowned scientists with lifetime appointments as professors, with no teaching responsibilities. There would additionally be a regular flow of short-term visiting scientists to the institute.

Flexner, the founding director of the IAS, was a strong proponent of curiosity-driven research; he published an essay with the provocative title "The Usefulness of Useless Knowledge."[7] He persuaded a wealthy New Jersey family, the Bambergers, to contribute the equivalent of $200 million today to his project. A permanent building was to be constructed for the IAS using these funds, about a mile from the main campus of Princeton University. In the meantime, though, Flexner needed to find a temporary office for the scholars who would comprise the institute. Fine Hall was ideal for this purpose; the new members of the IAS could interact with the mathematicians already there, as well as with the physicists in the adjacent Palmer Physical Laboratory (which is now the Frist Campus Center).

All was settled, and the IAS began operations in Fine Hall in 1933. Among its early recruits was the world-famous physicist, Albert Einstein. Another was a young mathematician named John von Neumann, who published his first mathematics paper before the age of eighteen. Von Neumann, the son of a wealthy banker, was born in 1903 in Budapest, Hungary. He was a prodigy who reportedly had the ability to recall entire books; before he joined the IAS, he had already written an influential book on the *Mathematical Foundations of Quantum Mechanics.* In 1935, von Neumann visited the University of Cambridge in England. There he met a young mathematician named Alan Turing, who was working on fundamental problems relating to the mathematics of computability.[8] Intrigued by Turing's research, von Neumann encouraged him to come to Princeton.

After receiving his degree from the University of Cambridge, Turing arrived in Princeton in 1936 to start his Ph.D. under Alonzo Church, a professor in the Department of Mathematics.[9] Between 1936 and 1938, Fine Hall housed Einstein, von Neumann, and Turing, three of the most legendary names in physics, mathematics, and computer science – the last being a field which did not even exist at the time. These three fields would come together in Princeton to birth the fields of numerical weather prediction and numerical climate prediction.

In a landmark 1936 paper, "On Computable Numbers, with an Application to the Entscheidungsproblem," Turing conceived of a universal computing machine with an infinite tape – a tape that the computing machine could read symbols from, write symbols on, or erase repeatedly. The tape could move forward or backward under the direction of the machine. This conceptual machine, now known as a Turing Machine, could carry out any computation which involved well-defined steps. Turing used this hypothetical construct to show that there were numbers that could not be computed using this machine. This led to the conclusion that there was no general computing procedure to prove if a computer program would eventually halt. In essence, Turing had invented a universal computing machine to prove the limitations of computing. This paper forms one of the foundations of the field of computer science.

In Princeton, Turing continued to work on problems related to computation. However, with the offices of luminaries such as von Neumann and Einstein just down the corridor from him, he felt that his own work would not be recognized.[10] After Turing completed his Ph.D. in 1938, von Neumann tried to persuade him to stay by offering him a research position at the IAS, but Turing decided to return to England. In the years following, von Neumann continued to be fascinated not only by the idea of building computing machines but also by the possibility of their practical utility.

Among the mathematicians of the IAS, von Neumann was somewhat of an oddity. He could hold his own among theoreticians like Einstein and Kurt Gödel, the great mathematical logician who had visited and later been recruited to the IAS. Gödel's work is about as theoretical as it gets: He is famous for proving the Incompleteness Theorem, which demonstrated the limits of mathematics and inspired Turing's work on the limits of computation. But, unlike either Einstein or Gödel, von Neumann was very much interested in useful applications of "useless knowledge," that is, the practical applications of science[11] such as hydrodynamics, meteorology, and the design of ballistic and nuclear weapons.

Being both a physicist and a newly minted computer scientist, von Neumann realized computers could be used to predict weather from basic physical principles.[12] But the Turing Machine, while a beautiful concept, was not a practical computer design. (It required an infinite tape, for instance.) So, von Neumann set out to design and build an electronic computer at the IAS. Due to such a computer's potential military applications, von Neumann was able to convince the US Atomic Energy Commission and various military agencies to fund this expensive endeavor.[13] While he waited for the IAS machine to be built, von Neumann would use the world's first general-purpose computer, the Electronic Numerical Integrator and Computer (ENIAC) – which was designed by another scientist interested in weather prediction.

1.1 From Sunspot Cycles to Compute Cycles

The cycles of heavenly objects have long fascinated humans. Inherent in cycles is their predictability: A peak in a ten-year cycle will be followed by a trough five years later, and by another peak ten years later. The English astronomer Edmond Halley was able to calculate the orbit of the comet eventually named after him and to predict its next appearance – fifty-three years in advance!

In 1933, John Mauchly was the head (and sole member) of the physics department at Ursinus College[14] in Collegeville, Pennsylvania, about 50 miles west of Princeton. He had tinkered with electronics throughout his youth and enrolled in the engineering school at Johns Hopkins University, but, turned off by the "cookbook style" of engineering courses, he ended up switching fields to physics and was awarded a Ph.D. in 1932.[15] Mauchly's research involved harmonic analysis – looking for periodic oscillations – in weather data. He built an analog harmonic analyzer machine, and he published a paper on oscillations in rainfall. The number of sunspots – dark areas on the surface of the sun – also exhibited 11-year oscillations, and Mauchly was looking for evidence that solar variations could be used to predict weather years in advance. (Recall that sunspots were also part of the secret formula used by Robert B. Thomas, the founder of the *Old Farmer's Almanac.*)

Mauchly wanted to analyze weather data to scientifically demonstrate a relationship between weather and the number of sunspots. The data was too voluminous for Mauchly to manually calculate the necessary statistical correlations; he figured that he could build an electronic computing device to speed up the calculations, but he needed support. Ursinus College was too small an institution to host such an endeavor. Mauchly persisted, and, in 1942, during the Second World War, he moved to Philadelphia to join the faculty of the Moore School of Engineering at the University of Pennsylvania.[16] During the war, the Moore School worked closely with the US Army, which required extensive computations for firing tables to find, for instance, the range of artillery shells. Using humans to perform these computations had proved too slow. Mauchly's proposed electronic computing device could solve the Army's problem. With an initial grant of $61,700 from the Army,[17] Mauchly partnered with J. Presper Eckert, an engineer, to design the ENIAC in 1943. The ENIAC was the first general-purpose electronic computer in the world, meaning it could be used for any type of calculation. Its predecessors were all computers custom-built for specific types of calculations.

Mauchly and von Neumann soon crossed paths. Purely by chance, a member of the ENIAC team recognized von Neumann on a railway platform in 1944 and invited him to meet with the team.[18] This led to von Neumann joining the team as a consultant. The ENIAC project itself ran behind schedule and was only completed in December 1945. The ENIAC was a behemoth, weighing 30 tons and containing 18,000 vacuum tubes.[19] Its final cost was about $500,000 (the equivalent of $7 million in 2019). It occupied an area of 30×60 feet and consumed about 160 kW. It had a memory of 40 bytes and could perform about 400 floating-point (or arithmetic) operations per second (flops). For comparison, a modern desktop computer can have almost a billion times more memory and can run more than a billion times faster.

The war was over by the time ENIAC was operational, and there was no longer an urgent need for ballistics computations. But the ENIAC would still be put to military use: A new type of weapon, the hydrogen (or fusion) bomb, was now in development, and its design would require complex calculations of shockwaves and explosions. Von Neumann used the ENIAC to carry out these top-secret calculations. Mauchly and Eckert went on to form a private company to commercialize their invention,[16] although Mauchly continued to work on the statistical analysis of sunspot cycles.[20] Later in this book, we will revisit the possible role of sunspot cycles in climate. But first, we turn to a problem that motivated both Mauchly and von Neumann: weather prediction.

1.2 Philosophy Break: Inductivism versus Deductivism

Computer pioneers von Neumann and Mauchly (Figure 1.1) were both interested in the accurate prediction of weather, but they approached the problem from diametrically opposite directions. Mauchly, the former engineer, wanted to analyze large volumes of weather data to reveal relationships between

Figure 1.1 (a) John von Neumann (Photo: US Dept. of Energy). (b) John W. Mauchly. (Photo: Charles Babbage Institute Libraries, University of Minnesota Libraries)

weather cycles and sunspot cycles, and then use those relationships to predict weather. Von Neumann, the grand theoretician and genius mathematician, hypothesized that the atmosphere could be divided into a grid of points and that weather could be predicted by applying the laws of physics at each point. Both approaches required an immense number of calculations, a number beyond human capability – hence the need for the electronic computer.

Mauchly and von Neumann exemplify two contrasting approaches to scientific progress: the *data-driven* approach and the *hypothesis-driven* approach. The data-driven approach to science may be referred to as *inductivism*, a branch of empiricism, which is the study of knowledge from experience. The hypothesis-driven approach may be referred to as *deductivism*, a branch of rationalism, which is the study of knowledge from reasoning.

A classic example of inductive reasoning is as follows: We watch one hundred swans pass in sequence, note that they are all white, and conclude therefore that all swans are white. What we might call the "theory of white swans" was indeed the accepted wisdom in the Western world regarding swans through the end of the seventeenth century[21] – until the exploration of Australia, when a black swan was spotted. The observation of a black swan falsified the theory of white swans. The Austrian–British philosopher Karl Popper, in his 1935 work *The Logic of Scientific Discovery*, emphasized that the falsifiability of theories is a key requirement of science.[22] Unlike mathematical theorems, scientific theories cannot be proven true, but they can be falsified. More and more observations of white swans can confirm the theory of white swans, but they cannot prove it.

The black swan has now become a metaphor for an unprecedented cataclysmic event, such as a stock market crash. Inductive reasoning cannot predict these kinds of events. This is a shortcoming known as the "problem of induction."[23] As discussed in the Introduction, the Great Galveston Hurricane of 1900 was not well predicted because it took a path across the Gulf of Mexico that went against the conventional wisdom at the time. But, unlike inductive reasoning, deductive reasoning can predict unprecedented events. The rainfall over Houston due to Hurricane Harvey in 2017 was (literally) off the charts, but the weather forecasts based on deductive computer models were able to predict it by solving the equations governing air motions.

Science progresses by creating new theories or models. In this book, the terms "theory" and "model" are used almost interchangeably. There is no real difference between the two. A model is an abstraction of reality; so is a theory. It is true that models are more commonly quantitative abstractions of reality, while theories are more commonly qualitative descriptions of reality. However, this distinction is far from absolute: Einstein's special theory of relativity involves rather complicated mathematical equations!

Discussions of scientific theories or models often emphasize their *predictive power*. In 1915, Einstein developed his general theory of relativity, hypothesizing that the sun's gravity could bend light and thus alter the apparent position of a star in the sky. This theory was verified by British astronomer Arthur Eddington's measurements during an eclipse in 1919.[24] The hypothesis-driven deductive approach emphasizes bold predictions. We can separate good theories (or models) from bad theories (or models) by looking at how accurate these bold predictions turn out to be.

The data-driven inductive approach, on the other hand, emphasizes the *explanatory power* of theories or models. The British naturalist Charles Darwin collected thousands of specimens and filled notebooks with careful observations and sketches of plants and animals during his five-year voyage on the ship the HMS *Beagle*. After studying the data he had amassed to understand how organisms had changed over time, Darwin (along with another British naturalist, Alfred Russel Wallace) proposed the theory of evolution through natural selection. The theory provides an elegant explanation of the observed characteristics of existing as well as extinct animals, and serves as a fundamental principle of modern biology.

In this book, we will frequently make distinctions for the purposes of analysis. We will use the concept of the analytic knife, also known as Phaedrus's knife (after the character in Robert M. Pirsig's 1974 book *Zen and the Art of Motorcycle Maintenance*). Our analytic knife[25] is the distinction between inductivism and deductivism. With it, we cut through the tangled web of multifarious approaches to scientific research.[26] It is admittedly a simplistic distinction, black and white without shades of gray. Much of real science involves a combination of both inductive and deductive approaches:[27] Theoreticians use old data to build scientific models, make predictions to test them, and refine them as new data are obtained from experimentalists. Asking whether the theory or the observation came first is like asking whether the chicken or the egg came first.

An ancient example of an inductive discipline is meteorology, the study of weather. The name itself is derived from the ancient Greek word *meteora*, referring to that which is high in the sky or the heavens. In the fourth century BCE, the Greek philosopher Aristotle wrote a treatise called *Meteorologica*, which remained the definitive work in the field for two millennia.[28] At the start of the twentieth century, meteorology was largely data driven, and more an art than a science: Meteorologists gathered observations of weather, mapped pressure and temperature, and made forecasts based on their previous experience.[29] This inductive approach to making forecasts was not very reliable. For instance, weather systems are often thought of as moving roughly with the prevailing winds. But prevailing winds near the surface can be quite different from prevailing winds higher up and can affect weather systems in different ways. Furthermore, weather systems can themselves alter the prevailing winds as they move. Human experience alone could not account for the variety of possible effects.

During this time, the physical sciences were making remarkable progress in explaining the fundamental behavior of nature, with the formulation of both quantum mechanics and the theory of relativity in the early twentieth century. The predictive power associated with some of these developments was spectacular: The theory of relativity, for example, accurately predicted small perturbations in the orbit of Mercury. Many physical scientists began to wonder why the same hypothesis-driven approach could not be applied to weather.[30] After all, weather was nothing more than air flow, driven by pressure forces and subject to Newton's laws of motion.

The mathematical equations governing the physics of weather were understood at this time; but solving these equations to forecast weather for the entire world was too much work for a single person. One British scientist, Lewis Fry Richardson, decided to try anyway.[31] He used a simple equation, called the *continuity equation*, to make a six-hour weather forecast for a day in the past (May 20, 1910). Richardson spent more than two years completing a long and tedious set of hand calculations. The results he obtained were, in the end, completely unrealistic. In fact, his forecast was so bad that he published a book about it, titled *Weather Prediction by Numerical Process*, in 1922. After his failed one-man effort, Richardson concluded that the computations he had attempted on his own would need to be farmed out to a large team of about 64,000 people in order to make a timely forecast (Figure 1.2a). This proposed

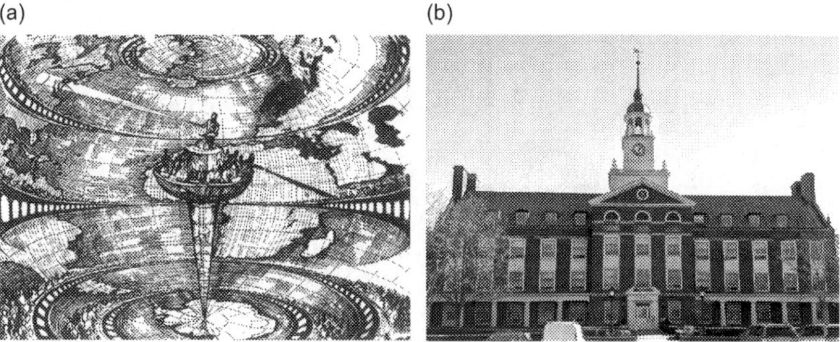

Figure 1.2 (a) A depiction of Richardson's proposed "forecast factory" (A. Lannerbäck, Dagens Nyheter, Stockholm). (b) Fuld Hall, Institute of Advanced Study, Princeton, New Jersey. (Photo: Shiva Saravanan)

"forecast factory" was impractical, but it set the stage for the use of digital computers in weather prediction. Richardson's ideas would play an influential role in the evolution of meteorology into a science.

1.3 The Weatherperson and the Computer

Let us return to Fine Hall (now Jones Hall) in Princeton. We walk a mile and a half southwest along College Road, past the Graduate College, turning left on Springdale Road, and then turning right on Ober Road, which becomes Einstein Drive. We have reached Fuld Hall, the permanent home of the IAS (Figure 1.2b).[7] The IAS faculty and staff moved into this Georgian-style building in 1939, soon after Turing's departure. While the ENIAC was being completed in 1945, von Neumann initiated the Electronic Computer Project to build a computer at the IAS, hiring many scientists and engineers for that purpose. Soon after that, in 1946, von Neumann formed the Princeton Meteorology Group, which would use the new computer to predict weather by solving the equations that govern atmospheric motions.

Since he himself had little detailed knowledge of atmospheric flows, von Neumann enlisted experts in that area.[32] The first such expert was Phil Thompson,[12] an Air Force meteorologist who learned about von Neumann's computer project from an article in the *New York Times Magazine*. Thompson joined von Neumann in 1946 and remained with the project for two years, during which time he learned about numerical analysis, the technique of solving mathematical equations using a digital computer. Before Thompson returned to the Air Force in 1948, he recruited one of his academic acquaintances, Jule Charney, to the Princeton Meteorology Group. Charney had recently derived a simple set of mathematical equations that could be very useful for computing weather forecasts of the sort envisioned by von Neumann.

Charney served as the anchor of the Princeton Meteorology Group, and he would go on to become a major figure in weather and climate science. His doctoral thesis on wavelike motions in the atmosphere was a groundbreaking work in the field, occupying the entire October 1947 issue of the *Journal of Meteorology*.[33] Upon completing his doctorate, Charney visited the University of Chicago to work with Carl-Gustaf Rossby, the preeminent meteorologist of the time. Rossby was one of the leading architects of the midcentury transformation of meteorology from an uncertain art into a certain science, one based on mathematics and physics. After Charney arrived in Princeton during the summer of 1948, he and von Neumann recruited several other atmospheric scientists from the United States and Europe to the Princeton Meteorology Group.[34]

The mathematical equations governing the motion of air, known as the Navier-Stokes equations, are derived from Newton's laws of motion applied to a fluid. Given the state of the atmosphere at a given time, these equations predict how the air will move under the action of the various forces. In principle, we can use a computer to solve these equations and forecast the weather for the next 24 hours. However, we cannot do this in a single step; we need to subdivide the problem into smaller steps. Say we make a forecast for the next hour. Then we can use that forecast to make a forecast for the following hour, and so on, until we have forecast 24 hours into the future. Choosing the number of steps for a forecast is like choosing the frame rate for a video. If there are too few frames per second, the motion will become increasingly jerky and inaccurate, and eventually cause the program to crash. If there are too many frames per second, the motion will be smooth, but it may overload the computer.

The time interval between the frames of a forecast, known as the *time step*, needs to be as large as possible, to minimize the total number of steps and the associated computational work. But for the computational work to be stable and accurate, the time step needs to be short enough to capture the evolution of each of a number of processes in the atmosphere. The processes governed by the Navier-Stokes equations include weather systems, fronts, and even sound waves. Sound waves have periods of much less than a second. This means that to solve the full Navier-Stokes equations directly on the computer and make a forecast, we need to use a time step of a fraction of a second.

With the computing power available in 1948, it was not feasible to make a 24-hour weather forecast with such a short time step. There would be far too many frames for the computer to handle. (Actually, even modern computers are not capable of such a short time step!) But sound waves are not important for weather forecasts. Charney analogized the atmosphere to a musical instrument that can play many tunes.[35] Sound waves are like the high notes, and weather systems are like the low notes. To forecast weather, the atmosphere only needs to play the low notes, because the high notes do not affect the weather.

The Navier-Stokes equations had to be simplified to eliminate "extraneous" processes, which would allow a longer time step for the forecast and a lower frame rate for the computer. The Princeton Meteorology Group considered two ways to accomplish this simplification. They could (1) use a two-dimensional model that took the drastic step of ignoring the vertical dimension of the atmosphere, thus eliminating the extraneous processes or (2) use a more

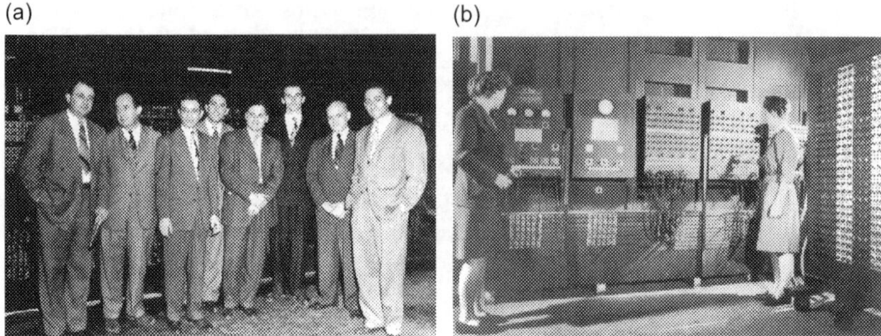

Figure 1.3 (a) Visitors and some participants in the 1950 ENIAC computations: (left to right) Harry Wexler, John von Neumann, M. H. Frankel, Jerome Namias, John Freeman, Ragnar Fjørtoft, Francis Reichelderfer, and Jule Charney, standing in front of the ENIAC. Wexler, Frankel, Namias, and Reichelderfer were visitors from the Weather Bureau. G. Platzman and J. Smagorinsky are absent. (Photo: Courtesy of John Lewis, reproduced from the Collections, Library of Congress). (b) Programmers Jean Jennings (left) and Frances Bilas operating the ENIAC (US Army photo)

elaborate three-dimensional model that incorporated a mathematical approximation (called quasi-geostrophy) that would also eliminate the extraneous processes. The two-dimensional model was already in use and therefore familiar to meteorologists, and Charney had just derived the equations of quasi-geostrophy necessary for the three-dimensional model.[12]

In 1950, the group decided to make an initial set of weather forecasts using the two-dimensional model. But the IAS computer was not ready yet. Von Neumann arranged to borrow some computer time from the ENIAC, which had by then been moved from Philadelphia to Aberdeen, Maryland. The group traveled to Aberdeen in March of that year to make the world's first set of numerical weather forecasts (Figure 1.3a). Three members of the Princeton team supervised the programming of the ENIAC for the forecasts: George Platzman, John Freeman, and Joseph Smagorinsky. Harry Wexler of the US Weather Bureau also worked with the team.

Group photos of scientists from the early days of computing and weather prediction (e.g., Figure 1.3a) may give the impression that all of the work was carried out by men. But photos of early computers like the ENIAC, with the first programmers standing next to the computers, often feature women (Figure 1.3b).[36] Despite the institutional barriers that inhibited their advancement in science, women often played important roles behind the scenes. While the engineers and scientists who worked on the computer hardware were almost exclusively men, the computer programmers were frequently women.

The wives of many scientists in the Princeton group[37] – including Klara von Neumann, Marj Freeman, and Margaret Smagorinsky – worked with computers as part of the research endeavor. Their contributions were significant but received little public recognition[38] aside from the occasional "thank you" in the acknowledgments section at the end of a paper. Klara von Neumann, in particular, played a crucial role in the project. Klara had previously worked with her husband, John, on classified atomic bomb simulations using the ENIAC.[39] She was a skilled programmer who taught other scientists how to code and was the one to check the final version of the program.

Programming in 1950 was a very different experience than it is today. Computers had neither keyboards nor monitors. Instead, there were massive banks of switches and large plug boards with connecting wires.[40] Platzman describes the ENIAC programming technique as follows:

> the programmer had available only about a half-dozen or at most 10 words of high-speed read/write memory. An intermediate direct-access but read-only memory of 624 six-digit words was provided by three so-called "function tables," on which decimal numbers were set manually by means of 10-pole rotary switches, a tedious and lengthy procedure.[41]

To transmit and receive data from the ENIAC, the well-established technology of punch cards was used. Each punch card, somewhat larger than a postcard, represented one 80-character line of input or output. For each character in the line, a unique combination of holes, readable by a machine, was punched in the card. While printing and magnetic tape were later invented for computer output, punch cards remained the preferred way to input data well into the 1970s.

The Princeton Meteorology Group worked intensely over a five-week period in Aberdeen. Since the ENIAC had very limited memory, about 100,000 punch cards were used to record the details of the weather forecasts.[42] The programming required round-the clock effort, with programmers working in shifts.[12] In the end, four 24-hour forecasts were made for selected days in the previous year. The forecasts had some skill, although it took ENIAC 36 hours to compute a 24-hour "forecast"[43] – too slow to be of practical use!

A paper describing these forecasts – which were the first numerical forecasts ever – was published, coauthored by Charney, von Neumann, and Ragnar Fjørtoft, a visiting scientist from Norway. Since the two-dimensional model was considered too drastic a simplification of the atmosphere, the Princeton group decided to use the three-dimensional model for numerical prediction, expecting that it would improve the skill of the forecasts. The vertical dimension in the three-dimensional model was represented using either two or three grid points, or pressure levels.[44] The IAS computer, completed in 1952, was used to carry out three-dimensional forecasts to predict the famously intense storm that had occurred on Thanksgiving Day 1950.[45] Although the results initially appeared promising, further analysis indicated that three-dimensional forecasts were less skillful than two-dimensional forecasts, and the approach was abandoned for the time being.

The forecasts made by the Princeton group in 1950 using the ENIAC were for research purposes only, but the field of meteorology was never the same again. The mathematical approach pioneered by the Princeton group provided the previously subjective discipline with an objective foundation. Meteorologists around the world read the papers published by the group and improved upon them, using faster digital computers and better algorithms. Today, this technique of making forecasts by solving the equations of motion using a computer is called *numerical weather prediction*.

Although the United States was the first country to research numerical weather prediction, it was not the first country to use it for operational forecasts. A Swedish scientist named Bert Bolin visited the IAS in 1950 to work with Charney to study the effect of mountains on weather.[46] When Bolin returned home, he joined a Swedish weather prediction effort led by Charney's Swedish-born mentor, Rossby. The first operational weather forecast in Sweden was made in late 1954, using a locally built computer called the Binary Electronic Sequence Calculator (BESK).[47]

The Princeton Meteorology Group continued for several years to carry out research on numerical weather prediction using the new computer. It eventually disbanded in 1956,[29] and its former members went on to have illustrious research careers elsewhere, many playing important roles in future scientific developments. Unfortunately, Von Neumann had been diagnosed with cancer by 1956. We will discuss his final days in Chapter 19.

1.4 The Dark Side of Weather Prediction

All stable processes we shall predict.
All unstable processes we shall control.

John von Neumann[48]

Predicting weather seems like a harmless application of science, but the motives behind some early efforts at weather prediction were not always benign. The military supported weather prediction because it could provide valuable intelligence to troops in the field, but that was not the only reason for its interest. The early- to mid-twentieth century was the heyday of hypothesis-driven physical science, and many scientists, including von Neumann, believed that if the weather could be predicted, it could also be controlled for civilian and military purposes.[49] The *New York Times* wrote in 1946 that "some scientists even wonder whether the new discovery of atomic energy might provide a means of diverting, by its explosive force, a hurricane before it could strike a populated place."[50]

Like the hydrogen bomb, weather modification could prove to be the ultimate weapon if the enemy were to acquire it first during the Cold War. The early computers used overtly to make weather forecasts were also used covertly to design hydrogen bombs. Von Neumann himself had helped design these bombs, as part of the top-secret Manhattan Project. He is even credited with using his expertise in game theory to develop the doctrine of Mutual Assured Destruction (MAD) that maintained the uneasy peace between the United States and the Soviet Union. Mutual Assured Destruction relied upon the immense destructive power of nuclear bombs. The very first atomic bomb dropped on Hiroshima, codenamed "Little Boy," yielded the equivalent of 15 kilotons of TNT explosive; the first hydrogen bomb tested, codenamed "Ivy Mike," yielded 10 megatons; and the most powerful nuclear weapon ever tested, the Soviet weapon Tsar Bomba, yielded 50 megatons. (Most nuclear weapons currently mounted on ballistic missiles have yields of less than one megaton.)

As powerful as nuclear weapons are, nature can be much more so. The average Atlantic hurricane generates about 100 megatons of mechanical energy per day,[51] the equivalent of 100 ballistic missile detonations per day! Von Neumann was aware of the immense power of these natural processes, but he believed that they could either be predicted or controlled in the same way that an unruly horse could be steered or reined in, through the judicious use of small amounts of force – or, as he put it, "the release of perfectly practical amounts of energy."[52] Von Neumann envisioned a committee of experts who would decide how to release this energy at the right points in space and time: They could ensure that there was no rain during national celebrations such as the Fourth of July,[53] for example – but they could also ruin Soviet harvests with an artificially induced drought.

***

Weather prediction was an important driver of, and one of the first applications of, digital computing. To this day, weather and climate scientists are among the first customers for the latest and greatest computers. Two pioneers of digital computing, John von Neumann and John Mauchly, were both motivated by a desire to forecast weather. The philosophical battle for weather prediction between von Neumann's deductive approach and Mauchly's inductivist approach was, in simplistic terms, won by the deductive approach. (Ironically, the first deductive weather forecast was achieved using the machine created by Mauchly!) Von Neumann sought to use this approach not only to predict but also to control weather; thankfully, his dream of weather control would turn out to be impossible, sparing us the nightmare of weather warfare.

In Chapter 2, we describe how a very simple model of a very complex system was used to demonstrate the limits of weather prediction. In subsequent chapters, we chronicle how deductivism extended its reach to climate prediction. However, inductivism hasn't gone away: Data-driven approaches have continued to play a role in model calibration and may come to play a more prominent role in weather and climate prediction as well through machine learning.

2
Predicting Weather
The Butterfly and the Tornado

It all started with a mistake – a rounding error.[1] A programmer and a professor were working in a university office, trying to solve a set of 12 mathematical equations that described how 12 variables changed over a certain time period.[2] This undertaking required a computer that occupied the entire room next door. Given 12 initial values for the 12 variables, the computer printed out the variables' intermediate values in the middle of the calculation, and then the final result. The professor decided to repeat the second portion of the calculation using the intermediate values that appeared on the printout. He plugged these numbers into the computer and stepped out for a long coffee. (Computers were quite slow at the time.)

When the professor returned an hour later, he found that this final result was completely unlike the final result he had obtained previously. It turned out that the computer program had been designed to round the 12 intermediate values to only three decimal places on the printout, in order to fit them all on one line – even though, internally, the computer used six decimal places for calculations. The small difference in the intermediate values that were used to restart the calculation led to a large difference in the final results! This demonstrated that the computed final values of variables, governed by certain types of mathematical equations, could be incredibly sensitive to small changes in the specified initial values of those variables.

The year was 1960. The location was the Massachusetts Institute of Technology (MIT). The computer was a Royal McBee LGP-30, a desk-sized vacuum tube machine. (The LGP-30's arithmetic performance, 400 operations per second, was about the same as the ENIAC's, but the LGP-30 had 100 times more memory for programs and data.) Its programmer was Margaret Hamilton, a mathematics graduate who had just been hired at MIT. The professor was Edward "Ed" Lorenz of the MIT Department of Meteorology (Figure 2.1a).

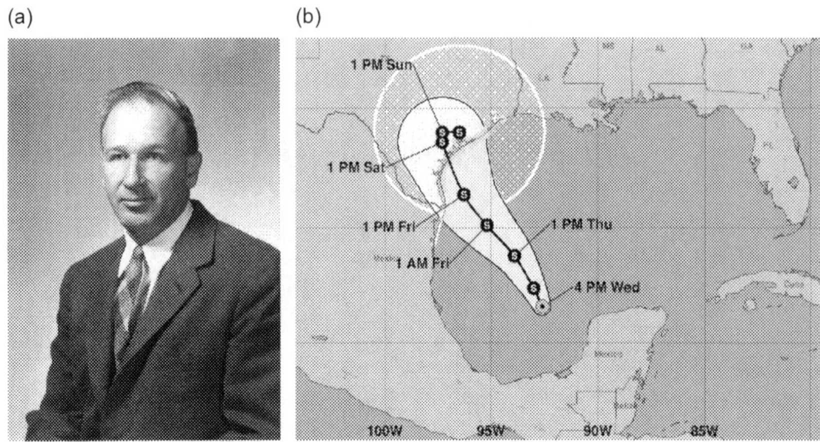

Figure 2.1 (a) Edward Lorenz (Photo: MIT Museum). (b) Five-day forecast of Hurricane Harvey's track over the Gulf of Mexico, starting at 4 p.m. on Wednesday, August 23, 2017, issued by the US National Hurricane Center. The thick black solid line is the day 1–5 track forecast; the white area demarcated by the thin solid line represents the "cone of uncertainty" in the day 1–3 track forecast. The stippled circular area bounded by the white line represents the uncertainty for the day 4–5 forecast.

The set of 12 mathematical equations that Lorenz was trying to solve represented (in a highly simplified manner) weather phenomena like storms. The 12 initial values, one for each variable, are called the *initial conditions*. In the atmosphere, the initial conditions correspond to the weather conditions for a given day, such as temperature, humidity, and windspeed. Solving the equations is like predicting the following day's weather, with the solution corresponding to the following day's forecast. What Lorenz had stumbled upon was the fact that a small error in an observation of today's weather could dramatically alter tomorrow's forecast.[3] This phenomenon would later become known as the Butterfly Effect. Since there will always be errors in observing today's weather, this effect limits how far into the future we can predict weather.

Lorenz had been pursuing a doctorate in mathematics at Harvard University when the United States entered the Second World War.[4] While working as a forecaster for the military, Lorenz realized that he could make a greater impact in the young science of meteorology than he could in mathematics. He changed his plans accordingly, and, in 1948, he received his doctorate in meteorology from the MIT Department of Meteorology and joined the department as a research scientist. One of Lorenz's eventual colleagues was Jule

Charney, the pioneer of numerical weather prediction, who moved to MIT after leaving the Princeton Meteorology Group in 1956.[5]

Lorenz, like John Mauchly, was interested in the idea of statistical weather prediction.[1] Statistical weather prediction relies on a mathematical property of physical systems called *linearity*. Many problems in fundamental physics are linear, in the sense that the equations describing them do not involve nonlinear terms. (Nonlinear terms are products of two or more variables, such as X times Y.) Linear equations lead to predictable solutions, which can be either monotonic (steadily increasing or decreasing) or periodic (cyclical). Statistical approaches typically predict the behavior of linear systems using a regression technique, which involves fitting a line to data points in a graph and then extrapolating that line into the future.

Lorenz knew that the full equations governing global weather were complex and nonlinear, involving millions of variables and products of pairs of variables. So, he constructed a conceptual model of weather with a highly simplified set of equations. This model used just 12 variables, but it retained some of the nonlinearity of the full equations. When Lorenz solved his simplified equations on the computer, the solution exhibited complicated noncyclical behavior. That is to say, it did not steadily increase or decrease, nor did it oscillate like a sine wave; it appeared to behave randomly. But the behavior was not truly random, because it was determined by the initial conditions. Such a solution is said to be *chaotic*. The inadvertent small change in the initial values led to a completely different realization of the chaotic behavior, which led in turn to a large change in the final results. Mathematically, he had discovered the extreme sensitivity of the final results to the initial conditions in nonlinear equations. This discovery formed the foundation of *chaos theory*, a field of science that studies unpredictable phenomena in nonlinear systems.

Lorenz did not publish this result immediately, but when he happened to mention it at a weather conference, he realized that people were interested in this issue of extreme sensitivity to initial conditions.[2] To better explain what was going on, he constructed an even simpler model for weather. One of Lorenz's colleagues, Barry Saltzman, had developed a seven-variable nonlinear model.[1] Lorenz borrowed it and simplified it to just three variables, X, Y, and Z, where X represented air motion, and Y and Z represented air temperature variations, in a very approximate sense.

The three-variable system of equations was sufficient to capture the essence of an atmospheric process known as Rayleigh-Bénard convection.[6] This process is similar to what happens when you boil a pot of water. The heat at the bottom of the pot warms the water there and makes it lighter, causing it to rise. The (relatively) cooler and heavier water at the top of the pot then sinks to replace it, setting up a rolling motion that redistributes heat from the bottom to the top. As you may know from your own experience, this leads to chaotic motion of the water.

Margaret Hamilton, the programmer who worked on the 12-variable model with Lorenz, moved on to a different project in 1961.[2] The new programmer for the LGP-30 was Ellen Fetter, another recent mathematics graduate. The computations carried out by Fetter and Lorenz using the three-variable nonlinear model were also very sensitive to the initial conditions, displaying the same behavior as the 12-variable model. Small errors in the initial conditions, like the third-decimal-place rounding error, led to large errors in the final results. The error in the initial conditions grew exponentially, meaning that it kept doubling over fixed intervals of time. This means that, where the doubling interval was one hour, an initial error of one unit became two units after one hour, four units after two hours, eight units after three hours, and about 1,000 units after just 10 hours!

The process of rapid error growth identified by Lorenz is illustrated by the five-day forecast for the track of Hurricane Harvey, which devastated the city of Houston in 2017 (Figure 2.1b). A small error in the initial location of the hurricane at 4 p.m. on Wednesday, August 23, when the hurricane was over the Gulf of Mexico, grew to an error of hundreds of miles in predicting the location of the hurricane's landfall along the Texas coast on Friday, August 25. As the track error widens into the future, it forms a cone, known as the error cone or the "cone of uncertainty." The average track error for all Atlantic hurricane forecasts in 2017 was 37 miles for a 1-day forecast, 101 miles after 3 days, and 179 miles after 5 days, which corresponded to a doubling interval of about 40 hours. During the past 15 years, the cone of uncertainty has narrowed as track errors have declined by a factor of two, thanks to improvements in weather models and better initial condition data.

Lorenz did eventually publish his results about exponential error growth in chaotic systems, in a now-classic 1963 paper entitled "Deterministic Nonperiodic Flow."[7] This paper did not attract much attention outside of meteorology when it first appeared, but, for the next two decades, the ideas presented in it found application in many other disciplines that dealt with complex systems. It is now considered one of the foundational papers of chaos theory.

In this and other papers, Lorenz acknowledged the important roles played by Hamilton and Fetter. However, public narratives surrounding Lorenz's discovery typically failed to mention either woman.[2] Two factors contributed to this – that they were women and that they were programmers. In the 1960s, women had few opportunities to advance in the scientific profession. And programming has always been considered somewhat of a rote task. The role of programmers in scientific research is not always appreciated, but today Hamilton and Fetter would likely have been recognized as coauthors of their respective papers. Hamilton went on to have a particularly groundbreaking career of her own, developing software for the Apollo lunar missions and founding multiple software companies. She is credited with inventing the term "software engineering."

The adjective *deterministic*, which appears in the title of Lorenz's landmark 1963 paper, denotes a mathematical system whose future behavior is determined by its initial condition. This contrasts with a *stochastic* system, whose future behavior is explicitly random and unpredictable for individual realizations. The three-variable system used by Lorenz was deterministic, but *nonperiodic*, which meant that its future behavior was very hard to predict beyond a certain time. Lorenz's results were surprising, not least because, at the time, deterministic systems were expected to be indefinitely predictable. In Section 2.1, we will learn more about the principle of determinism, which dominated scientific thought for two millennia prior to the twentieth century.

2.1 Philosophy Break: Laplace's Demon, Determinism, and Paradigm Shifts

Luke, you can destroy the Emperor. He has foreseen this. It is your destiny!

Darth Vader, *Star Wars, Episode V: The Empire Strikes Back*

On January 3, 2004, the NASA rover *Spirit* landed in the impact crater Gusev on Mars, within six miles of its target area[8] – after a journey of more than 300 million miles that took almost seven months. This 1-in-50-million accuracy is like throwing a dart from New York and hitting a bullseye in Los Angeles! How did NASA manage this astonishing feat of interplanetary navigation? To avoid corrective maneuvers that use up precious propellant, the navigation team at the NASA Jet Propulsion Laboratory had to consider even the tiniest of forces, including the gravitational effect of the churning of molten iron in the Earth's core, the plate tectonic motion of the continents, and the solar radiation pressure on the spacecraft. Taking these effects and more into account, the team was able to compute the trajectory of the spacecraft from its launch and stick the landing without having to make last-minute corrections. This was a stunning demonstration of the predictive power of mechanics, the branch of physics concerned with Newton's laws of motion.

The eighteenth-century French scientist Pierre-Simon Laplace wondered what would happen if a "vast intelligence" were to use this power to compute the trajectory of every atom in the universe, knowing the initial location and velocity of each[9] – akin to anticipating every collision in a cosmic billiards game. Such a "vast intelligence" would possess perfect knowledge of all past and future events in the universe, and of the causal linkages between them. This hypothetical all-knowing intelligence is referred to in philosophy as *Laplace's Demon*. The notion, championed by Laplace and articulated by many philosophers preceding him, that there is an identifiable preexisting cause for every event, is known as *determinism*.

Determinism prods us humans to seek a cause for every event, from hurricanes to heat waves. It also motivates much scientific and nonscientific inquiry. Even the *Old Farmer's Almanac*, as we have seen, operates on the principle that "there is a cause-and-effect pattern to all phenomena."[10] The power of prophecies, and the battle between fate and free will, are recurring themes in religion, literature, and cinema. The concept of the "chosen one," for instance, drives the storylines of the popular *Harry Potter* and *Star Wars* franchises. *Star Wars'* villain Darth Vader is in fact making a deterministic prophecy when he warns hero Luke Skywalker about his destiny.

The principle of determinism and the concept of Laplace's Demon came under attack following the major scientific breakthroughs of the nineteenth and twentieth centuries. Collisions of atoms, like collisions of billiard balls, can be exactly reversed in classical mechanics and traced back into the past. However, the irreversibility associated with the principle of increasing entropy or disorder – the Second Law of Thermodynamics – does not allow reversal in time, thus contradicting the exact reversibility of trajectories and invalidating the premise of an all-knowing Laplace's Demon.

The probabilistic nature of quantum mechanics, as embodied in German physicist Werner Heisenberg's uncertainty principle, is also incompatible with determinism. In a famous thought experiment, Austrian physicist Erwin Schrödinger imagined a cat trapped in a steel chamber along with a radioactive substance, a Geiger counter, and a poison release mechanism.[11] If a single atom in the radioactive substance decays, it is detected by the Geiger counter, which triggers the release of a poison to kill the cat. Since radioactive decay is probabilistic, according to quantum mechanics, we have to assume that the cat is both alive and dead – technically in a "superposition" of alive and dead states – until the chamber is opened. Laplace's Demon cannot tell whether Schrödinger's cat is dead or alive and is thus unable to know the future.

Laplace envisioned his "vast intelligence" as encompassing all domains of knowledge, writing, "it would embrace in the same formula the movements of the greatest bodies of the universe and those of the lightest atom."[9] For the purposes of this book, we envision several lesser demons toiling away in Laplace's Dungeon, each responsible for a particular domain.[12] Consider a Weather Demon: Given perfect knowledge of current weather conditions, the Weather Demon should be able to predict weather indefinitely into the future by following the trajectories of air masses using Newton's laws. An Economy Demon could predict economic growth or the lack thereof; then there could be an analogous Politics Demon, an Ecology Demon, and so on. These lesser Demons would work together to assist Laplace's Demon in making all-encompassing predictions about every aspect of the universe.

The concept of Laplace's Demon motivated early pioneers of mechanical computing like Charles Babbage in the nineteenth century, as described by historian Jimena Canales in her comprehensive survey of conceptual demons in science.[13] But it was the invention of the electronic computer in the late 1940s that made feasible the basic computational premise behind Laplace's Demon; ENIAC, the first electronic computer, was invented to compute trajectories (of artillery shells). Computer modeling opened up a portal for the deterministic demons to escape from their philosophical realm into the real world. The computations of Jule Charney and John von Neumann breathed life into the Weather Demon, allowing weather to be predicted scientifically. And those interested in weather control, including von Neumann, expected one day to treat the Weather Demon as they would Aladdin's genie, to ask for the granting of wishes such as wreaking floods or drought on the enemy.[14] But the work of Lorenz on deterministic chaos shackled this newly awakened demon, limiting how far into the future it could predict. Weather prediction would survive within its shackles, but, due to the unpredictability introduced by chaos, weather control was pretty much doomed.

Despite being under attack from the principle of entropy and quantum uncertainty, classical determinism was still alive and kicking in mid-twentieth-century science. Recall that John Mauchly, coinventor of the first digital computer, sought to predict weather using the sunspot cycle. Even Lorenz initially tried to use statistical techniques like linear regression to predict weather. We might think the discovery of deterministic chaos by Lorenz and others would have put the final nail in the coffin of determinism, but, like a zombie, it continues to live on in human thought and discourse.

***

Scientific developments relating to determinism provide an opportunity to discuss important concepts in the philosophy of science introduced by

American philosopher Thomas Kuhn in his 1962 book, *The Structure of Scientific Revolutions.*[15] Kuhn applied induction to the historical practices of science itself and analyzed scientists as a group rather than as individuals.[16] He observed that day-to-day science is generally incremental: If we glance at the titles of papers in scientific journals, for example, we find that they often discuss minutiae of established theories. This is completely routine because established theories have gaps that need filling or small extensions that expand their domain of application. Kuhn refers to this as *normal science,* in which scientists essentially solve puzzles. Every now and then, though, this normal routine is broken and punctuated by a major scientific advance that upends the established order. This is a *scientific revolution.*

The two theories we discussed in our previous philosophy break (Section 1.2), Darwin's evolution and Einstein's relativity, are both examples of scientific revolutions. Darwin's theory challenged the existing theory of creationism, showing that the origin of new species could be explained through natural selection. Einstein's theory replaced the existing absolute notions of space and time with a relativistic worldview. Quantum mechanics was another scientific revolution, overturning deterministic classical mechanics. Interestingly, Einstein – himself no stranger to bucking conventional wisdom – reacted to the probabilistic framework of quantum mechanics by exclaiming, "God does not play dice with the universe!" – an example of the powerful hold determinism exerts on human thought.

At the conclusion of a scientific revolution, there is a "new normal" or new paradigm in that area of science, which Kuhn refers to as a *paradigm shift.* Unlike political revolutions, which are often cataclysmic, scientific revolutions can take place in slow motion and may not always be recognized as such when they are occurring. Quite often it takes some time for the scientific community to switch to a new paradigm. Darwin's theory was not generally accepted for a long while due to scientific and societal resistance. Einstein's theory was accepted more quickly, thanks to quantitative predictions that were spectacularly verified (Section 1.2).

Chaos theory is often considered a scientific revolution, in the sense that it "overthrew" classical determinism. Although Lorenz is perhaps the most famous name associated with chaos theory, he had many scientific forebears – most notably the French mathematician Henri Poincaré, who discovered non-periodic orbits in his studies of celestial mechanics in the late nineteenth century.[17] Another was Phil Thompson, formerly of the Princeton Meteorology Group. In 1957, Thompson published a paper about the sensitivity of weather forecasts to initial conditions.[18] Recall that even Lorenz's work on deterministic chaos, published in 1963, was not widely appreciated until decades later, underscoring the slowness of scientific revolutions.

The system of mathematical equations used by Lorenz was a heuristic representation of the atmosphere, involving only a handful of variables. The chaotic behavior of such systems with few variables is referred to as low-order chaos, and its mathematical properties have been studied in detail. Low-order chaos has found applications in diverse fields such as astronomy, chemistry, and ecology. It has also had a profound influence on meteorology, but mostly in a qualitative sense. Weather in the atmosphere requires millions of variables to describe, and many of the quantitative insights of low-order chaos theory do not easily translate to the very high-order chaos in the weather system.

In Section 2.2, we describe how we measure global initial conditions and actually make forecasts for the high-order chaotic weather system every day.

2.2 Atmospheric Data: Sharing Selfies on the WWW

The advent of numerical weather prediction in the early 1950s was a scientific revolution in the field of meteorology. It would change weather forecasting forever, but it took some time for the paradigm shift to occur. The forecasts made by Charney and von Neumann were not true forecasts; they were research forecasts, made long after the weather events actually happened. The numerical forecast of the famous 1950 Thanksgiving Day storm was made more than a year later, in 1952.[19] This is common in meteorological research, where new forecasting techniques are tested against hard-to-predict weather events that occurred in the past.

Early computer models for numerical weather prediction, derived from Charney's experiments, did not perform very well when used to make true operational forecasts for public consumption.[20] Although these models captured winds in the upper levels of the atmosphere, they had trouble predicting temperatures and the behavior of surface winds, which were of the most practical value in forecasts. Therefore, computer models served mostly an advisory role for human forecasters in the early years. Over the following decades, these models became much more elaborate; the computers grew much more powerful; and, by the 1980s, numerical weather prediction using computers had become the primary mode of weather forecasting.

Today, weather forecasts are considered a routine service – a public utility, like electricity or the Internet. But behind the scenes, there is a vast machinery that enables the magic of forecasting to happen. As Lorenz demonstrated, accurate initial conditions are key for a skillful forecast. Central to generating these initial conditions is an information-exchange network called the World Weather Watch (WWW).[21] This WWW was established in 1963, long before the World Wide Web, at a time when digital computing was still in its infancy. An international consortium, the World Meteorological Organization (WMO), coordinates this activity among 187 member countries.

This WWW is a network of observing instruments, telecommunication facilities, and data protocols that enables the near-instantaneous exchange of weather information across the globe. It is an example of genuine international cooperation, born of the truism that "weather knows no boundaries." For example, we cannot predict weather in the United States simply by observing it in the United States; since weather systems move eastward across the country, we must observe weather coming from upstream, from the Pacific Ocean, and even all the way from China. Similarly, European weather is affected by weather from North America and the Atlantic Ocean.

The primary data feeding the WWW is a three-dimensional "selfie," a snapshot of the atmosphere taken from within the atmosphere. Weather balloons with instruments called *radiosondes* are launched twice a day at more than 800 stations across the globe.[22] The launches are tightly synchronized, happening at exactly 00:00 UTC and 12:00 UTC, corresponding to midnight and noon in Greenwich Mean Time. Radiosonde balloons, filled with helium or hydrogen, carry instruments that measure pressure, temperature, and humidity as they rise up in the atmosphere.[23] In most cases, the drift of the balloon is also monitored to track its motion and calculate the winds.

Weather data from the radiosondes is immediately transmitted to a central clearinghouse managed by the WMO. It is then shared among weather modeling centers around the world. Just as multiple photos of a landscape from different angles can be stitched together to create a 360-degree panorama, the data from the 800 stations is combined to create a global three-dimensional snapshot of weather. Recall that Thompson and Lorenz showed that a weather forecast is very sensitive to errors in the initial conditions. A "selfie" of the atmosphere with only about 800 "pixels" in the horizontal plane provides a rather fuzzy view of the atmospheric state. Radiosonde "pixels" are located mostly over continents, except for some over islands. This means that the marine atmosphere is mostly unobserved, which leaves a data void. For forecasts made using just data from the radiosonde network, initial errors from the data void regions grow rapidly and degrade the skill of the prediction. This was indeed the case with early weather forecasts before the advent of satellites.

Meteorology's satellite era began in 1960 with the launch of the TIROS-I satellite. This satellite incorporated instruments that could measure data related to weather.[24] Unlike the coarse "selfie" provided to the WWW by the radiosonde network, satellites could provide a very high-resolution image of the atmosphere taken from space. Modern weather satellites, such as the GOES-16 launched in 2016, can provide images with more than 100 million pixels of resolution. Unlike the radiosonde network, satellite images provide global coverage, over land and ocean alike. These satellite images capture temperature, clouds, rain, and other atmospheric properties.

In addition to providing finer spatial resolution, satellites also send data more frequently to weather modeling centers. Geostationary weather satellites orbit 36,000 km above the Earth's surface and take photos of a whole hemisphere every 15 minutes. But satellite data have one drawback: The images are two-dimensional. Although some three-dimensional information is retrievable from satellite images, they do not provide the same accuracy in the vertical dimension as the radiosonde "selfie." For example, a satellite can see the fine horizontal structure of a cloud, but it cannot see below the cloud. A radiosonde can ascend through the cloud, making continuous measurements of the fine vertical structure. Radiosondes are therefore used alongside satellites to provide three-dimensional data coverage over land, but, due to the lack of radiosondes over the ocean, there is still a three-dimensional data void over the ocean.

2.3 The Assimilation of Data

We are Borg. You will be assimilated. Resistance is futile.
Third of Five in *Star Trek: The Next Generation*, Season 5, Episode 23 ("I, Borg")

Unlike the *Star Wars* franchise, which is preoccupied with determinism, entries in the *Star Trek* franchise have a decidedly nondeterministic flavor. *Star Trek: The Next Generation* features an alien civilization called the Borg. The Borg are organized as a collective of cybernetic organisms, rather like a beehive. Individual members of the species are essentially drones, obeying rules and carrying out duties assigned by the collective mind. The Borg civilization grows by integrating captured members of other species through a process called *assimilation,* during which the body of each captive is physically altered. Once assimilated, new members obey Borg rules, losing their individual identities and receiving numeric names like Third of Five or Seven of Nine. A single member disobeying the rules could seriously disrupt the balance of the Borg collective, which relies on orderly execution of the will of the collective mind.

The Borg were successful in assimilating many different species, but they were singularly unsuccessful in assimilating one important *Star Trek* character – the android named Data. The Borg tried to alter Data to assimilate him, but (as chronicled in the movie *Star Trek: First Contact*) he was strong enough to resist them. Yet what the fictional Borg could never do, meteorologists do every day: They assimilate data (measured by radiosondes and satellites) into weather models to generate consistent initial conditions, which are the initial values of weather variables at each altitude and location across the globe. This mathematical procedure is called *data assimilation*.

The values of variables in weather models, like wind and temperature, must follow strict rules and balances that come from the mathematical equations for fluid flows. But observed values of winds and temperatures from the radiosondes inevitably have measurement errors and may not obey those rules. This means that new measurements cannot be simply inserted into the weather model's grid to create a new initial condition; they must be altered to conform to the rules. Errant measurements may even be rejected if they seriously violate the rules, because they could degrade forecasts. One reason for the spectacular failure of the numerical weather forecast chronicled by Lewis Fry Richardson in 1922 was that the initial weather data Richardson used were not properly "assimilated" into the mathematical framework he was using.[25]

Data assimilation operates on a continuous cycle. First, a short "assimilation forecast" is made, say for six hours, starting at 00:00 UTC. At 06:00 UTC, all newly available radiosonde and satellite measurements are collected from the globe. Recall that there are regions of the globe with very few on-site radiosonde measurements, such as the areas over the oceans. These "data void" regions in the measurements are filled with predicted values from the six-hour forecast. The blend of new measurements and forecast weather is assimilated, or altered, to satisfy the balance requirements of the weather model. This generates a new initial condition at 06:00 UTC, which is used to make another six-hour "assimilation forecast." At 12:00 UTC, the whole assimilation process is repeated using the new forecast and the new measurements that have become available.

The initial conditions generated by data assimilation are also used to make regular weather forecasts, extending out to 10 days and beyond. The forecasts are calculated using a three-dimensional computer model of the atmosphere, a much more sophisticated version of the one used by Charney and von Neumann in 1950 (Section 1.3). The weather model is like a face-aging app on a smartphone. Given a selfie, the face-aging app shows us what our faces will look like decades into the future. Given the initial weather conditions, the weather model shows you what the weather will be like days into the future. The better the quality of the initial "selfie" the better the final forecast will be. Some of the world's largest computers run these weather models to generate the weather forecasts that we see on television and hear on the radio.

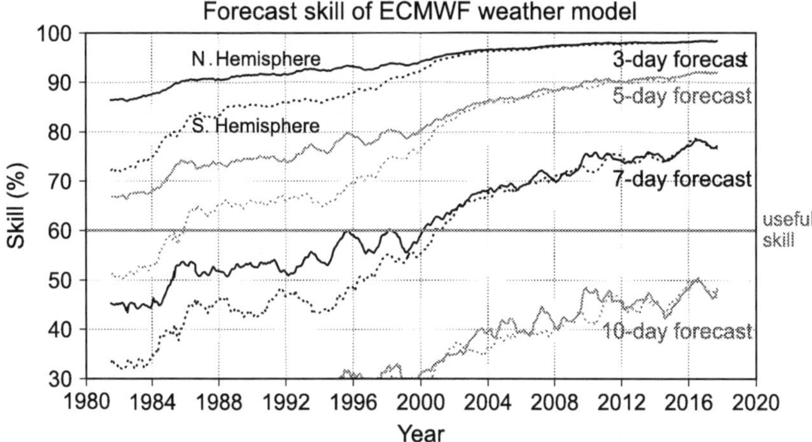

Figure 2.2 The chart shows the evolution of the anomaly correlation, a measure of forecast skill (expressed in %), for upper-air forecasts from 1982 through 2018 using the ECMWF model. The skill of 3-day, 5-day, 7-day, and 10-day forecasts (as indicated) is averaged outside the tropics in the Northern Hemisphere (solid) and the Southern Hemisphere (dashed). The horizontal line at 60 percent indicates the useful skill level. (Bauer et al., 2015; adapted from ECMWF chart/cc-by 4.0)

Weather models are evaluated on how skillful they are in forecasting weather patterns. Meteorologists commonly use a metric called "anomaly correlation" to measure the skill in forecasting large-scale atmospheric flow patterns. This metric, which is expressed in percentage units, computes the spatial statistical correlation between the predicted weather pattern and the observed weather pattern in the upper atmosphere. An anomaly correlation of 100 percent implies perfect prediction of the weather pattern, a value of more than 80 percent indicates high accuracy, and any value of more than 60 percent indicates a useful forecast.

The forecast skill for one of the premier weather prediction models in the world, from the European Center for Medium-Range Weather Forecasts (ECMWF), is shown in Figure 2.2. This model's skill has increased as the model has become more elaborate over the years. During the initial years, there was a gap in forecast skill between the data-rich Northern Hemisphere and the data-poor Southern Hemisphere, which has large oceanic regions without radiosonde observations. Since 2001, the gap has virtually disappeared due to better Southern Hemisphere initial conditions obtained through the assimilation of satellite data: A five-day forecast is now as good as a three-day forecast was 20 years ago, and the seven-day forecast is almost as good as the five-day forecast was. Like the precise landings of Mars missions, this "quiet revolution" in numerical weather prediction represents a triumph of determinism.[26]

Notice that the skill of three-day and perhaps even five-day forecasts has flattened over the last decade, after increasing steadily since 1982. There appears to be scope for improvement in the longer forecasts – that is, 7–10 days – but even they show indications of eventual flattening at lower skill levels. Overall, there appears to be a *predictability limit* of about two weeks beyond which weather forecasts will have little skill. This means that even with a "perfect weather model," a 14-day forecast would not be able to reach the useful skill level. The predictability limit arises from the sensitive dependence on initial conditions identified by Lorenz. Simply put, exponential error growth makes it harder to predict further into the future: There is a practical limit to how accurately we can measure the initial conditions, based on the number of radiosonde stations and the resolution of satellite data.

2.4 The Butterfly Effect

For want of a nail, the shoe was lost.
For want of a shoe, the horse was lost.
For want of a horse, the rider was lost.
For want of a rider, the message was lost.
For want of a message, the battle was lost.
For want of a battle, the kingdom was lost.
And all for the want of a horseshoe nail.

Fourteenth-Century Nursery Rhyme

The mathematical concept of the sensitive dependence of events on prior conditions, presaged by the nursery rhyme above, is now commonly known as the Butterfly Effect. This metaphor itself came into existence through a series of odd coincidences.[27] When Lorenz started to write about this mathematical concept,[28] he used the phrase "flap of a sea gull's wings" to denote the uncertainty in the initial conditions; he never mentioned a butterfly. The idea was that even a tiny disturbance such as a sea gull flapping its wings could be exponentially amplified over time and eventually lead to a much bigger weather phenomenon, such as a tornado or a hurricane.

In December 1972, Lorenz was scheduled to give a talk about his chaos research at the American Association for the Advancement of Science meeting in Washington, DC.[29] He had not provided a title for the talk, so the event organizer, a scientist named Philip Merilees, suggested a title along the lines of

"Does the Flap of a Butterfly's Wings in X Set Off a Tornado in Y?" where X and Y were place names. In the end, the talk was called "Does the Flap of a Butterfly's Wings in Brazil Set Off a Tornado in Texas?" Lorenz had chosen Brazil and Texas for X and Y because they made the title more alliterative, but the butterfly had been Merilees's idea. When science historian James Gleick used the same metaphor in his bestselling 1987 book *Chaos: Making a New Science*, the Butterfly Effect caught the public's imagination and became an enduring emblem of unpredictability.

Where did the butterfly metaphor come from? The most likely explanation attributes it to Joe Smagorinsky of the Princeton Meteorology Group, which had carried out the first numerical weather forecast in 1950 (Section 1.3). Smagorinsky continued to work on weather modeling after he left Princeton, and in a 1969 paper he wrote about the "flap of a butterfly's wings" altering a weather forecast. Both Lorenz and Merilees were aware of Smagorinsky's work, and it is plausible that this led Merilees to use the metaphor for the title of Lorenz's talk.

A less likely, but more dramatic, explanation traces the butterfly metaphor to the science fiction writer Ray Bradbury.[27] His 1952 short story "A Sound of Thunder" tells of a time-travel safari's unintended consequences, wherein a butterfly plays a key role. The Butterfly Effect has important implications for the notion of time travel. Lorenz himself was once asked whether he would live his life differently if he could start over again.[30] He responded by saying that he might have pursued different research interests, but he added, "The one thing I can be sure of is that I would marry the same girl – if I met her!" Presumably the conditional clause alluded to the Butterfly Effect. Ed Lorenz, who passed away in 2008, had a long and happy marriage to his wife Jane Loban.

There is yet another twist in the tale of the Butterfly Effect, identified in a 2014 paper by British scientist Tim Palmer and others.[31] Lorenz's 1972 talk was not about the claim he had made in his landmark 1963 paper of how the sensitive dependence on initial conditions leads to the exponential growth of errors. Lorenz was actually addressing a more profound claim in a 1969 paper, about the even stronger dependence of error growth on initial conditions.[32] Errors at larger scales associated with weather patterns doubled in matters of days. But errors at smaller scales, such as the locations of individual clouds, grew much faster, doubling in hours or less. These errors would eventually affect the larger scale and could lead to the much faster, or *super*-exponential, growth of errors.

To understand the implications of exponential error growth, let us consider the use of models with increasingly finer spatial grids to make longer forecasts.

These require more and more expensive observing systems to provide finer initial condition data, and more and more powerful computers to calculate forecasts. Say the current computer, with a performance of 10 petaflops (10 quadrillion floating-point, or arithmetic, operations per second), predicts weather for up to 9 days with useful skill (Figure 2.2). A 100-petaflop computer may extend the prediction to 10 days with better initial conditions, and a 1,000-petaflop computer may extend it to 11 days. The time period of the useful forecast grows linearly, but the computational requirement goes up exponentially. Basically, the Weather Demon can provide us a forecast that is valid a day longer, but it will need better data and a much more powerful computer to do that. This means that there is a practical predictability limit for weather forecasts since we have only finite resources available with which to build computers and observing systems. But it is not a theoretical or absolute limit because we could hypothetically have infinitesimally accurate initial conditions and infinitely powerful computers.

However, if a finer spatial grid leads to even faster exponential growth, or super-exponential error growth, it will mean that there is not only a practical limit but also an absolute limit to how far into the future forecasting can extend. To illustrate this, consider a 1,000-petaflop computer that predicts weather with useful skill for 11 days. The 10,000-petaflop computer would extend the same prediction skill to 11.5 days, a 100,000-petaflop machine would extend it to 11.75 days, a 1,000,000-petaflop machine to 11.875 days, and so on. Even a centillion-flop computer could not push the limit beyond 12 days. This is similar to what is known as Zeno's Paradox: The additive increments get exponentially smaller each time, so the limit can never be crossed or even reached. If such a predictability limit existed, it would be an absolute limit – just like Einstein's relativistic limit, which says that even spending a centillion dollars cannot make a spaceship fly faster than the speed of light. The Weather Demon would be truly unable to see beyond an absolute predictability limit.

There is clear evidence for exponential error growth in weather forecasts, corresponding to the Butterfly Effect demonstrated in Lorenz's 1963 paper. This is the accepted wisdom. The evidence for super-exponential error growth is not so clear. In their 2014 paper, Palmer and others argued that there are certain weather situations where error growth is faster than exponential, because errors much finer in scale grow much more quickly.[31] They suggest that the concept of this much faster error growth be referred to as "the real butterfly effect," since that was the actual subject of Lorenz's 1972 talk that inspired the metaphor. They also argue that this effect can be mitigated using ensembles, a forecasting methodology that is described in Section 2.5.

2.5 An Ensemble of Demons

The work of Lorenz and others showed that there is a deterministic predictability limit. The skill of weather forecasts starts to decline as this limit is approached. But there is a way to stretch this limit – if we relax the need for an exact, or deterministic, forecast. Consider the following scenario. We watch a butterfly flap its wings in Brazil and try to count the number of flaps. We don't have an exact number, but we estimate that there were between 11 and 15 flaps. Say we have access to a Weather Demon, which makes perfect weather predictions. We create an "ensemble" of five identical Weather Demons, each corresponding to the same weather model, rather like a quintet of five identical musical instruments.[33] But each of those identical clones is given a different initial condition. The first demon of the quintet is told that the butterfly flapped its wings 11 times, the second is told 12 times, and so on, until the fifth demon is told that the butterfly flapped its wings 15 times. Each of the five Weather Demons then makes a deterministic forecast for the next 30 days.

Due to sensitive dependence on initial conditions, the five forecasts from the demon quintet will typically be different. You take all five forecasts and average them to create what is called an *ensemble forecast*. If two of the forecasts predict rain in Texas, and three don't, then you predict a 40 percent chance of rain in Texas. This is no longer a deterministic forecast; it is a *probabilistic forecast*. In this way, ensemble forecasting extends the skill of weather forecasts beyond the deterministic predictability limit. The impact of the error in the initial condition is mitigated by averaging out some of the uncertainty.

We can get even more information out of the ensemble forecast. Meteorologists love to hold forecasting contests: Each weather aficionado makes a forecast for a particular day, and the one who gets it right gets to take home a prize. We could treat the forecasts made by the demon quintet as a contest, declaring the demon that makes the most accurate forecast the winner. But all the demons in the quintet are identical; each was just given a slightly different set of initial conditions. The winning demon is likely the one that received the best set of initial conditions, purely by chance. This realization can actually be used to improve the initial conditions for the next forecast.

Recall that the estimation of the initial conditions for a forecast starts with a fuzzy 800-pixel "selfie" of the atmosphere taken with weather balloons. Satellites add horizontal, but not vertical, resolution. It would be nice if we could touch up and sharpen this fuzzy "selfie." To do that, we use the demon quintet to make five forecasts for a short period in the past, say, six hours. We pick the demon that made the most accurate forecast for the past six hours, and we ask that demon to improve the current selfie and add detail to the data void regions, because we expect that demon to have received the best estimate of the initial conditions (the number of butterfly wing flaps) that happened six hours earlier. More sophisticated versions of techniques like this are used to fill gaps in the data and generate better initial conditions, as part of the data assimilation process.

Uncertainty in the initial conditions is not the only factor that can degrade a forecast. Structural imperfections in the construction of the weather model, such as poor formulas used to approximate the behavior of clouds, can also cause the quality of a forecast to deteriorate over time. These are known as *model errors*, and averaging over an ensemble of forecasts will not mitigate the impact of these errors. If the five identical Weather Demons all make a similar error in their forecasts, such as overestimating rain, averaging their forecasts will not eliminate that error. This problem with model errors gets worse as the forecast period becomes longer, because these errors grow slowly with time.

Even ensemble forecasting loses predictive skill at some point. For the current generation of weather models, this happens about a month into the forecast. At this point, the memory of the atmospheric initial condition is essentially lost. As the chaotic error growth saturates, weather evolves randomly (or stochastically). We then move into the realm of climate, the long-term behavior of weather.

2.6 The Infinite Forecast

Having achieved a measure of success in numerical weather prediction, von Neumann wanted to extend forecasts beyond a day or two, to a much longer period. He called this the "infinite forecast."[34] Imagine making a very long forecast using the Weather Demon, starting from the fuzzy initial conditions and running the computer weather model thousands of years into the future. This "infinite forecast" is an alternate realization of the real world: Beyond a few weeks, weather events such as hurricanes will occur at different times in the alternate realization than in the real world, but they will occur at the same frequency. This means that if we calculate the weather statistics of this infinite forecast – such as the number of heat waves, severe rainstorms, or droughts – they will match the weather statistics of the real world.

In other words, the infinite forecast captures the climate of the real world, because *climate is the average of weather.*[35] The average temperature of a region for a particular month, the average time of onset for the rainy season, the average number of hurricanes making landfall – these are all different kinds of climate information. Does this mean that if you run a weather model long enough, it becomes a climate model? In a sense, yes, but this glosses over some important details. External factors that affect weather, like the angle of the midday sun or the surface temperature of the ocean, do not change much over the week or two-week duration of a weather forecast. However, these factors become quite important when we forecast for more than a month or so in the future.

Does climate exhibit nonlinear chaotic behavior, like the three-variable Lorenz equations introduced at the beginning of this chapter? For a simple chaotic system, histograms of the frequency distribution of variables typically exhibit skewed shapes or multiple peaks, deviating from the bell (or Gaussian) shape. Such non-bell-shaped frequency distributions are a common signature of chaotic behavior, as opposed to random or stochastic behavior, which is characterized by bell-shaped distributions. But for data from the last 150 years, when we have direct measurements of climate variables, frequency histograms tend to stay close to a bell-shaped distribution, indicative of stochastic behavior. (Note that the climate system has exhibited other kinds of behavior over the past hundreds of thousands to millions of years, during the ice ages and beyond.)

The ubiquity of approximately bell-shaped frequency distributions in climate can be intuitively explained by a mathematical theorem known as the Central Limit Theorem. This theorem states that if you add together a large number of variables, each of which may have a non-bell-shaped frequency distribution, the sum will tend to have a bell-shaped or Gaussian distribution. The more variables that are added together, the closer the distribution is to a bell shape. Climate is the long-term average of many chaotic weather events, so even if the weather frequency distributions are not bell-shaped, the climate frequency distributions will tend to be bell-shaped. Predicting climate therefore means predicting the average and standard deviation of these bell-shaped distributions using the "infinite forecast."

***

The success of weather prediction is a triumph of deductive scientific modeling.[36] The forecasts made by the Weather Demon are verified every day to a high degree of accuracy. But the Weather Demon is shackled by chaos and cannot make such forecasts beyond two weeks. Is there a Climate Demon that can predict beyond that limit – years, decades, and centuries into the future?

We will address the issue of climate prediction in later chapters, but we can indulge in some speculation about the properties of the lesser demons. The Weather Demon is a solitary beast. It toils alone to compute the movement of air without ever talking to, say, the Economy Demon or the Politics Demon, because economics and politics don't affect weather forecasts for the next week or two. However, as we shall see in subsequent chapters, the Climate Demon needs to be a social demon that frequently talks to many other demons, because economic growth and political decisions over the next years and decades will profoundly affect climate predictions. In Chapter 3, we introduce the basics of climate, the scientific underpinnings of the Climate Demon.

3

The Greenhouse Effect

Goldilocks and the Three Planets

What makes the Earth habitable? Why is its climate so hospitable to plant and animal life, with an acceptable range of temperatures, sufficient availability of water, and so on? We might argue that each planet supports life that is uniquely adapted to the local conditions on that planet. This would imply that other planets in the solar system support other kinds of life. But none of the other planets in the solar system appear to support any life at all, that we know of. Astrobiologists, who study the possibility of life on distant planets, use a well-known children's story to illustrate this conundrum: "Goldilocks and the Three Bears."

In case you haven't read the story, here's a brief retelling. Three bears – a big papa bear, a medium mama bear, and a little bear cub – lived in a house. Every day they made porridge for breakfast and poured it into three bowls – a big bowl, a medium bowl, and a little bowl. While the porridge was cooling, the three bears would go out for a walk. One day, when the bears were away, a mischievous little girl named Goldilocks snuck into the house. She sampled the porridge in the big bowl, but it was *too hot*. She sampled the porridge in the medium bowl, but it was *too cold*. Then she sampled the porridge in the little bowl – and the temperature was *just right!* She ate all that porridge, performed other acts of mischief around the house, and eventually fell asleep on the little bear's bed. When the bears returned, they were shocked by the mess and promptly kicked Goldilocks out. (The full story includes many more details and is more fun to read.)

The fact that some planetary temperatures are just right for life, neither too hot nor too cold, is known as the "Goldilocks principle" in astrobiology.[1] Some planets are too hot, like Mercury, and some too cold, like the outer gas giants. There is believed to be a zone for planets in which they are just the right distance from the sun and therefore receive just enough sunlight to be neither too hot nor too cold for life. Venus, Earth, and Mars are roughly in this zone, which is known as the *circumstellar habitable zone*, or simply the Goldilocks zone. Venus might have once supported life: It is about the same size as the Earth and has a thick atmosphere. However, it currently has clouds of sulfuric acid and is extremely hot. Mars, on the other hand, has very little atmosphere and is rather cold.

The climate over much of the Earth at this time is conducive to life, but Earth's climate has naturally changed over millions of years. There were times when it was colder, when ice sheets covered large portions of the continents. More recently, over the last 150 years, the climate has been warming, and the pace of warming has accelerated in the last few decades. Almost everyone now agrees on the reality of this warming. The vast majority of scientists attribute it to rapid increases in the concentrations of carbon dioxide and certain other gases emitted by human activities into the atmosphere. In the rest of this chapter, we try to understand why the temperature of the Earth happens to be "just right," what balances help keep it that way, and what forces can disrupt these balances. We start by identifying some of the basic concepts of climate science, using lessons from everyday life.

3.1 Lessons from a Hot Summer Day: Radiation, Albedo, and Convection

Consider a hot summer day. If you park a car outside, you know that it will become very hot, almost scalding, to the touch: The car's metal roof gets much hotter than the surrounding air. This is because the metal roof, unlike the surrounding air, is heated directly by the sun. The surrounding air is mostly transparent to the sun's rays, meaning that sunlight passes through air without being absorbed by it. The metal roof absorbs the sunlight and heats up.

Light and heat are forms of energy. Specifically, they are forms of electromagnetic radiation, which is often just referred to as *radiation*.[2] Light is the visible form of radiation, and heat is one of the many invisible forms of radiation. Each type of electromagnetic radiation has a property called *wavelength*. The visible light from the sun has a wavelength of about half a micron. (A micron, or micrometer, is one-millionth of a meter.) The car's roof emits heat as invisible *infrared radiation*. (If you hold your hand over the roof of the car, without touching it, you can feel the heat.) Infrared radiation has much longer wavelengths than the visible light from the sun, typically around 10 microns. Therefore, infrared radiation is also known as *longwave radiation*. Accordingly, light from the sun is known as *solar radiation* or *shortwave radiation*.

You might have noticed that a black car left out in the sun is hotter than a white or silver car left out in the sun. Scientific studies have confirmed that a white or silver surface reflects more of the sunlight falling on it, whereas a dark surface absorbs more of the sunlight falling on it. The reflective property of a surface is called *albedo*, which means "whiteness" in Latin. The greater the albedo of a surface, the more reflective it is. Clouds also have albedo. When a cloud passes overhead, you feel cooler because the cloud reflects some sunlight, reducing the amount that reaches you.

You might also have noticed that when you park the car with the windows closed, the inside of the car gets much hotter than it does when you park the car with the windows slightly open. As we know, the air around the car is not heated by sunlight and is therefore much cooler than the surface of the car. If the windows are even slightly open, air from the inside of the car circulates outside and mixes with the cooler surrounding air, which cools the interior of the car. If the window is tightly shut, the interior surfaces get much hotter, because this cooling option is not available. This is similar to what happens in a greenhouse. We shall discuss this process again in Section 3.2.

Shortwave radiation from the sun heats the surface of the Earth directly, without heating the air in the atmosphere, because the molecular properties of air make it mostly transparent to direct sunlight. The hot surface then warms the atmosphere above it. This means that the atmosphere is heated from below, rather like water in a pot placed on a stove. This heating generates an overturning circulation called *convection*, which becomes chaotic when the water begins to boil (Chapter 2). Convective motions transfer heat from the bottom to the top of the pot.

Solar heating of the surface causes a similar process of convection to occur in the atmosphere, because warm air is lighter than cold air. The resulting ascent of warm air and descent of cold air transfers heat from the hot surface to the atmosphere. Additionally, the convection of moist air cools the surface through evaporation and leads to the formation of clouds. When the moist air rises, it cools due to a process called *adiabatic expansion*, and the moisture in it condenses to form clouds and rain droplets. Condensation of water vapor releases heat, resulting in more heat being transferred to the atmosphere.

While the air is mostly transparent to solar radiation, it is not transparent to infrared radiation. The air can block and absorb the infrared radiation emitted by the surface, capturing its energy. Therefore, the two processes that warm the atmosphere are heat transfer through convection and direct infrared radiation from the surface. This explains why the air is colder on top of a mountain, even though the top of the mountain is closer to the sun. The air surrounding the peak is further away from the hot surface below, so it receives less of the heat from the surface.

But what process causes the air to block the infrared radiation? That process is obscured by solar heating during a hot summer day. We can experience it best during a different time of the day at another time of the year.

3.2 Lessons from a Cold Winter Night: Greenhouse Effect and Climate Sensitivity

Consider now a cold winter night. When the sun goes down at the start of the night, it is no longer heating the ground. But the ground still retains the energy from the earlier daytime solar heating, and, throughout the night, it sends that energy back to space as infrared radiation via the atmosphere. You may have noticed that a clear winter night is colder than a cloudy winter night. That's because clouds impede this radiation of heat back to space; they act like a blanket that traps the infrared radiation and sends it back to the ground. Water vapor in the air traps heat in the same way, which is why humid nights are warmer than dry nights (when other conditions are the same). Due to lack of humidity, nights can be quite cold in dry desert climates, even if the days are extremely hot.

In addition to clouds, which you can see, and humidity, which you can feel, there is one other component of the atmosphere – a component you can neither see nor feel – that also traps infrared radiation. That component is carbon dioxide, which, like water vapor, is an invisible gas that prevents infrared radiation from escaping to space and that warms the Earth's surface. Carbon dioxide is present in tiny amounts in the atmosphere – currently at a level of 400 parts per million, which means that 400 out of every one million molecules of air are carbon dioxide molecules. Despite only being present in this minuscule amount, carbon dioxide is key to making the Earth habitable. Water vapor is present in larger amounts than carbon dioxide, but it still amounts to only a few percent of the total molecules in the air. Without any water vapor or carbon dioxide in the atmosphere, global temperatures would drop by about 25°C (45°F; see Box) from their present values.[3] Such a dramatic cooling would likely result in the entire Earth becoming covered in ice.

Temperature units: We mostly use the Celsius (°C) temperature scale, as is commonly done in science books. To approximately convert changes in temperature from °C to °F, the Fahrenheit scale, you can double the original number: A temperature change of 1°C corresponds to roughly 2°F (or 1.8°C, to be precise). But absolute temperature values are a bit more complicated to convert: 0°C corresponds to 32°F, or freezing, and 100°C corresponds to 212°F, the boiling point of water.

Distance units: For distances, we use kilometers (km): 1 km = 0.62 miles, or 1 mile = 1.6 km. For shorter distances, we use meters (m): 1 m = 3.3 feet, or 1 foot = 0.3 m

The effect of gases like water vapor and carbon dioxide that keep the Earth warm is called the *greenhouse effect.* The name itself arose from a misunderstanding about how actual greenhouses work. A greenhouse, typically used to grow tropical plants in colder climates, is essentially a glass hut. The glass allows sunlight to pass through and heat the interior, but it also prevents the heat from escaping, thus keeping the interior warm. Originally, it was thought that this was primarily because the glass blocked the infrared radiation from leaving the greenhouse. That is why the trapping effect of carbon dioxide and water vapor was labelled the greenhouse effect. Later studies of greenhouses showed that the main reason for their warm interiors was that the glass prevented convection, which is to say that it prevented the interior air from mixing with the exterior air. (Recall the car parked outside in the summer; lack of convection is why it becomes warm when the windows are closed as compared to when the windows are open.) Glass does actually block infrared radiation, but that turns out to be a secondary factor in keeping greenhouses (or car interiors) warm.

Even though it is a misnomer, the "greenhouse effect" has stuck, and everyone uses it. Gases like water vapor and carbon dioxide that trap infrared radiation are called *greenhouse gases.* The other greenhouse gases, such as methane and nitrous oxide, are present in even smaller concentrations than carbon dioxide, but they are more powerful on a per-molecule basis and make an important contribution to the overall greenhouse effect.[4]

Let us return to the cold winter night. You are home in bed and trying to fall asleep. Your body typically generates about 100 watts of heat, the equivalent of about six LED light bulbs. To maintain normal body temperature at equilibrium, your skin must transfer this energy away through convection and infrared radiation. If you lose more heat than you generate, you start to feel colder; if you don't lose enough heat, you start to feel warmer.

Suppose you haven't put on a blanket. Your skin feels cold because it loses heat to the colder surrounding air. You decide that you are uncomfortable, and you cover yourself with a blanket. The blanket impedes the transfer of heat away from your skin and sends some of it back, slowly warming your skin. The heat loss outside the blanket temporarily falls below 100 watts. Once your skin reaches a new warmer temperature, the amount of heat loss is back to 100 watts, to maintain the equilibrium. The difference is that there is a temperature gradient between your warmer skin and the cooler outer surface of the blanket. The outer surface of the blanket will have approximately the same cold temperature your skin had without the blanket on.

You still feel cold. You put on a second blanket. The second blanket impedes heat loss from the inner blanket and sends some of it back, again

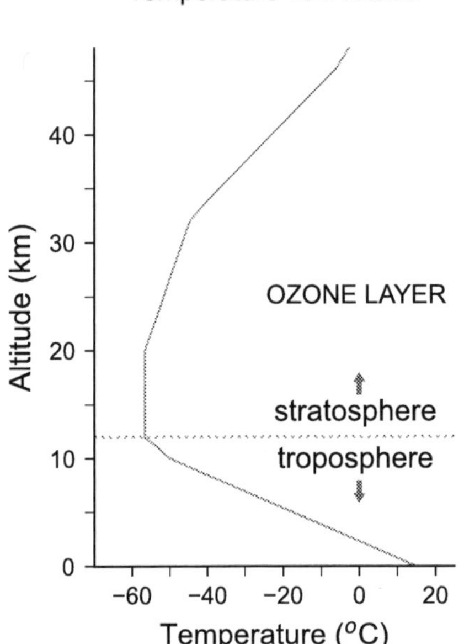

Figure 3.1 Variation of temperature with altitude in the atmosphere (US Standard Atmosphere [middle latitudes]).

temporarily reducing the heat loss below 100 watts. Your skin starts to warm again. After some time, the temperature of your skin settles into a new, warmer value. We can call the difference between the two-blanket skin temperature and the one-blanket skin temperature the *equilibrium response to doubling the number of blankets*.

At the new equilibrium, the outer surface of the second blanket will have approximately the same cold temperature your skin had without any blankets, and the heat loss will once again be back to 100 watts, as required by the law of conservation of energy. But there is a temperature gradient between your warm skin, the cooler exterior of the first blanket, and the cold exterior of the second blanket. We can refer to this temperature gradient between the skin and the outermost blanket as the "lapse rate," that is, the rate at which the temperature "lapses."

A similar effect happens in the atmosphere: Since it is heated from below, the air temperature decreases as the altitude increases in the lower part of the atmosphere, known as the *troposphere,* which extends to an altitude of about 10–16 km (Figure 3.1). (The region above the troposphere is known as the *stratosphere*, and it is heated from above, as discussed later in Section 6.1.) The temperature gradient between the warm surface of the Earth and the cool upper regions of the troposphere is known as the tropospheric *lapse rate*. It has an average value of 6.5°C per kilometer, meaning that for each kilometer gained in altitude, the air temperature drops by 6.5°C. At the top of Mt. Everest, which is 8.8 kilometers high, the air temperature will be 57°C colder than it would be at sea level.

In the case of the blankets, the temperature gradient between the skin and the outside is determined by how insulating each blanket is: The greater the insulation, the greater the warmth and the stronger the gradient. The tropospheric lapse rate is determined by a more complex balance between infrared radiation from the surface, infrared absorption by the greenhouse gases, and convection, which also transfers heat from the surface to the atmosphere.

The heating of your body is analogous to the heating of the Earth's surface by solar radiation. Even the numbers are fairly close. After accounting for sunlight reflected and absorbed by the atmosphere, the Earth's surface receives an average of about 160 watts of energy for every square meter of its area.[5] Increases in the concentration of carbon dioxide are like additional blankets on the planet, which trap heat to make the surface progressively warmer. An important metric in climate science is the *equilibrium response to carbon dioxide doubling*, which is what the change in the surface temperature would be if the concentration of carbon dioxide in the atmosphere were doubled instantaneously. Since this is a measure of how sensitive the planet is to increases in the concentration of carbon dioxide, it is known as *equilibrium climate sensitivity* or, commonly, as just *climate sensitivity*. This metric provides a simple way to compare estimates of the greenhouse effect from different sources without worrying about details like how long it will take for the doubling to occur and when the climate will reach equilibrium.

After you put on a new blanket, you reach a warmer equilibrium several minutes later. But the Earth would take several millennia to reach a warmer equilibrium, because the ocean has the capacity to absorb a lot of the excess heat (unlike the human body). This means that for our currently observed warming, we are nowhere near equilibrium. Also, carbon dioxide concentrations do not double instantaneously in the real world. So climate scientists consider a more realistic scenario of climate change, one in which the concentration of carbon dioxide in the atmosphere is slowly doubled over a period of about 70 years, rather than at once. The rate at which the Earth warms in this scenario, called the *transient climate response* (TCR), is another commonly used metric in climate science.

3.3 Eunice Foote and the Three Glass Tube Experiments

In the early nineteenth century, encouraged by contemporary advances in physics and chemistry, scientists turned their attention to climate. In 1822, French physicist Joseph Fourier explained how heat was transported through materials[6] such as the blankets considered in Section 3.2. To study heat transport, Fourier invented the famous trigonometric series that bears his name. A few years, later, Fourier was the first to articulate the energy balance maintaining the climate of the Earth, noting that the energy from space reaching the surface was balanced by infrared radiation back to space.[7] He also proposed the analogy between the infrared absorbing properties of the atmosphere and a glass enclosure, which would later become known as the *greenhouse effect.*

The next milestone in climate research features an unlikely intersection with the story of the women's rights movement in the United States.[8] It involves Eunice Newton Foote, a scientist and women's rights activist. Foote attended the 1848 Woman's Rights Convention in Seneca Falls, New York, where she became the fifth signatory to the Declaration of Sentiments (Figure 3.2). This declaration was a major milestone in the fight for women's right to equal treatment, including the right to vote. It was modeled after the US Declaration of Independence but included the modified phrase "that all men and women are created equal."

Foote was the first person to assert a link between carbon dioxide and the warmth of the Earth's atmosphere.[9] In the 1850s, she conducted three experiments using pairs of glass cylinders, each with a thermometer attached. In the first experiment, one cylinder had air removed from it, and the other had air pumped into it. When exposed to sunlight, the latter cylinder warmed more. In the second case, Foote compared a cylinder filled with dry air to another filled with moist air. The latter warmed more, leading her to conclude that moisture warms the atmosphere, consistent with our sensory experience. Finally, Foote compared a cylinder with plain air to another filled with carbon dioxide; the latter warmed more when exposed to sunlight. She concluded that an atmosphere of carbon dioxide would "give to our earth a high temperature."

Foote's paper briefly describing her findings was presented at the annual meeting of the American Association for the Advancement of Science in 1856. The paper was read not by Foote but by a male scientist, Joseph Henry, although Foote may have been in attendance, indicative of the unequal gender norms at the time.[10] The September 1856 issue of *Scientific American*[11] mentioned Foote's work in an article entitled "Scientific Ladies – Experiments with Condensed Gases," which began:

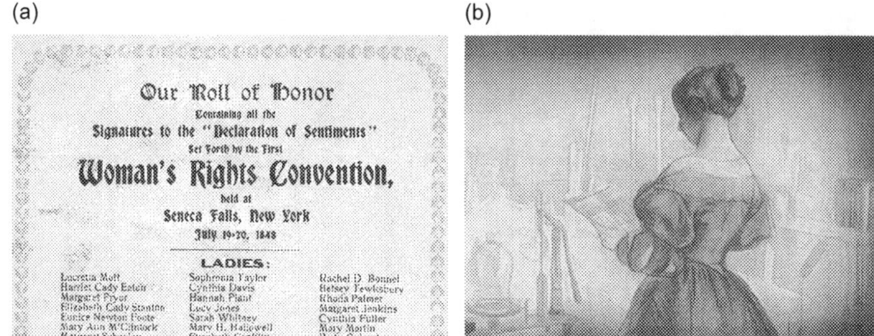

Our Roll of Honor
Containing all the
Signatures to the "Declaration of Sentiments"
Set Forth by the First
Woman's Rights Convention,
held at
Seneca Falls, New York
July 19-20, 1848

LADIES:

Lucretia Mott
Harriet Cady Eaton
Margaret Pryor
Elizabeth Cady Stanton
Eunice Newton Foote
Mary Ann M'Clintock
Margaret Schooley

Sophronia Taylor
Cynthia Davis
Hannah Plant
Lucy Jones
Sarah Whitney
Mary H. Hallowell
Elizabeth Conklin

Rachel D. Bonnel
Betsey Tewksbury
Rhoda Palmer
Margaret Jenkins
Cynthia Fuller
Mary Martin
P. A. Culvert

Figure 3.2 (a) Eunice Foote's signature in the Declaration of Sentiments at the first Women's Rights Convention in Seneca Falls, New York, July 19–20, 1848 (Image: US Library of Congress). (b) Eunice Foote's gas tube experiments. (Drawing by Carlyn Iverson, NOAA/Climate.gov; from Huddleston, 2019)

> Some have not only entertained, but expressed the mean idea that women do not possess the strength of mind necessary for scientific investigation. Owing to the nature of woman's duties, few of them have had the leisure or the opportunities to pursue science experimentally, but those of them who have had the taste and the opportunity to do so, have shown as much power and ability to investigate and observe correctly as men.

In her brief paper, Foote even speculated that increased concentrations of carbon dioxide in the distant past could explain the warm climates that the Earth had experienced millions of years ago.[10] However, Foote was limited by the equipment to which she had access, and her experiments were unable to distinguish between solar heating and infrared heating. The heating she observed could not, therefore, be attributed to the greenhouse effect. In 1859, the Irish physicist John Tyndall carried out laboratory measurements that identified water vapor and carbon dioxide as the most significant absorbers of infrared radiation,[9] acting as the "blankets" responsible for the greenhouse effect. Tyndall's work on this topic has long been recognized. Foote's prior research on the role of carbon dioxide in climate was carried out when women were not allowed to achieve their full potential in fields of science; it was overlooked for a century and a half, and was rediscovered only in 2011.[12]

3.4 Arrhenius and the First Climate Model

In the first half of the nineteenth century, geologists discovered there were periods in Earth's history where large continental areas had been covered by massive ice sheets. These periods, when the Earth's climate was much colder, are referred to as *glacial periods* or *ice ages*. Periods with warm climates, like the one we are currently in, are known as *interglacials*. By the second half of the nineteenth century, explaining the origin of ice ages became a grand challenge for scientists around the world. One of the scientists to propose an explanation was Swedish chemist Svante Arrhenius.[13] He reasoned that the greenhouse effect could offer an explanation: If we could somehow lower the amount of carbon dioxide in the atmosphere, this lowering could cool the Earth's surface and lead to an ice age.[14]

Arrhenius decided to construct a simple model of the atmosphere, conceiving of it as a slab of air with constant temperature. But he still needed a way to calculate the greenhouse effect. In 1890, an American astronomer, Samuel Langley, had measured moonlight's absorption by the atmosphere at different wavelengths, using a prism and an instrument called a bolometer. Arrhenius used Langley's data to estimate the infrared absorption properties of water vapor and carbon dioxide. After tens of thousands of hand calculations, he was able to approximate how varying carbon dioxide would change surface temperature at different latitudes for different seasons. In 1896, he published a paper showing that reducing the concentration of carbon dioxide to half its present value could explain the temperature changes associated with the ice age. To explain warm periods in Earth's history that occurred millions of years ago, he also estimated that doubling the concentration of carbon dioxide would increase the global temperature by about 5.5°C. This was the very first scientific estimate of Earth's climate sensitivity, that is, the temperature change associated with doubling of carbon dioxide. Arrhenius had constructed the first quantitative model of the Earth's climate.

Arrhenius' paper is for this reason considered a milestone in the history of climate prediction. But his main goal was to try to explain naturally occurring (not human-induced) cold and warm periods in Earth's past (not Earth's future). The paper discussed the release of carbon dioxide from human coal consumption but dismissed it as a small effect. Taking into account the absorption of carbon dioxide by the ocean, Arrhenius estimated that it would take 3,000 years for the burning of coal to increase the carbon dioxide concentration in the atmosphere by 50 percent and warm the Earth.[15] A resident of chilly Scandinavia, he thought that such warmth might actually be beneficial to mankind, writing that "[b]y the influence of the increasing percentage of carbonic acid [carbon dioxide] in the atmosphere, we may hope to enjoy ages with more equable and better climate."[16]

Arrhenius used a value of 300 parts-per-million (ppm) for the carbon dioxide concentration in 1896. In 2016, the measured carbon dioxide concentration exceeded 400 ppm, representing a 33 percent increase in 120 years, and on course for a 50 percent increase within a few more decades. Arrhenius's estimate of the impact of humans burning coal was off by a factor of 16! His estimate of Earth's climate sensitivity – the warming due to doubling of carbon dioxide concentrations – was closer to the mark: Modern studies suggest a value for climate sensitivity of around 3°C, not far from his estimate of 5.5°C.

Arrhenius was lucky to get close to the correct answer.[17] His estimate of infrared absorption, which he based on Langley's moonlight study, had an error that would have resulted in a much higher estimate of climate sensitivity. However, his neglect of the vertical dimension of temperature variation would have led to a considerably lower estimate of climate sensitivity. The errors happened to offset each other, resulting in an answer in the right ballpark. This is not to detract from his accomplishment: He made reasonable assumptions, used the best data that he could lay his hands on, and carried out as many calculations as humanly possible without a computer. But the fortuitous cancellation of errors in the first climate model also marks the first occurrence of *compensating errors* in climate modeling – a problem that persists to this day.

Arrhenius' estimate of climate sensitivity was a scientific prediction, but his estimate of human-induced carbon dioxide increase was more of an economic prediction. As we shall see, uncertainties in the latter are typically far greater than those in the former. The overall prediction for the actual change in climate is a combination of scientific and economic predictions. Improving the accuracy of the scientific prediction alone is not sufficient to improve the accuracy of the overall prediction.

3.5 Climate Feedbacks and Forcings

Until now, we have focused on how increases in the concentration of carbon dioxide in the atmosphere can lead to a warmer planet. But water vapor actually contributes much more to the greenhouse effect than carbon dioxide. If that is the case, why don't we talk about variations in the concentration of water vapor as causing a warmer or cooler Earth? Well, we don't directly control the amount of water vapor in the atmosphere. It is regulated by the planet's surface temperature, which, as we have seen, determines the atmospheric temperature. The Clausius-Clapeyron relation specifies the maximum amount of moisture allowed in the air for each value of temperature. Warmer air can hold more moisture, and cooler air can hold less moisture. The ratio of the air's actual moisture content to its maximum possible moisture content is called *relative humidity*. This ratio, which can range from zero to one, is usually expressed as a percentage. A relative humidity of 100 percent means that the air holds as much moisture as it can possibly hold at that temperature.

As Earth's climate warms or cools, the relative humidity of the atmosphere tends to stay roughly the same – even though the total amount of moisture, known as absolute humidity, can vary quite a bit.[18] The ocean is essentially an "infinite" reservoir of water. Any extra moisture that we "emit" into the atmosphere, say by extracting groundwater and evaporating it, is a drop in the bucket compared to the total amount of water in the ocean. The atmospheric temperature determines how much sea water evaporates and remains in the atmosphere. Once sufficient water vapor evaporates to make the relative humidity 100 percent, the atmosphere becomes saturated, and evaporation will cease.

There is a corresponding reservoir of carbon dioxide in the ocean, as well as in the rocks on Earth's surface. Carbon dioxide in the atmosphere can dissolve in water to form carbonic acid, which reacts with rocks and changes their composition. This process, called *weathering*, allows atmospheric carbon dioxide to be stored in rocks.[19] Weathering happens slowly, over many millions of years. Carbon dioxide absorption by the ocean is somewhat faster, occurring over many centuries rather than many millions of years. However, the exchange of water vapor between the atmosphere and the ocean is almost instantaneous, occurring through evaporation and rainfall.

In discussions of global warming, we talk about the climate changing over time periods of many decades. Over this timescale, the concentration of carbon dioxide is affected by human activities, but the concentration of water vapor is regulated by the Earth's surface temperature. This does not mean that water vapor is unimportant. In fact, it plays a very important role in amplifying the greenhouse effect of carbon dioxide changes through a process called *climate feedback*.

The best example of a feedback in everyday life is perhaps a public address audio system – an amplifier hooked up to a microphone and a speaker. Any sound picked up by the microphone will be amplified by the electronics and broadcast by the speaker. If the speaker is too close to the microphone, the amplified sound will be picked up again by the microphone and then amplified again, repeatedly. This will overload the amplifier and the speaker, resulting in a loud screeching sound. Usually, moving the speaker away from the microphone will break this "feedback loop" and fix the problem. A similar situation can arise when two people on the same conference call are seated close enough for each of their phones or computers to hear the other's phone or computer.

An analogous feedback occurs in the climate system. If the concentration of carbon dioxide in the atmosphere increases slightly, the greenhouse effect will strengthen and cause atmospheric temperatures to warm slightly. Relative humidity tends to remain constant; so the warmer the atmosphere gets, the more water will evaporate to maintain relative humidity, causing the amount of water vapor in the atmosphere to increase. Since water vapor is itself a powerful greenhouse gas, this will compound the greenhouse effect and cause additional warming. The additional warming will lead to further evaporation, and so on. The net result will be a warming of the atmosphere much greater than that due to the carbon dioxide effect alone. In climate science, this is referred to as the *water vapor feedback.*

Like a megaphone amplifying your voice, the water vapor feedback amplifies the warming due to increased carbon dioxide, making it much stronger. To distinguish it from a climate feedback, the increase or decrease in carbon dioxide that drives temperature changes is referred to as a *climate forcing*. There are also other types of climate forcings – those arising from volcanic eruptions, changes in the sun's energy output, variations in the Earth's orbit, and so on.[20] Returning to the audio analogy, the sound that you speak into a megaphone would be the forcing. The forcing precedes the feedback, or alternatively, the feedback lags the forcing.

Although carbon dioxide currently acts as a forcing for global warming, that wasn't always the case in past climates.[21] At the end of the ice ages, the data show that there was an increase in carbon dioxide concentrations, but it wasn't a climate forcing, because it *lagged* the warming of temperature by centuries. The climate forcings that drove the ice age cycles were the slow variations in the Earth's orbit that occur on timescales of many tens of thousands of years. These cause the polar regions to receive more summer sunlight at the end of an ice age. The resulting warmer temperatures release the carbon dioxide stored in the reservoirs of land and ocean, further amplifying the warming. The greenhouse effect of carbon dioxide therefore acted as a feedback during the ice age cycles, not as a forcing.

There is one other important feedback that plays an important role in climate. Recall the albedo, a measure of the reflective property of a surface to incoming solar radiation. Snow and ice tend to reflect solar radiation; they have high albedos. Land and water, on the other hand, tend to absorb solar radiation; they have lower albedos. Suppose that the Earth's tilt changed slightly, so as to decrease the amount of solar radiation directed at the polar regions. This would cool the Earth's polar climate and increase the amount of snow and ice in the polar regions, which would lead the polar regions to reflect more sunlight, cooling the Earth further. The net effect would be a much stronger cooling than could be explained by orbital tilt changes alone. This is referred to as the *ice-albedo feedback.*

The two feedbacks discussed so far are positive or *amplifying feedbacks* – the feedback amplifies the original perturbation, which can destabilize the system. However, there are also many negative or *stabilizing feedbacks*, in which the feedback weakens the original perturbation. Had Goldilocks not entered the bears' house, all three bowls of porridge would have been at the same temperature when the bears returned from their walk: Hot porridge loses heat and cools to room temperature. This heat exchange with the air is a stabilizing feedback that dampens the initial temperature perturbation. Similarly, if the amount of sunlight reaching the Earth increases, the planet will get warmer, and, as the planet warms, it will emit more and more infrared radiation. This will mitigate the impact of increased solar heating.

Because they can lead to rapid changes in a system, amplifying feedbacks receive a great deal of attention in climate research. Stabilizing feedbacks are less glamorous but far more ubiquitous – they are crucial to the Goldilocks effect, keeping the temperature "just right" for us. By definition, for any system in a stable equilibrium, the stabilizing feedbacks must dominate the amplifying feedbacks to keep the system stable and resistant to small perturbations.

3.6 Too Hot, Too Cold, and Just Right

The three terrestrial planets in the solar system with atmospheres are Venus, Earth, and Mars. They could all have potentially been habitable, but only Earth is actually habitable. We have the greenhouse effect to thank for that. Without it, Earth would probably be covered in ice.

Venus is about the same size as the Earth and, like Earth, has a thick atmosphere. But it has a very different climate; at 470°C, its surface is hot enough to melt lead! Why is Venus so much hotter than Earth? It is believed that Venus has a *runaway greenhouse effect.* Venus receives considerably more solar heating than the Earth, being closer to the sun. Due to the extra heating, the greenhouse effect became so strong that the water vapor feedback greatly amplified the warming and evaporated all of Venus's water.

Venus is sometimes held up as a cautionary tale of what could go wrong if a similarly uncontrolled amplifying feedback were to occur in the Earth's climate system. However, no climate models project this kind of feedback for the next century. There is a theoretical minimum amount of sunlight required for the runaway greenhouse effect, known as the *Komabayashi-Ingersoll limit.*

This limit applies to planets where there is a large reservoir of a condensable greenhouse gas (that is, water). This threshold is about 10 percent higher than the amount of sunlight currently received by the Earth, so there is no need to worry about a runaway greenhouse effect at this time. However, stellar models predict that as the sun burns up more of its hydrogen fuel through nuclear fusion, it will need to burn hotter to maintain its gravitational balance, and it will increase its energy output, or luminosity, accordingly. About a billion years from now, the amount of sunlight reaching the Earth will exceed the Komabayashi-Ingersoll limit, and a runaway greenhouse effect may become inevitable.

What about Mars? It has an atmosphere, but the mass of the atmosphere is only about 1/100 of Earth's atmosphere. This means the atmosphere is too thin to provide a strong greenhouse effect, even though it is almost all made up of carbon dioxide. There is also no water vapor in the Martian atmosphere to amplify the greenhouse effect, although there may be water underground. Mars also receives much less sunlight because it is further away from the sun. The combination of these factors results in a very cold planetary surface, inhospitable to life.

Since it is possible to land probes on Mars, and since it will soon be possible to land humans on Mars, there is talk of making Mars more Earth-like by triggering a greenhouse effect to warm the atmosphere. But it is not clear that there is enough greenhouse gas material stored in either solid or liquid form to create an atmosphere thick enough to sustain a strong greenhouse effect,[22] even if enough energy could be found to release that material. There is no such talk of terraforming Venus, because it is so inhospitable even to metallic probes. It seems that there may indeed only be one planet in the solar system that is "just right" for humans.

***

Although the work of scientists like Fourier and Tyndall was driven by the limited data about Earth's climate available at the time, it was more of a deduction about how climate *should* work, rather than an induction about how climate *did* work. Granted, Arrhenius was trying to explain the then-recently acquired data about past climate change, but his statements about the future of climate were deductive. At the start of the twentieth century, there was no data available to motivate the idea that the climate was warming, or that the concentration of carbon dioxide in the atmosphere was increasing rapidly. As such, Arrhenius' work on the greenhouse effect did not receive due recognition until many decades later – although he won a Nobel prize for his other research in chemistry.

Arrhenius's first attempt at numerical climate modeling used tedious hand calculations, as did Lewis Fry Richardson's first attempt at numerical weather modeling in 1922. Although the former's errors canceled out and the latter's did not, they were both pioneers who used radically new, physics-based approaches to solve their respective problems. Nevertheless, in narratives of global warming, Arrhenius is frequently hailed as a genius partly because he got Earth's climate sensitivity "just right" in 1896. There is a reluctance to attribute success to chance, but Arrhenius's "lucky" estimate of climate sensitivity was perhaps just that – lucky. We should remember that a correct prediction does not prove that the underlying model is also correct. Compensating errors can lead to the right answer for the wrong reasons. Recall our discussion of the white swans and the black swan: Theories and models can be falsified, but they cannot be proven true.

In their research, Arrhenius and Richardson both pushed the limits of human calculating ability. They had to carry out tens of thousands of calculations by hand to solve even the simple equations of their models. During the second half of the twentieth century, the twin streams of climate research and weather research found a confluence in a common tool, the computer, that could be used to solve more complex equations. In Chapter 4, we describe how computer models of the atmosphere developed for weather forecasting were adapted for use in climate prediction.

4

Deducing Climate

Smagorinsky's Laboratory

For the next chapter in the story of climate prediction, we return to Princeton University. Driving southeast out of the main campus, we reach Route 1, the north–south US highway that runs all the way from Florida to Maine. After traveling about three miles northeast along this road, we reach the Forrestal Campus of Princeton University. This sprawling campus was created to host government-sponsored research labs affiliated with the university.

Near the entrance to Forrestal Campus, we see a nondescript, rectangular, two-story office building, vaguely reminiscent of a greenhouse (Figure 4.1a). Within this building is a US government laboratory called the Geophysical Fluid Dynamics Laboratory, or GFDL, which carries out research on fluid dynamics in the atmosphere and the ocean, studying the motions of air and water that control weather and climate. There is a bunker-like annex behind the main GFDL building, which until recently housed state-of-the-art computers used for weather and climate modeling. As computers became more and more power-hungry, they were moved offsite. Now GFDL's computers are hosted remotely, at Oak Ridge National Laboratory (ORNL) in Tennessee.

The Geophysical Fluid Dynamics Laboratory originated with a young scientist named Joseph Smagorinsky (Figure 4.1b), formerly of the Princeton Meteorology Project. Smagorinsky, or "Smag" as he was known to his friends, was a forceful and influential figure in meteorology during the second half of the twentieth century.[1] The lab that he founded and nurtured – sometimes known as Smag's Laboratory – would play a pioneering role in weather and climate prediction.

4.1 General Circulation of the Atmosphere

Joseph Smagorinsky grew up in New York City, in a family of Belarusian refugees,[2] and worked in his father's paint store. His dream was to become a

Figure 4.1 (a) Geophysical Fluid Dynamics Laboratory (GFDL) in Princeton's Forrestal Campus. (b) Kirk Bryan (left) and Suki Manabe, with GFDL Director Joseph Smagorinsky (right) in 1969. (Photos: NOAA/GFDL)

naval architect, but he failed to gain admission to the prestigious Webb Institute near his home. Smagorinsky settled for his second choice of career, meteorology, and began attending New York University (NYU) in 1941. When the United States entered the Second World War, Smagorinsky joined the Air Force and was selected for its meteorology training program.

After the war, Smagorinsky completed his master's degree in dynamic meteorology at NYU and joined the US Weather Bureau as an assistant to Harry Wexler, Chief of the Special Scientific Services Division.[3] Wexler was involved with the Princeton Meteorology Group, and, in 1950, Smagorinsky himself joined the group. He worked as a scientific programmer, while simultaneously carrying out his doctoral research in atmospheric weather patterns at NYU.

Smagorinsky returned to the US Weather Bureau in 1953, after receiving his Ph.D. from NYU. There he continued to work on computational weather forecasting. In 1955, another member of the Princeton Meteorology Group, Norman Phillips, carried out an innovative calculation using the IAS computer. Phillips used Charney's simplified three-dimensional equations for the atmosphere, with just two grid levels in the vertical direction, to carry out a 30-day forecast, well beyond the weather prediction horizon. For the first time, the boundary between weather and climate prediction had been crossed.

Phillips's calculation immediately caught the attention of von Neumann. It was a step toward the infinite forecast he had envisioned, which fitted into his grand vision of modifying weather and climate through better understanding of the atmosphere. The infinite forecast would describe the long-term average of atmospheric flow patterns, referred to as the *general circulation of the atmosphere*. For this reason, the earliest atmospheric models were known as *general circulation models* or GCMs. (Today, GCM is more commonly used to denote *global climate model.*)

In August 1955, von Neumann proposed the establishment of a research unit to study the dynamics of the general circulation using state-of-the-art computers.[4] He requested $262,000 per year, or about $2.5 million in today's money. The United States government signed off within a month. The General Circulation Research Section (GCRS) was established soon afterwards in Suitland, Maryland, as a joint Weather Bureau–Air Force–Navy venture, with Joe Smagorinsky as the director.

In 1959, the GCRS unit changed its name to the General Circulation Research Laboratory and moved to Washington, DC.[1] It acquired its current moniker in 1963, and, five years later, it moved to Princeton University's Forrestal Campus, where it continues to operate today. Through the Department of Geosciences, scientists at GFDL were provided with affiliate appointments at Princeton University, enabling them to teach courses and advise graduate students. The management structure of GFDL, though, is quite different from that of a university department; it operates more like the IAS, where Smagorinsky had worked in the early 1950s. The emphasis is more on long-term research than on the short-term publication of papers.

Princeton Meteorology Group was organized around the ENIAC and then around the IAS computer. Von Neumann wanted Smagorinsky's lab to likewise have as its centerpiece the most advanced computer available. After the success of the ENIAC, private companies had entered the business of building computers. One of them was the Remington Rand corporation, which had acquired a smaller computer company started by Mauchly and Eckert, the designers of ENIAC. In 1951, it released a computer called the UNIVAC, for "universal automatic computer." Around the same time, a competing corporation, International Business Machines (IBM), was creating its own computer, having hired von Neumann as a consultant.[5] This computer, named the IBM 701, followed the design of the IAS computer.

The UNIVAC and the IBM 701 were the first computers that were sold commercially; they were not just one-of-a-kind research computers. Both were large and were built using vacuum tubes, like the ENIAC. Although they were not much faster than their predecessor, they were more reliable and had far more memory for data and programs. These larger computers were referred to as "mainframes," to distinguish them from the smaller computers that followed. Later, the most advanced computers of the day would become known as "supercomputers," as mainframes became commoditized for business uses.

GFDL started out with access to an existing IBM 701 in 1955.[6] Smagorinsky rapidly assembled a team of scientists and programmers. The scientists came up with ideas and algorithms, and the programmers implemented these in code. As the team began to build a comprehensive three-dimensional model of the atmosphere to simulate weather and climate, Smagorinsky recognized the need for additional expertise, beyond fluid dynamics. He was looking for a scientist with a strong background in atmospheric physics to work with him in his new lab.

4.2 From Tokyo with Meteorology

Akira Kasahara was looking for a job. While Smagorinsky was building computer models of weather in Princeton, Kasahara was a graduate student working under Shigekata Syono, an eminent professor of meteorology at the University of Tokyo.[7] Professor Syono was an excellent researcher, but he did not provide many employment opportunities for his students. In 1952, a scientist from Syono's group, Kanzaburo Gambo, left to work with Jule Charney in Princeton, and Kasahara was able to obtain Gambo's former research position. Gambo wrote excitedly from Princeton about how computers were being used to predict weather. Kasahara and others in his research group closely followed these developments, which were revolutionizing their field. In 1954, Gambo returned from Princeton, and Kasahara found himself out of work.[8]

Kasahara wrote to one of the members of the Princeton Meteorology Group, John Freeman, who had recently moved to Texas A&M University in College Station, Texas, and was in search of a research assistant. Freeman hired Kasahara, and, in 1954, Kasahara took a cargo ship to the United States. Kasahara was among the first of many Japanese scientists who emigrated to the United States in the postwar period and made lasting contributions to weather and climate science.

In 1956, Kasahara took a new position at the University of Chicago, working on hurricane prediction with George Platzman, another former member of the Princeton Meteorology Group.[9] Kasahara needed a computer for his research, but the University of Chicago did not have one at the time, so Kasahara had to travel regularly to locations with computers. Some of the computers he used, such as those at national laboratories, were sensitive installations operated by the US Atomic Energy Commission. Since Kasahara was not a US citizen, and these were restricted facilities, he was not allowed to enter the building with the computer. So, he would leave the deck of punched cards containing his computer program at the guard's office. The deck would then be sent to the computer operators, who would run the program and send the output back to the guard's office. Kasahara would come back later to pick it up.

One of the computers that Kasahara was able to use more easily, through Platzman's Princeton connections, was the IBM 701 at GFDL. During one of Kasahara's frequent visits, Smagorinsky mentioned to him that he was looking for a scientist with a strong background in atmospheric physics to help him build a climate model.[10] Kasahara suggested Syukuro Manabe, a colleague from Professor Syono's group at the University of Tokyo. Impressed with Manabe's work on rainfall prediction, Smagorinsky hired him in 1958.

We will encounter Kasahara once more in a later chapter: He changed jobs again in 1963, at the invitation of another former member of the Princeton Meteorology Group.

4.3 Three-Dimensional Climate Modeling

Syukuro "Suki" Manabe was born in rural Japan, the son and grandson of medical doctors (Figure 4.1b).[11] When he went to the University of Tokyo, it was assumed that he would follow in their footsteps. But Manabe despised biology and soon switched fields to physics. Figuring that he was not smart enough to be a theoretical physicist, and not handy enough to be an experimental physicist, he decided to study geophysics. After obtaining his bachelor's degree, he couldn't find a job. He went on to complete a master's degree and then a Ph.D. in Professor Syono's group. There still were no well-paying jobs.

At the time, Manabe was working 18-hour days doing rainfall prediction with manual calculations. When Smagorinsky offered Manabe the chance to work on computer modeling in the United States, Manabe immediately accepted it.[12] In the fall of 1958, he moved to Washington, DC, where GFDL was then located. Smagorinsky also ended up hiring two of Manabe's colleagues from Syono's group, Kikuro Miyakoda and Yoshio Kurihara, who would go on to become pioneers in extended weather forecasting and hurricane modeling, respectively.

Manabe's primary task was to work with a team of meteorologists and programmers to build a three-dimensional climate model. Initially, Manabe worked under Smagorinsky's direction, improving various components of the model. As Smagorinsky became progressively more involved over the years with administrative activities and international projects, Manabe took charge of the team.[13]

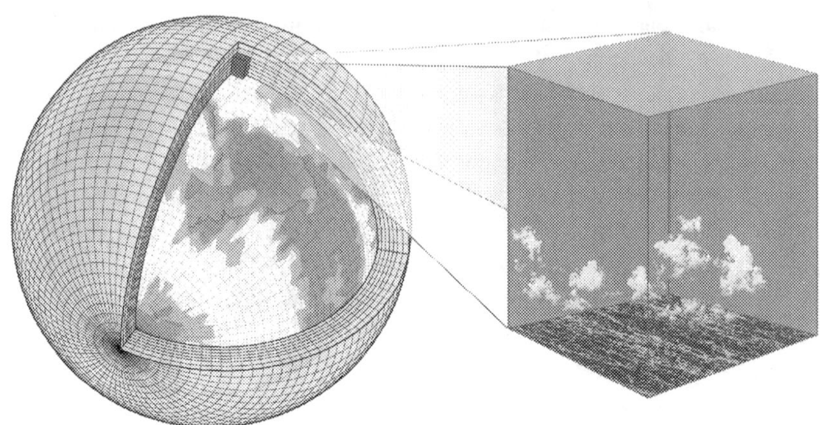

Figure 4.2 Schematic illustration of the three-dimensional spherical grid of a global climate model. The blow-up on the right shows a single column of air over a horizontal grid cell, which is too coarse to capture the details of fine-scale clouds inside the column. (From Schneider et al., 2017a; reprinted by permission from Springer/Nature Climate Change, © 2017)

The process of constructing three-dimensional weather and climate models starts from the Navier-Stokes equations that govern air motions (Section 1.3), along with thermodynamic equations for air and water vapor.[14] The Earth's atmosphere is a spherical shell (Figure 4.2). To solve the equations on the computer, the spherical shell is first divided into a two-dimensional grid of cells covering the entire globe in the horizontal (latitude–longitude) dimensions. Then the column of air over each horizontal grid cell is further subdivided into a grid of levels in the vertical (or altitude) dimension, extending from the ground to the stratosphere, to form a three-dimensional grid spanning the entire globe. The horizontal grid cell is typically about 10×10 km for weather models and about 100×100 km for climate models. The number of levels in the vertical is about 50. This means that a typical climate model has about three million elements in its three-dimensional grid, and weather models can have a hundred times more grid elements. At each grid element, air temperature, humidity, and winds are represented by numbers that change over time, as determined by the equations.

The size of each horizontal grid cell is too coarse, even in weather models, to capture the details of individual clouds and other fine-scale processes (Figure 4.2). Instead, approximate mathematical formulas known as *parameterizations* are used to represent clouds and other small-scale processes in the computer model. The equations governing the large-scale air motions represented in the grid, and the parameterization formulas for fine-scale processes not captured by the grid, are solved using a powerful computer, starting from an initial condition and marching forward one time step at a time (Section 1.3).

Although weather and climate prediction share the same computational framework, there are additional challenges when it comes to climate prediction, because many more processes need to be considered. The early weather models used by Charney and von Neumann were focused on simulating the movement of air, a process known as fluid dynamics. But air is warmed by infrared radiation from the Earth's surface, not directly from the sun (Section 3.1). Water vapor also heats and cools. When water evaporates from the surface, it cools the surface. When water vapor condenses to form clouds, rain, or snow, it heats the air. Like infrared radiation, water vapor also transfers heat from the surface to the air. Radiation and rainfall are two important processes that need to be represented well in climate models.

In 1965, Manabe and Smagorinsky published papers[15] that described their success in building one of the earliest three-dimensional climate models.[16] The model solved the fluid equations in a hemispherical shell, meaning that only the Northern Hemisphere was represented. The horizontal grid cell size was 500 × 500 km, and there were nine levels in the vertical. Even though it had no geography – it was all land and no ocean – the model managed to capture the basic properties of weather and climate. But this state-of-the-art three-dimensional model was complex and computationally expensive, making it hard to tweak it in different ways to understand its behavior.

4.4 Philosophy Break: Reductionism and Emergentism

Little drops of water,
Little grains of sand,
Make the mighty ocean
And the pleasant land.

From the nineteenth-century poem by Julia Carney

When scientifically analyzing a complex system, we might find it simpler to reduce the system to a collection of smaller interacting subsystems and study each of these subsystems separately. This approach to science is known as *reductionism*. An early practitioner of this approach was the Greek philosopher Democritus, who posited in the fifth century BCE that the physical world was composed of indivisible particles called atoms. In the seventeenth century, the French philosopher René Descartes,[17] the namesake of Cartesian geometry, introduced the related concept of *mechanism*. Descartes argued that the world was like a machine, driven by parts that acted as a clockwork mechanism.

One can think of reductionism as a repeated wielding of the analytic knife. Pirsig's Phaedrus used the analytic knife to "split the whole world into parts of his own choosing, split the parts and split the fragments of the parts, finer and finer and finer until he had reduced it to what he wanted it to be."[18] The idea is that once the parts of the whole are small enough, they are easy to study and understand.

The opposite of reductionism is *emergentism*. Emergentism says that new properties in a complex system emerge through interactions between other properties. An early proponent of emergentism was the nineteenth-century British philosopher John Stuart Mill, who used properties of chemical compounds as an example. The related concept of *holism* literally means that the whole is more than the mere sum of its parts. In other words, an emergent or holistic complex system exhibits behavior that cannot be understood by simply studying each of its parts.

The distinction between reductionism and emergentism, both of which deal with the physical world, is somewhat orthogonal to the distinction between deductivism and inductivism, which are both concerned with scientific thought. Typically, reductionism is associated with deductivism, and inductivism with emergentism, but other combinations are possible. Among the sciences, physics is more reductionist: Matter is composed of molecules, molecules are composed of atoms, and so on. Biology, on the other hand, is more emergent; the behavior of organs like the brain or the heart cannot be explained by simply studying the properties of individual cells and molecules.

Reductionism forms the basis for the construction of computer models of weather and climate. A computer model is a giant program that is made up of smaller programs, each of which deals with different components – such as atmosphere, ocean, and land – as well as different processes that are too small to be represented on the model grid – such as rain, clouds, and turbulence.

To use the Laplace's Demon analogy, the Weather Demon can be described as working with lesser demons such as the Air Demon, the Radiation Demon, and the Cloud Demon that interact with each other. Each lesser demon makes a prediction for its domain, and the Weather Demon combines them to make an overall prediction. A reductionist approach would study each of the lesser demons separately, under the assumption that the Weather Demon simply adds up the individual predictions. An emergent approach, on the other hand, would argue that the isolated behaviors of the lesser demons are less important to the overall prediction than are the interactions between these lesser demons.

Weather and climate prediction therefore depend upon the emergent behavior of the atmosphere simulated by models.[19] The models can predict complex phenomena like hurricanes and droughts, even though such phenomena are not explicitly represented in the individual model components. The collective interaction of different components such as air circulation, radiation, and clouds is responsible for generating these phenomena.

4.5 Models of Simplicity in the Garden of Complexity

Big models have little models
To delegate their deductivity
And little models have lesser models
And so on to reductivity

variation of Lewis Fry Richardson's 1922 poem,
itself a variation of Augustus De Morgan's *Siphonoptera* (1872)
derived from Jonathan Swift's satirical *On Poetry: A Rhapsody* (1733)

In the 1960s GFDL was a hive of research activity in weather and climate modeling. It welcomed a number of international visitors. One of these visitors was a German scientist named Fritz Möller, an expert on radiation.[20] Möller, like Arrhenius before him, was interested in the greenhouse effect that kept the Earth warm. In 1938, a British engineer named Guy Callendar claimed that the Earth had warmed in the previous 50 years and that the concentration of carbon dioxide in the atmosphere had increased in the same period – and that the former was attributable to the latter.[21] This was the first time such a link had been made based on data; Arrhenius's earlier work merely raised this link as a possibility. Callendar estimated climate sensitivity, long-term warming due to a doubling of the carbon dioxide concentration, to be about 2°C, which appeared to be consistent with the data. His analysis, although groundbreaking, was flawed; it only considered the role of radiation in heating the atmosphere, not the role of water vapor.

Möller revisited the greenhouse effect of carbon dioxide in 1963, using a much more accurate calculation of radiation. He found that estimates of climate sensitivity depended upon how water vapor was handled in the calculation. Manabe, who was working with Möller, was strongly influenced by this result.[22] At the time, Manabe was helping Smagorinsky improve the radiation formulas in the GFDL's comprehensive climate model. But tweaking the radiation formulas and running the full climate model each time to test it would be very expensive computationally – not to mention very slow because of the wait time between model runs.

What Manabe did, which is a common approach in climate science, was to narrowly focus on the vertical transfer of heat, reducing the complexity of the problem. He constructed a much simpler model of the atmosphere – a one-dimensional climate model that only considered the altitude dimension, ignoring the latitude and longitude dimensions. At first glance, this appears to be a giant leap of faith, approximating the spherical shell of the atmosphere as a single column (Figure 4.2). But it is not as drastic as it sounds. Temperature decreases with altitude at each location on the globe, and local lapse rates, the rates at which temperature falls off with height, do not diverge much from the global average value of 6.5°C/km (Figure 3.1). Mathematically, this means that there is a certain amount of symmetry along latitude and longitude, allowing those dimensions to be averaged out, leaving only the altitude dimension.

Manabe's one-dimensional model, known as the *radiative–convective model,* represents both radiation and moist convection, the turbulent transport of water vapor that causes rain. The effects of radiation were captured through tedious but doable computations using the basic equations of electromagnetic radiation. The model computed solar radiation passing through the atmosphere and reaching the surface, and infrared radiation emitted from the surface passing through the atmosphere back to space. It also calculated the effects of the two main greenhouse gases, water vapor and carbon dioxide, which impede the passage of infrared radiation.

Representing convection, though, was much trickier. Convection is a chaotic and turbulent process, especially when moisture and clouds are included in the mix. During convection, rising air cools due to an effect called *adiabatic expansion* (Section 3.1). If the air contains moisture and its temperature cools below a threshold temperature known as the dew point, the moisture will condense to form cloud droplets and rain. Since this happens at very fine spatial scales – of a few hundred meters – it cannot be captured by the coarse grid of a climate model.

Manabe came up with a clever shortcut to represent rain due to convection. The one-dimensional model followed a parcel of warm, moist air rising upward from the surface as it cooled due to adiabatic expansion. When the temperature of the parcel cooled below the dew point, it could no longer hold all its moisture. The excess moisture would then be condensed and fall as rain. The air parcel's temperature would be continuously readjusted to the dew point as it moved upward, until all its moisture was rained out.[23] This procedure, known as *moist convective adjustment*, is one of the earliest examples of a parameterization – a formula or algorithm used to represent small-scale processes like clouds that are too small to be captured by the spatial grid of the model. Manabe first tested it in his one-dimensional model before including it in the full climate model.

With this convective adjustment parameterization in place, Manabe had a simple, but complete, one-dimensional climate model that could be solved rapidly using a computer. With his colleagues Robert Strickler and Richard Wetherald at GFDL, he published papers in 1964 and 1967 using this model. The latter paper, titled "Thermal Equilibrium of the Atmosphere with a Given Distribution of Relative Humidity," is considered one of the most influential papers in climate science.

Manabe's model is a classic example of the power of the reductionist approach. The model encapsulated the essence of the problem that had previously been studied by Arrhenius, Callendar, and Möller. Building on this prior work, Manabe was able to fix many of its deficiencies. He used the one-dimensional model to study the global warming associated with carbon dioxide, including the amplifying effect of the water vapor feedback. He estimated climate sensitivity to be about 2.3°C, not far from today's estimate of about 3°C.

Manabe's model also showed that, while increasing carbon dioxide warms the lower part of the atmosphere, the troposphere, it actually cools the upper part of the atmosphere, the stratosphere (Figure 3.1).[24] This happens because carbon dioxide is both a good absorber and a good emitter of infrared radiation. In the lower atmosphere, the absorption of strong infrared radiation from the surface plays a bigger role and leads to warming. In the upper atmosphere, where there isn't that much infrared radiation from below to absorb, the emission effect dominates, leading to more cooling with increased carbon dioxide concentration. This counterintuitive stratospheric cooling effect, verified by observations, demonstrates the power of scientific modeling. Many results from Manabe's radiative–convective model have stood the test of time, despite its highly simplified nature – or perhaps because of it.

Like the atmosphere, the ocean is an important component of the climate system. Early climate modelers used a very simple representation of the ocean; they treated the upper ocean as a motionless slab of water, completely ignoring ocean currents. This may be acceptable if one is only interested in predicting the equilibrium climate, but it is a very poor representation if one is interested (as we are) in predicting the evolving climate. In Section 4.6, we discuss the development of a complex ocean model that is an analogue of the complex atmospheric model.

4.6 Flywheel of Climate: The Circulating Ocean

A flywheel is a very heavy wheel that spins, acting as a reservoir of kinetic energy. It's difficult to get it to spin, but, once it is spinning, it is equally difficult to get it to stop. Flywheels are commonly used in exercise machines, like stationary bikes and ellipticals, to maintain momentum.

The ocean is climate's equivalent of a flywheel, storing heat energy rather than kinetic energy. It can serve as a heat reservoir because of a property called *heat capacity*. Heat capacity is defined as the amount of energy it takes to warm an object by 1°C.

Say you put an empty metal saucepan on a heating stove. The saucepan will heat up rapidly, because the metal of the saucepan has a small heat capacity. Of course, the saucepan is not really "empty" – there is air inside of it – but as air has even smaller heat capacity, this makes little difference in how fast the saucepan heats up. When the saucepan is filled with water, it takes much longer to heat up, because the water has a much larger heat capacity than air. The heat capacity of a saucepan-sized amount of water is about 4,000 times the heat capacity of a saucepan-sized amount of air, accounting for the low mass of the air.[25] We can do a similar calculation for the atmosphere and ocean, taking into consideration the average depth of the ocean and the area it occupies. It turns out that the ocean has about 1,000 times the heat capacity of the entire atmosphere.[26] In other words, it takes 1,000 times more energy to warm the entire ocean by 1°C than it takes to warm the entire atmosphere by 1°C. The ratio is smaller if we only consider the warming of the upper ocean, which is what happens in the short term, over a few decades.

The enormous differential between the heat capacity of the atmosphere and the heat capacity of the ocean has important consequences for the global warming problem. Say the carbon dioxide concentration doubles instantaneously, and the greenhouse effect begins to trap more infrared radiation. As the atmosphere warms up, it begins to transfer excess heat to the upper ocean, through a process known as ocean *heat uptake*. This slows down global warming, resulting in the *transient climate response* that occurs over many decades (Section 3.2). The excess heat from the upper ocean eventually makes it way to the deep ocean, over many centuries, through a process called *thermohaline circulation*. Once the entire ocean has warmed up, the combined atmosphere–ocean system reaches a new equilibrium. The warming at this point is the *climate sensitivity* that we discussed previously. Of course, if the carbon dioxide concentration keeps increasing, the system will not reach equilibrium. (The land has about the same heat capacity as the atmosphere, and therefore does not take up much heat.)

To predict the short-term evolution of global warming, we need to consider the ocean as well as the atmosphere. When Manabe and his predecessors used simple atmosphere models to estimate the warming associated with doubling the amount of carbon dioxide, they worked around the lack of a proper ocean model by considering only the long-term equilibrium response, after the ocean has completely warmed up. It would take the entire ocean thousands of years to reach equilibrium. But since we want to mitigate climate change, we are more interested in knowing what is likely to happen in the next 30–100 years.

Smagorinsky was prescient enough to understand that any climate model used for practical prediction would have to include both the atmosphere and the ocean. His long-term vision was to realize Lewis Fry Richardson's dream of a "forecast factory" for the entire climate system (Figure 1.2a). In 1960, he hired an oceanographer named Kirk Bryan (Figure 4.1b) from the Woods Hole Oceanographic Institution to start building an ocean model at his lab.[27] Back then, oceanographers often had a very different scientific culture from meteorologists; they focused more on regional problems than on global problems, for instance. The ocean was much more poorly observed than the atmosphere, and oceanographers concentrated more on data gathering than on modeling. Due to lack of sufficient data to constrain the three-dimensional structure of the global ocean, efforts to model it were considered premature. But Bryan had originally trained as a meteorologist and had experience with numerical modeling, having completed his Ph.D. under the supervision of Ed Lorenz.[28] This background enabled him to transcend the disciplinary boundary and work more easily with the atmospheric modelers who dominated GFDL.

Bryan worked alongside a talented programmer named Michael Cox to build a model of the ocean that could represent ocean currents like the Gulf Stream or the Kuroshio. Bryan and Cox essentially had to start from scratch because ocean modeling was far less developed than atmospheric modeling: The ocean models then were quite simple and assumed the ocean was in steady state. To develop their three-dimensional model of the ocean, they borrowed computational design ideas from a parallel climate modeling effort underway at the University of California, Los Angeles (UCLA), led by Yale Mintz and Akio Arakawa – the latter being from Professor Syono's group in Tokyo, like Manabe and Kasahara. Later, Bryan and Cox shared the code for their model with scientists around the world, pioneering the open-source tradition that has become increasingly common in climate modeling.[29]

In 1969, Manabe and Bryan constructed a climate model that included both the atmosphere and the ocean, known as a "coupled model" – the first of its kind.[30] It was not a truly global model, given the state of modeling at the time, but it did allow atmospheric winds to interact with ocean currents for the first time ever. This collaboration between the oceanographer and the meteorologist would continue for several decades, as they built more complex models of the climate system.

To use the Laplace's Demon analogy, the Climate Demon needs to work with lesser demons like the Atmosphere Demon, Ocean Demon, Land Demon, and Sea Ice Demon to make predictions. Early versions of the Climate Demon relied mostly on the Atmosphere Demon, interacting with very crude caricatures of the other demons. Manabe and Bryan were the first to get the Atmosphere Demon to talk to a realistic Ocean Demon.

***

Many years later, Manabe said of his time at GFDL: "this is one of the fascinating things about this laboratory. We never, I never in my whole life – I never wrote [a] grant proposal for my own research."[31] The success of Smagorinsky's lab owes much to the management philosophy of the man who led it from its inception in 1955 until his retirement in 1983.[1] He hired young, talented scientists like Manabe and Bryan and provided them the resources and the freedom that they needed to grow to be world leaders in their disciplines. Scientists at GFDL were provided a small team of programmers to assist in their work and computer resources to run their model without ever having to write proposals so long as they carried out research broadly consistent with the mission of the laboratory. Apart from the pioneering climate modeling activities discussed in this book, GFDL scientists have also made many groundbreaking contributions in other areas of the atmospheric and oceanic sciences.

During the 1960s, computer models of climate were being built at research centers around the world. Manabe's model at GFDL was just one among them. Jule Charney and John von Neumann, along with the rest of the Princeton Meteorology Group, had already shown how we could use the basic principles of physics in conjunction with a computer to scientifically predict weather. It is then natural to ask whether we can also scientifically predict climate, which is simply the average of many weather events, using those same physical principles. The computer climate models of Manabe and others, as well as the simple model of Svante Arrhenius, essentially aimed to do just that. But recall that Lorenz had shown that we cannot predict weather beyond the next two weeks. How, then, can we hope to predict climate months and years into the future?

5

Predicting Climate

Butterflies in the Greenhouse

The next stop in our climate prediction story is the town of Punxsutawney, Pennsylvania, about 300 miles west of Princeton. We have timed our visit so that we arrive on the second day of February – Groundhog Day. A groundhog named Phil will emerge from his burrow in Gobbler's Knob, about two miles outside of the town (Figure 5.1).[1] If Phil sees his shadow on this day, he has predicted that winter will continue for six more weeks. If he doesn't see it, he has predicted an early spring. A group of men in top hats and tuxedos, known as the Inner Circle, orchestrate this whole exercise. Although Groundhog Day is now essentially a scripted media event, it is rooted in an old tradition that German settlers brought with them to rural Pennsylvania. February 2 is celebrated as Candlemas in many Christian traditions, and in Germany it was the native badger whose behavior foretold the arrival of spring.[2] The event was first celebrated in the United States during the early nineteenth century. Groundhog Day was immortalized in a movie of the same name starring Bill Murray. After the release of the movie, the event became immensely popular. You can now livestream Punxsutawney Phil's prognostication, if you can't make it to Punxsutawney in person.

The enduring tradition of Groundhog Day underscores our primeval fascination with prophecies. But it's not just groundhogs and badgers that predict climate. We humans do it all the time, even if we don't realize it. If you live north of the equator, you can say years in advance that January will be cold, and July will be warm. Then there are adages about April showers that bring May flowers. These are statements about the typical weather for those months, not about individual weather events like snowstorms or hurricanes. As meteorologist Marshall Shepherd puts it, "weather is your mood and climate is your personality."[3] In other words, climate is what you expect, and weather is what you get! You may expect the winter months to be cold, but you can still get a

Figure 5.1 Punxsutawney Phil predicting the arrival of spring at Gobbler's Knob on Groundhog Day, 2013. (Photo: Anthony Quintano/cc-by 2.0)

warm spell during these months. To put it more mathematically, climate is the average of weather.

The seasons are easy to predict because they repeat themselves after a fixed interval of time. This is referred to as *cyclical behavior*. When people see cyclical behavior, they often think of prediction. The 11-year cycle of sunspots is what motivated John Mauchly to try to predict weather, eventually leading him to build the first digital computer. One of the phenomena climate scientists try to predict is El Niño, which is quasi-cyclical with a three- to five-year period. It manifests itself as alternating warm and cold phases of ocean temperature in the tropical Pacific, but it affects weather and climate in many other parts of the globe, including the United States.

Apart from the seasons, weather is not very cyclical. Good and bad weather occur at random intervals. We can now predict weather rather well a few days in advance, using computer models rather than almanacs or groundhogs. Lorenz and others showed that weather cannot be predicted beyond about two weeks. But we can predict climate, or averaged weather, years into the future, because climate prediction is not just an extension of weather prediction. It is prediction of a very different kind.

5.1 Nature versus Nurture: Initial and Boundary Conditions

The philosopher Karl Popper distinguished between unconditional prophecies and conditional predictions.[4] An unconditional prophecy will come true no matter what. Most scientific predictions are not like that. They are almost always of the form "if X is true, then Y will happen." A weather forecast is conditional on our knowledge of initial properties of the atmosphere such as the temperature, humidity, and winds, collectively known as *initial conditions* (Chapter 2). A climate prediction, on the other hand, is conditional on our knowledge of time-dependent external factors known as *boundary conditions*. Predicting boundary conditions allows us to predict climate. (Ed Lorenz emphasized this distinction, referring to climate prediction as prediction of the second kind, with weather forecasting being prediction of the first kind.[5])

When every year we predict that the winter months will be cold, we use our knowledge of the zenith angle of the sun. In the winter, the sun tends to be lower in the sky, rather than directly overhead, so it heats the atmosphere less than it does in the summer, when it is higher in the sky. This is a boundary condition, and knowledge of it allows us to predict climate. To predict whether the Earth will continue to warm in the future, we need to consider other boundary conditions that affect atmospheric heating. Carbon dioxide and other greenhouse gases in the atmosphere also trap heat and warm the planet (Section 3.1). If we can predict how concentrations of these gases will change in the future due to human activity, we can predict how Earth's climate will respond to these changes.

Like the Weather Demon, who assists Laplace's Demon in predicting weather (Section 2.1), we can also conceive of a Climate Demon, who assists in predicting the climate. To make its predictions, the Weather Demon only needed scientific measurements of the initial conditions. But the Climate Demon needs boundary conditions from scientific demons like the Solar Demon, as well as nonscientific demons like the Economy Demon, to make its predictions.

Initial conditions are like the genetic code, or genotype, of an organism. In nature, a gene is a long sequence of As, Cs, Gs, and Ts, corresponding to the four nucleotide bases adenine, cytosine, guanine, and thymine in a DNA molecule. In a computer weather model, atmospheric initial condition values – such as winds, temperature, and humidity – are represented as a long sequence of binary numbers, in the form of 0s and 1s.

Boundary conditions are more like the environment that nurtures the growth of the organism. Genotype plays a bigger role in the early stages of growth, but the environment starts to play a bigger role later on. Loosely speaking, predicting weather is like predicting the early growth of an individual organism. Climate prediction is more like predicting how a group of organisms will develop later on, given certain environmental conditions that may evolve over time. The better we know how the environmental conditions are likely to change, the better we can try to predict their influence on the group of organisms.

Punxsutawney Phil can apparently make predictions without a complicated computer model. When Phil predicts an early spring, he is making a statement about the average weather over six weeks, not about individual weather events. This information may not be useful to individuals planning a family picnic, but it could be quite useful to farmers, who need to decide when to plant crops, as well as to corporations whose businesses are affected by the changing seasons.

How accurate is Phil? If we define an early spring as above-average temperatures during the six weeks after February 2, meteorologists who track Phil's predictions estimate that he is right about 40 percent of the time.[6] That is a bit suspicious. Even if Phil were acting randomly, tossing a coin, he should be right about 50 percent of the time. If we are confident that Phil is wrong 60 percent of the time, we can simply use the opposite of his forecast, and we will be right 60 percent of the time! We can even use this information to make money in the commodity markets. Perhaps Phil is cunning enough to deliberately mislead humans and is secretly buying heating oil futures. The serious point, though, is that climate predictions are inherently probabilistic. They are not exact.

For the Climate Demon to predict the climate over a six-week period, the boundary conditions that it needs to consider are the snow or ice cover in cold regions, or the soil moisture in warm regions, as these can affect atmospheric air flow and rainfall. Human activity will not increase carbon dioxide all that much in six weeks, so it is not an important boundary condition for a seasonal prediction. But if the Climate Demon is predicting climate 50–100 years ahead, concentrations of carbon dioxide and other greenhouse gases – affected by human activity – become crucial boundary conditions. Recall that Callendar had tried to raise interest in the problem of increasing carbon dioxide concentrations in 1938, but his estimates of increases in carbon dioxide were suspect due to the quality of data that were available to him. It turns out that carbon dioxide concentration can be quite difficult to measure accurately.

5.2 Far from the Madding Crowd: The Keeling Curve

Around the time that Manabe received a job offer from Smagorinsky, a young scientist named Charles David Keeling was building an instrument to make climate measurements as far from inhabited areas as possible.[7] He chose the remote north slope of a volcano, Mauna Loa, on the island of Hawaii.

Keeling wanted to make accurate measurements of how carbon dioxide concentration in the atmosphere changes over time. Carbon dioxide is what is known as a well-mixed atmospheric gas, which is to say that its concentrations are roughly uniform above the turbulent lower layer of the atmosphere close to the ground, known as the boundary layer. Within the boundary layer, though, the local emission of carbon dioxide from human activities or plant growth can cause its concentration to deviate from the well-mixed value, adding noise to the measurements.[8] Moving away from a continental location reduces the noise – which is why Keeling located his primary instrument in Hawaii, even though his lab was based in the Scripps Institution of Oceanography in San Diego, California. Another instrument was in an even more pristine location, Antarctica. (Measurements in Antarctica used a continuous monitoring instrument at a station near the coast, as well as air samples collected in glass flasks from the South Pole.)

After receiving his Ph.D. from Northwestern University, Keeling was carrying out postdoctoral research at the California Institute of Technology, or Caltech, when he started to measure carbon dioxide in the atmosphere using a precise gas manometer and realized that prior measurements were not very accurate. During a trip to Washington, DC, in 1956, he boldly proposed a new technique to Harry Wexler of the US Weather Bureau for accurate measurements of carbon dioxide that used an infrared gas analyzer. Wexler agreed immediately and offered Keeling a job with the Weather Bureau. Although employed by Wexler, Keeling was convinced to base his lab at Scripps by its director, Roger Revelle, an oceanographer who was very interested in how the ocean absorbed carbon dioxide.[9] This was a very important issue – if it turned out that the ocean could absorb most of the carbon dioxide entering the atmosphere due to human activities, then there would be less need to worry about increasing carbon dioxide concentrations and the resultant global warming.

In 1958, Keeling began measurements of carbon dioxide concentrations at a Weather Bureau observatory on Mauna Loa as part of a global scientific project to carry out interdisciplinary research known as the International Geophysical Year. He soon found that carbon dioxide concentrations went up and down cyclically with the seasons, with a maximum in May and a minimum around September. As plants started growing in the spring, they absorbed carbon dioxide from the atmosphere, lowering the concentration; at the end of the growing season in the fall, they released it back, increasing the concentration again.[10] This formed Earth's "breathing cycle."

Another surprise was that the cycle did not repeat exactly from year to year. Each year there was a little more carbon dioxide than the year before, for the same season. This was true of measurements at Mauna Loa and in Antarctica, which is even further removed from contamination by local emissions. The graph of carbon dioxide concentration plotted against time started to look like an upward-sloping sawtooth (Figure 5.2a). As Keeling continued his

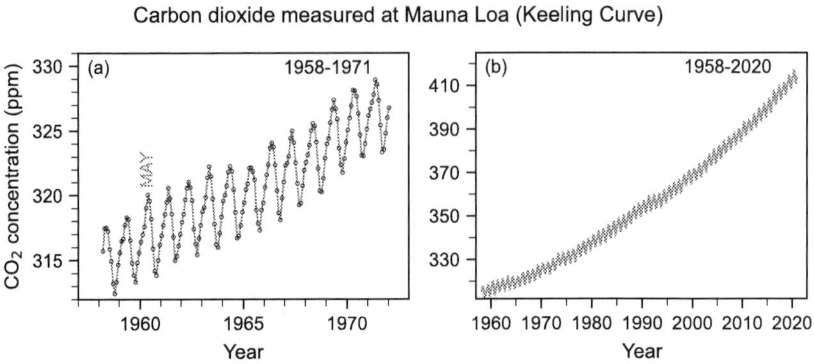

Figure 5.2 Long-term variation in the concentration of atmospheric carbon dioxide at Mauna Loa Observatory. (a) 1958–1971: The circles indicate the observed monthly average concentration (in parts per million). The "sawtooth" curve is a smoothed version of the monthly measurements. (b) The same smoothed carbon dioxide concentration, but for the period 1958–2020. (Adapted from Keeling, 1976; extended using data from Keeling et al., 2001; cc – by 4.0)

measurements, the rise continued year after year. In 1976, Keeling published his now-classic paper documenting the inexorable rise of carbon dioxide. The sawtooth graph in the paper would become famous as the Keeling Curve – a sloping signature of the human imprint on climate.

By locating his instruments far from inhabited continents and well above the surface of the Earth, Keeling had extracted the carbon dioxide signal from the noise. The rate of increase in carbon dioxide levels has been slowly accelerating in recent years (Figure 5.2b), consistent with the growing use by humans of carbon-based fuels like coal, petroleum, and natural gas. The long record of accurate measurements from the Mauna Loa Observatory, starting from 1958, is perhaps the clearest and most reliable evidence of our ability to change Earth's climate.

5.3 It Takes Two to Tango: Carbon Dioxide and Temperature

The interplay between carbon dioxide concentrations and global temperatures is like a partner dance. One leads, and the other follows. If we watch a couple dance several steps, we can usually figure out who is leading and who is following. But if we barely get to see one step of the dance, and we miss the beginning, it's much harder to tell. The global warming we observe currently is barely one step of this dance. Even if we see carbon dioxide increasing and

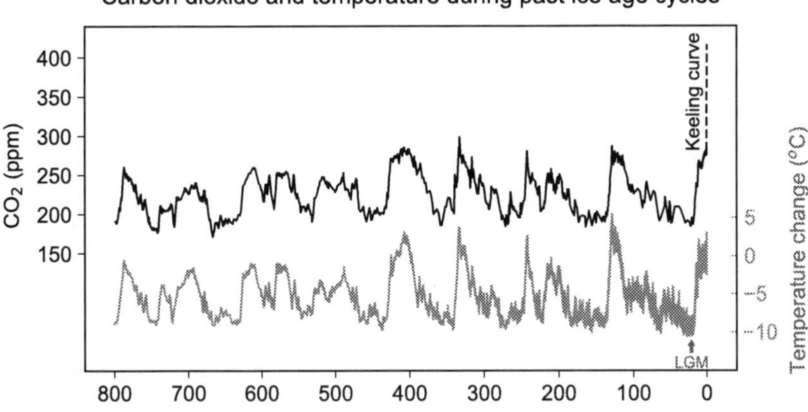

Figure 5.3 Atmospheric carbon dioxide concentration (black) and temperature (gray) data derived from Antarctic ice core measurements. The Keeling curve – the dashed vertical segment appended to the ice core data at the upper right – shows the unprecedented nature of the recent increase in carbon dioxide concentration. "LGM" denotes the Last Glacial Maximum, or the peak of the last ice age, which occurred about 20,000 years ago. (Adapted with permission from Harris, 2010, © 2010, American Chemical Society)

temperature increasing at the same time, we can't be sure which is leading simply by looking at the data.

Climate variations of the past tell a different story. We can find many steps to this partner dance. The field of paleoclimatology uses tree rings, sediments, fossils, and other geological evidence to study ancient climates. There is paleoclimatic evidence that global temperatures underwent cycles of cooling and warming of about 5°C roughly every 100,000 years during the last 800,000 years (Figure 5.3). The cold glacial periods were the ice ages, when much of Scandinavia and what is now Canada was covered with large ice sheets. We are now in a warm interglacial period of this cycle.

In 1958, Keeling used air samples from the South Pole trapped in flasks to measure the concentration of carbon dioxide in the atmosphere. It turned out that nature had been taking its own samples for millions of years. Every time the old snow in Antarctica was compacted by new snow, bubbles of air were trapped inside. By drilling ice cores through the Antarctic ice sheet and carefully slicing them, scientists analyzed these ancient bubbles for carbon dioxide concentrations. They also estimated past temperatures from isotopic analysis, that is, from the ratio of deuterium to hydrogen in the ancient ice.

This ratio is sensitive to atmospheric temperature, with more deuterium present for warmer temperatures.

A graph of past carbon dioxide concentrations and temperatures, from a recent paper, is shown in Figure 5.3. We see that both temperature and carbon dioxide concentrations changed during the ice age cycles, with low carbon dioxide values during the cold periods and high carbon dioxide values during the warm periods. Clearly human activities could not have been responsible for these changes, which occurred hundreds of thousands of years ago. A more careful analysis of data shows that increases in carbon dioxide happened after the temperature started increasing. Temperature was leading carbon dioxide during the eight steps of the ice age dance over the last 800,000 years.

The greenhouse theory of global warming assumes that carbon dioxide leads temperature. This means that the greenhouse theory cannot explain the ice ages. The accepted explanation for the ice ages was put forward by Serbian astronomer Milutin Milankovitch in the 1920s. He discovered cycles in solar radiation reaching the Earth's surface during different seasons, now known as *Milankovitch cycles*. The axis of Earth's rotation and the ellipticity of its orbital motion around the sun have cyclical variations, with periods ranging from 26,000 to 100,000 years. These variations redistribute the amount of solar heating received by the Earth during winter and summer seasons, which can lead to ice sheets growing and melting, the effect of which is amplified by the ice–albedo feedback (Section 3.5). When temperatures warm during the melting phase, carbon dioxide stored in other components of the climate system, such as the ocean, is released, leading to increased carbon dioxide concentrations and further amplification. Carbon dioxide acts as the feedback in this case, and temperature acts as the forcing. This explains why temperature leads carbon dioxide in the ice age dance.

The Keeling curve appended to the right of the ice core data in Figure 5.3 shows the dramatic increase in carbon dioxide since 1950, as compared to past values. Global temperature is also increasing – but we cannot attribute that to increases in carbon dioxide based on this figure alone, as we cannot tell which is leading. We will use models to address the attribution of global warming in Section 7.2. But an important lesson to learn from the figure is that the recent increase in carbon dioxide levels is unprecedented. That means that we cannot use purely inductive reasoning to predict what will happen next as carbon dioxide continues to increase.

Until the early 1970s, climate modelers were more interested in the ice age dance, for which there was strong evidence, than in the global warming dance, which had barely begun. Trying to explain global cooling during the ice ages had long been considered scientifically more interesting than trying to explain any future global warming. Like Arrhenius, Manabe carried out his famous doubled carbon dioxide calculations in 1967 out of scientific curiosity, not because of its practical implications. At the time, he was not worried about the societal impact of greenhouse warming; he was studying it simply because greenhouse gases were the second most important factor in determining the Earth's climate, next to solar heating.[11]

5.4 Academia Warms to Global Warming

Keeling's measurements were clear evidence that the carbon dioxide concentration in the atmosphere was increasing, reaching levels not seen in the last one million years. However, in the 1970s, evidence for any associated global warming was still weak. Global temperatures rose between 1890 and 1940, roughly the period analyzed by Callendar, but they decreased slightly during the next 20–30 years. A reflective haze, linked to industrial pollution, had cooled the globe. In fact, there were some scientists who were worried about global cooling, rather than global warming.[12]

As was the case with early carbon dioxide measurements, there is a great deal of noise in global temperature calculations. Natural phenomena like El Niño can warm the globe for a year or two, and volcanic eruptions can cool the globe for a year or two. Disentangling these effects, and that of industrial haze, from the underlying steady warming associated with increasing amounts of carbon dioxide and other greenhouse gases is a difficult statistical problem. There is no easy way to improve the warming signal in the observations. We cannot use spatial distancing (moving far from the sources of noise, as Keeling did) to improve the temperature signal; unlike the well-mixed carbon dioxide concentration in the atmosphere, the temperature signal is highly variable from place to place. But we can use temporal distancing. With the passage of time, the global warming signal, if real, should get stronger, while the noise stays roughly the same.

The development of climate models progressed apace. While observationalists waited to see more steps in the carbon dioxide–temperature dance, modelers simulated those steps in their virtual worlds. At GFDL, which had by then moved to Princeton, Manabe and his colleagues built a three-dimensional global climate model (Section 4.3).[13] When the team carried out doubled carbon dioxide climate sensitivity experiments using this model in 1975, they discovered two very important properties of global warming: Polar regions warm more than tropical regions, and the land warms more than the ocean.

Another parallel climate modeling effort was gathering steam in New York City in the 1970s, not far from Princeton, at the Goddard Institute for Space Sciences (GISS), a lab operated by the US National Aeronautics and Space

Administration (NASA). The GISS climate model was derived from the one developed at UCLA by Mintz and Arakawa.[14] Climate modeling at GISS was led by Jim Hansen, who started his career as a planetary scientist studying the atmosphere of Venus.[15] (Venus, as we noted in Section 3.6, has an extremely hot atmosphere due to its runaway greenhouse effect.)

Hansen, more so than most climate modelers at the time, was concerned about the warming of the Earth. Several scientists who shared his fears convinced the US government to look into the issue. When the government is concerned about a scientific problem, it asks the National Academy of Sciences to provide advice. In 1979, the academy convened a committee of distinguished scientists to study the greenhouse effect. Jule Charney, who had worked with von Neumann to carry out the first numerical weather forecast, chaired this committee, called the "Ad Hoc Study Group on Carbon Dioxide and Climate."

Two of the "star witnesses" who provided evidence to the committee were Suki Manabe of GFDL and Jim Hansen of GISS, the leading climate change modelers in the country at the time.[16] Manabe's estimate of climate sensitivity using the three-dimensional GFDL model was 2°C. Hansen had recently carried out a climate sensitivity calculation using the GISS climate model and estimated it to be 4°C. This discrepancy appeared to be related to how the two models handled ice and snow processes. The committee averaged the two estimates to come up with a climate sensitivity value of 3°C, and it indicated a range of 1.5 to 4.5°C to reflect the uncertainty. In the end, the committee produced a report titled "Carbon Dioxide and Climate: A Scientific Assessment," more commonly known as the Charney Report.[17] This was the first formal attempt by the scientific community to assess the risk posed by global warming. Its climate sensitivity estimates would more or less remain the same for the next 40 years.

***

We can predict climate if we can predict the boundary conditions that control it. The concentration of carbon dioxide in the atmosphere is one such boundary condition, and it was increasing steadily as humans continued to burn fossil fuels to generate energy. If this trend continued unchecked, climate models indicated that global temperatures could get a lot warmer. In the 1970s, there was clear observational evidence for rising carbon dioxide levels, but the evidence for a corresponding rise in global temperatures was still weak. There didn't appear to be much to worry about, yet.

In the meantime, a very different type of environmental problem emerged, rather unexpectedly, in the 1980s. It had to do with the concentration of a different atmospheric gas – ozone.

6

The Ozone Hole

Black Swan at the Polar Dawn

It was a remote and desolate continent in the South. The explorers came from the North. They came to observe local birds, especially the swans. They had only seen white swans before, day after day. They assumed all swans were white. But one early dawn they saw something they had never seen before: a black swan. Not trusting their eyes, they waited another day to be sure. The next day at sunrise, they saw it again. After seeing a black swan many mornings in a row, they decided to tell the world about their startling discovery.

This explorer story reads like a classic example of the failure of inductive reasoning from the late seventeenth century. Until that time, Europeans were convinced that all swans were white, based on experience and history – so much so that the phrase "black swan" was used dismissively to refer to impossibilities. This notion was dispelled in 1697, when Dutch sea captain Willem de Vlamingh observed an actual black swan, *Cygnus atratus*, during his exploration of West Australia. In the nineteenth century, British philosopher John Stuart Mill used this unexpected observation to illustrate the problem with induction, or purely data-driven reasoning, as discussed in Section 1.2.[1]

The explorer story we will narrate is actually about a different sequence of events, featuring a metaphorical swan, which began with the first sighting of Antarctica in 1820 by a Russian expedition. Soon after that, maritime powers like Britain and France, as well as other European and South American nations, started laying claims to its territory. Each nation tried to buttress its claims by promoting exploration of this desolate continent, through efforts such as British explorer Robert Falcon Scott's ill-fated expedition to the South Pole in 1912. The Antarctic Treaty of 1959, which designated the Antarctic as a scientific preserve, put these territorial claims in abeyance, but research and exploration has continued to this day – a scientific legacy of imperial entanglements.

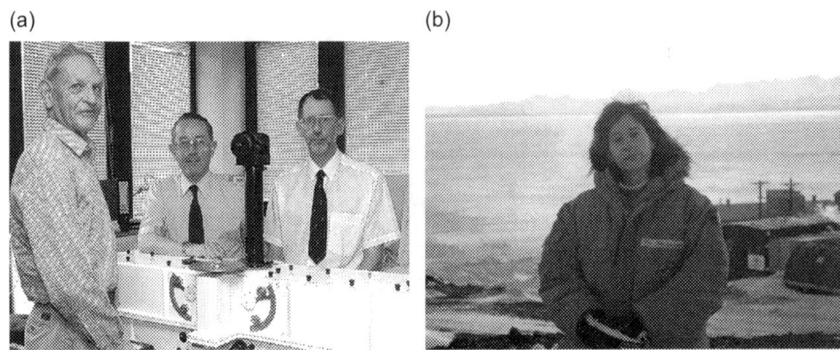

Figure 6.1 (a) Joe Farman (left), with his 1985 paper coauthors Brian Gardiner and Jon Shanklin, around a Dobson ozone spectrophotometer at the British Antarctic Survey in Cambridge, UK. (Photo: Chris Gilbert, British Antarctic Survey). (b) Susan Solomon at McMurdo Station, Antarctica (Photo: Courtesy of Susan Solomon)

The British Antarctic Survey (BAS) is the arm of the British government responsible for their country's exploration of the region. It is headquartered in the bustling university town of Cambridge in Britain, far from the icy and remote continent it is intended to explore. Many field personnel of the BAS – modern explorers – are like migratory birds. They travel to Antarctica each year, spending many months making measurements during the Antarctic summer. Polar days and nights, which are defined as periods during which the sun never sets or rises, respectively, can be very long in Antarctica – extending up to six months near the south pole. A small number of brave souls, known as "winterers," stay through the Antarctic night (April–August), making scientific observations.

BAS operates several research stations in Antarctica. Halley Station is one.[2] It was established in 1956 as part of the International Geophysical Year project, which also supported Charles David Keeling's carbon dioxide measurements starting around the same time. Halley Station has been measuring concentrations of a different atmospheric gas, ozone, since then. An ozone molecule is made up of three oxygen atoms, unlike the more familiar oxygen molecule we breathe, which has just two atoms. Ozone plays an important role in the Earth's ecology, shielding the planet from harmful ultraviolet radiation.[3] If ozone levels become too low, it can be a cause for worry, as more harmful radiation will reach the Earth's surface.

British scientist Joe Farman (Figure 6.1a) joined the BAS in 1956 and went to Antarctica to set up an instrument called the Dobson ozone spectrophotometer.[4] This instrument measures the total amount of ozone in a column of air from the surface to the top of the atmosphere, in units called Dobson Units or DUs. Normal values – the "white swan" values – of Halley Station's October ozone are about 300 DU.[5] Farman's duties shifted back to Cambridge a few years later, but the daily measurements of ozone continued. Handwritten ozone readings were sent back by ship and then typed up in the BAS headquarters in Cambridge. Year after year, ozone values hovered slightly above 300 DU in October, which is spring (or morning) at that latitude in Antarctica. By the late 1970s, Keeling's carbon dioxide measurements had already produced important insights into global warming, but only "white swans" (that is to say, nothing of interest) had shown up in Farman's ozone data for decades. When the BAS faced budget cuts under a new government in 1980, Antarctic ozone measurements were in danger.[6] Farman was asked why the Antarctic ozone measurements should be sustained, as no new results had emerged in 25 years. Somehow the measurements survived; subsequently, geopolitical considerations played a role in protecting BAS funding in the aftermath of the Falklands War of 1982.

By that point, the BAS had built up a backlog of several years of ozone data.[7] When Farman and his colleagues cleared the backlog and started analyzing the data in 1982,[4] they found a dramatic year-over-year decline in October ozone values starting in the late 1970s, from the normal value of 300 DU to below 200 DU.[8] This was unprecedented and totally unexpected – a "black swan" event. The scientists repeated their measurements using a new instrument, in case there had been a problem with the old instrument. When the new instrument confirmed the low ozone values, they published their results in 1985. Farman's perseverance had finally paid off, after nearly 30 years. The paper had a dramatic impact in shaping ozone science and policy. In Section 6.1, we step back to review the history of ozone science, to place Farman's discovery in context.

6.1 The Ozone Layer: Earth's Sunscreen

When you go out in the sun to partake in an outdoor activity or just to lie on the beach, you are supposed to put on sunscreen: a protective layer of a substance that blocks harmful ultraviolet (UV) radiation. UV radiation is part of the spectrum of sunlight that includes visible light, as well as the infrared radiation that we discussed in Sections 3.1 and 4.3. It has a shorter wavelength than visible light and is invisible. For light, shorter wavelength means greater energy, so photons of UV radiation are more energetic than photons of visible light. One type of shorter-wavelength UV radiation, known as UV-B, is particularly harmful to life.[9] If UV-B radiation falls on skin, its energetic photons can damage the DNA in the cells, leading to sunburn and an increased likelihood of skin cancer.

Much of the UV radiation from the sun is blocked by a layer of ozone gas high in the atmosphere, acting as a sunscreen for all life on the planet. The sunscreen that we put on our skin only protects us from the remnants of UV radiation that manage to get past the ozone layer. If the ozone layer were weakened, much more UV radiation would get through to the surface, leading to increased incidence of skin cancer, cataracts, and other damaging health effects.

The protective ozone layer is present in the part of the atmosphere known as the stratosphere (Figure 3.1).[10] Extreme UV radiation from the sun absorbed by oxygen molecules in the stratosphere causes them to break up and recombine with other oxygen molecules to generate ozone, through a process known as *photochemistry*.[11] The ozone thus created absorbs much of the harmful UV-B radiation, preventing it from reaching the surface. Absorption of UV radiation also heats the stratosphere, reversing the drop in temperature with increasing altitude.

Normally, little ozone is found in the lower part of the atmosphere, the troposphere. But when pollution levels are high, automobile and industrial emissions can interact with sunlight to form ozone in the troposphere. Ozone formed this way, tropospheric ozone, is harmful to human health – unlike stratospheric ozone, which humans don't breathe. Once ozone's properties were recognized, scientists began monitoring its global distribution, beginning with routine measurements in the 1920s after the invention of the Dobson spectrophotometer,[12] the instrument that Farman used to discover ozone depletion over Antarctica in 1985.

6.2 The Road to Ozone Depletion

The story of global warming begins with good intentions, with our desire to improve our living standards and provide ourselves with creature comforts. The discovery of electricity, and the invention and mass production of automobiles, improved the lives of billions of people. But we require affordable sources of energy to power cars and factories. The cheapest sources of fuel have historically been coal, petroleum, and natural gas, which are all carbon based: Burning them produces the greenhouse gases that lead to global warming.

The story of ozone depletion also began with good intentions. The air conditioner, invented at the start of the twentieth century, has tubes that circulate a gas called a "refrigerant." An electric motor compresses the gas into a liquid, while releasing heat to the exterior in the process. Hot interior air flows over the tubes with the liquid refrigerant, cooling down as the refrigerant evaporates and flows back to the compressor. A refrigerator also works on the same principle. Key to this process is the volatile refrigerant, which should be easily transformable from gas to liquid and back at the operating temperatures.

The earliest refrigerants were gases like ammonia and sulfur dioxide.[10] They were volatile but also toxic, and ammonia could cause explosions. Any leaks would be dangerous to life and property. What was needed was a refrigerant that was harmless or inert. Chemists worked hard to develop such a compound. In 1928, the Frigidaire division of General Motors created such a gas, which was later commercialized under the name Freon. This was a major industrial breakthrough, and its inventor, Thomas Midgley, received the top award of the American Chemical Society. To demonstrate how harmless the gas was, Midgley once inhaled it and used it to blow out a candle at a conference.

Freon is one of a group of chemical compounds known as chlorofluorocarbons or CFCs. They are volatile and inert, excellent for use in spray cans, air conditioners, and refrigerators. But CFCs can leak from faulty or poorly disposed equipment (or during normal use, in the case of spray cans). The fact that they are inert means that any leaked CFCs stay around in the atmosphere for a long time, unlike reactive gases. Although these gases begin in the troposphere, over time, air circulation carries them into the stratosphere and causes them to accumulate there.

The accumulation of CFCs and its possible side effects started to worry a chemist named F. Sherwood Rowland at the University of California, Irvine.[13] In the early 1970s, he and a postdoctoral researcher, a young Mexican scientist named Mario Molina, began studying CFCs. They found that there was no way to destroy CFCs in the troposphere. But they could be destroyed in the stratosphere. Once CFCs get through to the ozone layer, UV radiation can separate chlorine atoms from the CFC molecules.[14] Chlorine is in turn capable of destroying ozone through powerful catalytic reactions; one chlorine atom can destroy thousands of molecules of ozone.[10] Over time, this would weaken the protective ozone layer and expose the population to a greater and greater risk of skin cancer and cataracts.

Molina and Rowland published their results in a 1974 paper that didn't initially attract much attention. A few months later, they held a press conference at the American Chemical Society annual meeting to announce their estimate that, if CFC emissions were to continue at the current rate, the amount of atmospheric ozone would eventually drop by 7–13 percent.[15] This prediction caught the attention of other scientists and the public. The US National Academy of Sciences took up the problem in 1976 and issued a report outlining the harm that CFCs could cause to stratospheric ozone.[10] The reaction from the industry to this report was not very favorable, because there were no clear alternatives to CFCs available at that time. The use of CFCs in spray cans was banned by the US government in 1978, but its use in other equipment continued.

Scientists began to look for the depletion signal in stratospheric ozone, but the predicted trends were quite small, less than 1 percent per decade.[16] Trends varied from location to location, and few long-term measurements were available. The signal was too weak to be easily detectable against the background of noise, so scientists concluded that there was no measurable trend yet.

Then, in May 1985, the Farman paper on Antarctic ozone loss appeared. The paper created consternation at NASA, which was continuously monitoring ozone using satellites. How could they have missed such a large signal of ozone loss? The satellites did record the loss in ozone, but it seems that the data was flagged by a quality check because the numbers were too low, below 180 DU. The United States also had ground ozone measurements in Antarctica, but their instruments wrongly reported "normal" ozone values of 300 DU due to a calibration error. Perhaps the data analysis was backlogged due to this discrepancy and the possibility that the low values obtained by the satellites were due to instrument error. In any case, by 1984 NASA had begun processing the data.[17] Recall that the BAS also had backlogged data from the 1970s that they only started to process in 1982. Fortuitously, they managed to clear their backlog before NASA did, and thus scoop NASA on one of the most famous discoveries in climate science.

NASA did a complete analysis of their raw satellite data and published it soon after the BAS paper. Not only did the satellite data confirm Farman's ground measurements of ozone depletion, but they also provided a global picture of ozone values for the first time.[18] The low ozone values occurred in a roughly circular region centered near the South Pole, in some years covering an area larger than the continent of Antarctica.[19] The dramatic satellite images of ozone loss led to the coining of the name "ozone hole" to describe the phenomenon.[20]

The discovery of the unprecedented ozone loss stirred up excitement at the Aeronomy Laboratory, a government research institution in Boulder, Colorado, tasked with carrying out research on stratospheric ozone. It was crucial to understand why the loss occurred at a pace much faster than that estimated by the chlorine mechanism of Molina and Rowland a decade earlier. Farman's paper had also proposed a rudimentary mechanism involving chlorine, but it was not very convincing. In March 1986, scientists at the Aeronomy Laboratory decided to launch an expedition to Antarctica to make measurements.[21] They selected a young scientist named Susan Solomon (Figure 6.1b) to lead this expedition, which they called the National Ozone Expedition (NOZE).

Solomon had joined the Aeronomy Laboratory after completing her Ph.D. at the University of California, Berkeley, in 1981. She was an unusual choice to lead NOZE, because she was a theoretician, whereas the expedition was focused on making measurements.[22] But Solomon had volunteered to operate the Aeronomy Lab's visual absorption spectroscopy instrument for the expedition. The earliest the expedition could get to McMurdo Station in Antarctica, the American counterpart to the British Halley Station, was in August 1986. That was good timing, because it was the end of the polar winter, which meant that Solomon could watch the ozone hole form at the start of spring.

6.3 Cloudy Skies in the Polar Night

When Svante Arrhenius calculated in 1896 how the greenhouse effect could lead to climate change, he relied upon measurements of how moonlight was absorbed by the atmosphere at different wavelengths. In August 1986, when the NOZE expedition arrived at McMurdo Station in Antarctica, Susan Solomon was trying to understand a very different kind of climate change: the dramatic depletion of stratospheric ozone levels. Her instruments needed a good light source to measure how different wavelengths of light were absorbed by the atmosphere in order to calculate the atmospheric concentration of different chemical compounds. August was still polar night at McMurdo, but, 90 years after Arrhenius' calculations, moonlight once again came to the rescue of climate scientists.[23] Solomon had to stand up on the roof of a building in the cold Antarctic night, holding a mirror to catch the moonlight, to measure the concentration of a gas called chlorine dioxide. This was quite a turn of events for someone who had until recently been working on a computer model of atmospheric chemistry in the stratosphere.

Solomon was very familiar with Farman's 1985 paper because she had been one of its reviewers. As a theoretician and modeler, she was intrigued by the question of what had caused the unexpected ozone loss. Farman's paper mainly focused on the measurements, but it suggested that chlorine might be responsible. However, the chlorine-based ozone depletion model proposed by

Molina and Rowland could not destroy anywhere near as much ozone as was required to produce an ozone hole.[24] Other scientists quickly came up with mechanisms independent of chlorine. One suggested that the depletion was caused by oxides of nitrogen coming down into the stratosphere from above. Another suggested that it wasn't chemistry at all, but the upward motion of air from below the stratosphere that diluted ozone concentrations.

Solomon believed that chlorine was still key to explaining the ozone hole. Even before the start of the NOZE expedition, she and other scientists had come up with new chlorine-based mechanisms to explain the ozone depletion. These involved chemical reactions happening in clouds, which is somewhat unusual. The stratosphere is extremely dry, much drier than the troposphere below it, because very little of the water vapor in the troposphere makes it up to the stratosphere. But the Antarctic stratosphere gets extremely cold in the polar night, so cold that even its small amount of water vapor condenses to form clouds, called *polar stratospheric clouds*. Once cloud particles are formed, a whole new set of chemical reactions becomes possible on the surfaces of these particles. This surface chemistry can support new reactions, different from those considered by Molina and Rowland.

By June 1986, Solomon and other scientists had published papers proposing that surface chemistry associated with polar stratospheric clouds could enhance a compound called chlorine monoxide in the lower stratosphere and lead to accelerated ozone depletion. Now there were three different theories to explain the ozone hole: one based on chlorine surface chemistry, another based on oxides of nitrogen, and a third based purely upon air motions.[18] The next step was to use observations to select from among the competing theories.

Data from the 1986 NOZE expedition and subsequent expeditions provided a wealth of information to falsify theories. The ozone loss measured was in the lower stratosphere, which was inconsistent with the mechanism involving oxides of nitrogen coming down from above the stratosphere. There was also no evidence of upward air motions in the lower stratosphere, which ruled out that mechanism. But there was evidence for very large increases in amounts of chlorine monoxide and chlorine dioxide, confirming the chlorine-based chemical theory. This theory, based on polar stratospheric clouds and associated surface chemistry, is the currently accepted explanation for the ozone hole. Solomon was in the unique position of being involved both in proposing the theory and in making the initial set of observations that helped verify it.

In recognition of her discovery, six years later, Solomon became one of the youngest members elected to the National Academy of Sciences, among the highest honors accorded to US scientists. In 1999, she was awarded the

National Medal of Science. She remains connected to the continent of her discovery, with a glacier and a snow saddle in Antarctica named in her honor.

Earlier we discussed how unusual it was for a young theoretician to lead a high-profile observational expedition. It goes without saying that it was also rare for a female scientist to lead such an expedition. The media picked up on that.[25] When Solomon was in New Zealand leading a team of 16 men headed to Antarctica, a local reporter asked her, "How does it feel being a woman working with all these men?" Hiding her annoyance at the question, she looked around and said, "Wow, they are all men, aren't they?!," as if she'd just noticed for the first time.

6.4 Patching the Hole: Ozone Recovery and Climate Change

International efforts to curtail the use of CFCs were already underway in the late 1970s, through the United Nations Environment Program, but progress was slow, and the refrigerant industry was resistant.[26] In March 1985, even before the discovery of the ozone hole, the United Nations adopted the Vienna Convention for the Protection of the Ozone Layer, to coordinate action on CFCs. But this did not impose binding limits on emissions.[27] The dramatic discovery of the ozone hole in 1985, and the media coverage that followed, thrust the issue back into the spotlight. The public became aware of the direct health implications of ozone loss. This catalyzed action both in the United States and internationally.

A binding international agreement, known as the Montreal Protocol on Substances that Deplete the Ozone Layer, was signed in 1987 by 46 nations.[28] It came into force in 1989. When the new country of East Timor signed it in 2009, it became the first international environmental treaty to achieve ratification by all the parties, including 196 countries of the United Nations.

The Montreal Protocol has been very successful in reaching its objectives. Under it, the use of CFCs was phased out in many nations by the end of 1995.[29] That same year, three scientists were awarded the Nobel Prize in Chemistry for their work on the formation and decomposition of ozone: Mario Molina, F. Sherwood Rowland, and Paul Crutzen. This was the first time that a Nobel Prize had been awarded directly for environmental research. Crutzen had carried out research in the early 1970s showing that nitrogen chemistry could lead to ozone depletion in the stratosphere. He also mentored Susan Solomon's doctoral research.

The amount of chlorine in the stratosphere associated with CFCs peaked in the late 1990s and has been declining since then, as CFC emissions have decreased.[30] Springtime ozone values over Antarctica continued to decrease after 1985, into the beginning of the twenty-first century – but, in recent years, there has been indication of a recovery in ozone values,[31] amidst the strong noise of year-to-year variations. Due to the long lifetime of CFCs, a full recovery of ozone values, leading to a complete patch of the Antarctic ozone hole, will take many more decades.

The phasing out of CFCs required a switch to alternative refrigerants. As a transitional measure, a class of compounds called hydrochlorofluorocarbons (HCFCs) was allowed as an alternative to CFCs. These are not long-lived like CFCs, and they are much less harmful to the ozone layer. HCFCs have since been phased out in several countries and replaced with hydrofluorocarbons (HFCs), which have no chlorine at all and thus do not deplete ozone.

As tweaks to a complex system inevitably seem to do, the Montreal Protocol had some unintended consequences: some good, some bad. In addition to their ozone-depleting properties, CFCs are also powerful greenhouse gases. Their concentration is a million times lower than that of carbon dioxide, but CFCs are also many thousands of times stronger in their global warming potency on a per-molecule basis. Therefore, they make a significant contribution to global warming. Declining concentrations over the next few decades will eliminate their contribution to the warming.

The bad news is that HFCs, the current replacements for CFCs, are also powerful greenhouse gases. Due to their long lifetimes, they will accumulate over time and start contributing to global warming, which could add up to 0.5°C by the end of the century. Fortunately, the Montreal Protocol is amendable. In 2016, more than 170 countries adopted the Kigali Amendment with the goal of phasing out HFCs in 30 years. There are alternatives to HFCs that are under development, such as hydrofluoroolefins (HFOs). HFOs are expected to be harmless to the ozone layer and are also expected not to contribute to the greenhouse effect, which will help mitigate climate change.

Climate change has already caught up with Halley Station in Antarctica, though. The station was established on a floating ice shelf in the Weddell Sea, intended for year-round operation. But there were no winterers for three years in a row, 2017–2019, because the station was closed as a precautionary measure: The ice near Halley Station was no longer stable, and there was a danger of it breaking off into the sea.[32]

***

The discovery of the ozone hole is a stark example of the strengths and limitations of science. Nobody had foreseen it. Will models be able to predict the next dramatic climate phenomenon in the atmosphere involving complicated chemistry, or perhaps cloud microphysics? Reasoning inductively (albeit reasoning based on a limited sample), perhaps not. If models are fiction, as some critics would have it, the truth may well be stranger.

The deductive model of Molina and Rowland, although incomplete,[33] broadly predicted the properties of CFCs and their role in slow ozone depletion. However, it did not anticipate the rapid ozone loss in the polar stratosphere. This could be considered a limitation of deductive reasoning for complex systems, specifically the insidious problem of structural uncertainty. No amount of tweaking the parameters in an atmospheric chemistry model with only gas-phase chlorine chemistry could have made it capture the surface chemistry on polar stratospheric clouds.[34]

The initial delay in detecting the hole in the ozone layer, though, is a classic failure of inductive reasoning. The very low satellite-measured values of ozone were originally flagged as possibly erroneous because they fell outside reasonably expected values, based on past experience. There was no urgency to analyze the backlog of data because only a weak trend signal was expected. But this sort of inductive failure is perhaps unavoidable because monitoring resources are limited and need to be allocated based on past experience.[35]

Susan Solomon had experience with modeling ozone in the stratosphere before the ozone hole was discovered. Although her model could be considered "wrong" in the sense that it did not predict the hole, it was very useful in preparing her to process new information about ozone depletion, a perfect example of French biologist Louis Pasteur's dictum: "In the fields of observation chance favors only the prepared mind." Even the best models cannot predict phenomena that we know nothing about, but they can help us be better prepared to deal with such phenomena.

When observational scientists discovered the ozone hole, modelers quickly proposed multiple hypotheses to explain the new phenomenon. The existing model of chlorine chemistry provided important constraints, such as the angle of the sun, that allowed certain hypotheses to be ruled out immediately. Solomon's preparedness helped her decide what to measure to validate the remaining hypotheses, and where to measure it. New measurements allowed hypotheses to be falsified, leaving only the final accepted explanation. It is almost like a textbook example of Popperian falsification in the philosophy of science, although the actual falsification process was quite complex.[36]

The societal response to the ozone hole also has important policy lessons for the emerging problem of global warming. In the 1980s, industrialized nations had already made considerable progress in addressing local air pollution associated with automobile and industrial emissions. Like air pollution, ozone depletion posed a direct threat to human health that was easy to explain: increased incidence of skin cancer and cataracts. Unlike air pollution, though, ozone depletion required a globally coordinated response because it didn't matter where CFCs were emitted – they all ended up in the stratosphere. But only a narrow segment of the global industry, those using CFCs, would be affected, and affordable alternatives to CFCs could be found.

The harm due to global warming, though, is harder to explain to the public because the warming is happening very slowly, over the course of generations. Also, global temperatures being 2°C (or 3.6°F) warmer sounds more like a discomfort than a serious environmental issue. And the solutions to mitigate global warming would require a transformation of our entire energy infrastructure.

7

Global Warming

From Gown to Town

The summer of 1988 was an unusually hot one in the United States, with heat waves across the country. A drought that would extend into the next two years was just beginning. In Washington, DC, on June 23, it was a sweltering 98°F (37°C) – the warmest June 23 on record at the time. The Senate Committee on Energy and Natural Resources was holding a hearing that day on the greenhouse effect and global climate change. Two leading climate change modelers – Jim Hansen and Suki Manabe – testified, as they had before the Charney panel in 1979. This time, though, they were not speaking to scientists but to politicians and the public – in the full glare of the media.

Hansen spoke first. Showing the committee the global surface temperature data for the last 100 years, he said the four warmest years were all in the 1980s, and that 1988 was on track to set a new record.[1] He asserted that he was 99 percent confident that the "warming during this time period [was] a real warming trend." Hansen also showed recent predictions using the GISS model for three different scenarios for greenhouse gas emissions, each of which predicted different degrees of warming 30 years later, ranging from 0.6 to 1.5°C (Figure 7.1). Manabe was more measured when he spoke later, saying that the warming associated with increased carbon dioxide in the atmosphere would make droughts more likely. But, he continued, he could not be sure whether the current drought was caused by the warming trend because the background noise level of natural weather variability was high.

As one could guess, it was Hansen's assertion of 99 percent confidence that global warming had been detected that made the media headlines. It has been said that holding the hearing on one of the hottest days of one of the hottest summers on record amplified its impact. In a statistical note, as of the writing of this book, 1988 still holds the joint record for the hottest June 23 in Washington, DC, which means that no June 23 since then has been warmer, 31 years in a row and counting.

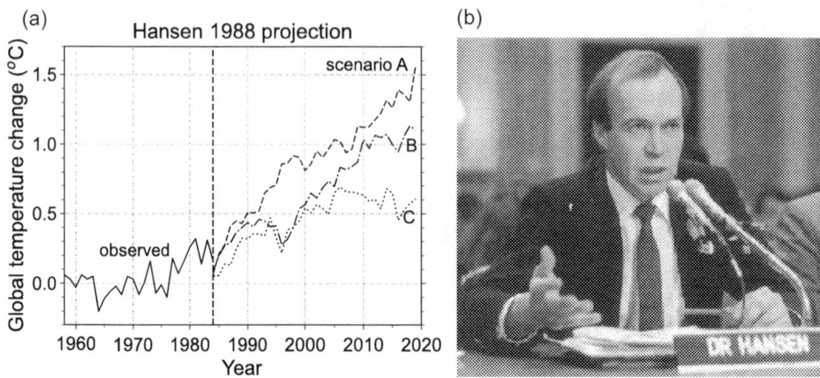

Figure 7.1 (a) Annual-average global surface air temperature change projected for 1984–2019 by the NASA/GISS climate model for greenhouse gas emission scenarios A, B, and C (as shown); observed annual-average temperature change (solid line), with reference to the 1951–1980 base period, is shown before 1984 (Adapted from Hansen et al.,1988). (b) Jim Hansen at a later Senate hearing (in 1989). (Photo: NASA)

After Hansen's 1988 testimony and follow-ups, the issue of climate change flew from its nest in the academic ivory tower to the front page of newspapers across the country.[2] The theoretical models of Manabe and other climate scientists had outgrown their academic playpens and entered the real world. It was a hostile world. From then on, climate science would become inextricably intertwined with economics, politics, energy policies, and, eventually, uncompromising ideologies.

7.1 Assessing Climate Change

As public awareness of global warming grew, so did the role of environmental activists who wished to prod governments into responding to this threat. Some worried that environmentalists would demand implementation of policies based on ideology, rather than scientific evidence. To avoid this, an international framework was created to obtain advice from the scientific community, but governments would have the last word on the recommendations.[3] The countries of the world had just come together to address the ozone depletion issue by agreeing to the Montreal Protocol in 1987. The hope was that a similar concerted global scientific effort would be able to address the global warming issue.

In 1988, two international organizations – the United Nations Environmental Program and the World Meteorological Organization – came together to create the Intergovernmental Panel on Climate Change (IPCC).[4] The IPCC would greatly influence climate science and policy in the years to come. Its first chairman was Swedish scientist Bert Bolin, who had worked with Jule Charney at Princeton in 1950 and who helped make the world's first operational weather forecast in Sweden in 1954.

The IPCC operates by consensus, rather than majority vote, and includes government and scientific representatives from 195 member countries around the world. Every five to six years, it produces an assessment report on climate change and climate impacts. Thousands of scientists are involved in producing this report, which is based on scientific results from published peer-reviewed papers. The Summary for Policymakers document, which needs to be approved line by line by the participating governments of the IPCC, is produced from the full report. (Other parts of the report also need to be approved, but the process for those is less strict.)

The IPCC assessment process considers all aspects of climate change and its societal impacts. It has three Working Groups (WG): WG1 deals with the physical basis of climate science, including observational evidence and climate model predictions; WG2 deals with the impacts of climate change; and WG3 deals with mitigation. For the rest of our discussion, we will focus on the conclusions reached by WG1, since they are the most relevant to climate modeling.

The IPCC reports released thus far are:

- 1990 First Assessment Report (FAR)
- 1995 Second Assessment Report (SAR)
- 2001 Third Assessment Report (TAR)
- 2007 Fourth Assessment Report (AR4)
- 2013 Fifth Assessment Report (AR5)

The Sixth Assessment Report (AR6) is expected sometime in 2021.

Climate models have gotten much more complex since the IPCC's first report in 1990; they now provide a much more comprehensive representation of important components and processes of the climate system. In the 1990 assessment, coupled models including the atmosphere and the ocean were still in the early stages of development. Today, they are routinely used. Processes like the carbon cycle were not part of the early models, but they are now incorporated in many of the newer models.

An important measure of model sophistication is the size of the grid box. The finer the grid box, the better certain processes such as the effects of mountains and clouds can be represented. The model grid size used by the IPCC for long simulations grew steadily finer over its first four assessments, from a horizontal grid of 500 × 500 km in 1990 to 100 × 100 km in 2007.[5] However, in the fifth assessment, released in 2013, the grid size stayed about the same because the model complexity had increased so much that finer grids were becoming prohibitively expensive computationally.[6]

One of the most important estimates related to climate predictions is climate sensitivity.[7] In the first assessment, the IPCC estimated climate sensitivity as having a range between 1.5°C and 4.5°C. This range estimate, which is the same as that in the 1979 Charney Report, remained unchanged in the second and third assessments. The fourth assessment narrowed the range somewhat, to 2.0°C–4.5°C. The fifth assessment widened it back to 1.5°C–4.5°C.[8] The needle on the climate sensitivity meter has barely moved in the past four decades, even though the meter itself has gotten a whole lot bulkier and more sophisticated. The lower estimates of climate sensitivity represent modest climate change that we could learn to live with, provided that our carbon emissions do not increase much over time. But the higher estimates imply a massive disruption of our environment by the end of this century if we do not drastically reduce our carbon emissions.

A simple inductive analysis of the evolution of the IPCC assessments over time leads us to some important conclusions about climate modeling:

1 The complexity of climate models continues to increase as more processes are included.
2 Computational costs increase as models become more complex and spatial grids get finer.
3 The uncertainty in our predictions has not decreased, despite the increasing sophistication of our models.

We will explore these conclusions further in Chapter 12.

Some of the most robust predictions in the IPCC reports come from the earliest climate models. Even at the time of the first IPCC assessment in 1990, the following features of global warming were anticipated by the work of Manabe, Hansen, and others, at times using fairly simple models:

1 The troposphere will warm and the stratosphere will cool due to the absorption and emission of radiation by carbon dioxide.[9]
2 Arctic surface temperatures will warm more than the global average, especially in the winter, due to local amplifying feedbacks (*polar amplification*).
3 The land surface will warm faster than the ocean surface.
4 There will be more heat waves, and fewer cold spells.

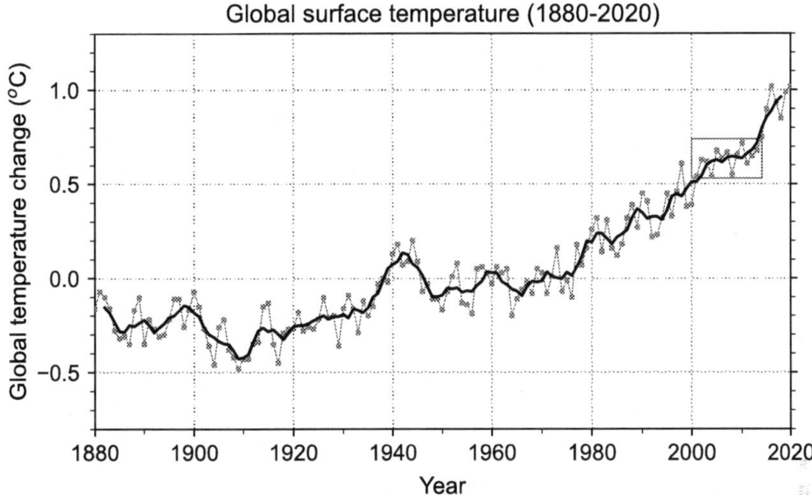

Figure 7.2 Annual global-average surface temperature change from 1880 to 2020 (gray line with square markers), and five-year moving average (black line), with reference to the 1951–1980 base period. Gray box marks the 2001–2014 "hiatus" period. (Plotted using NASA GISTEMP v4 dataset, Lenssen et al., 2019; GISTEMP Team, 2021)

5 The atmospheric water cycle will be more vigorous, with increased precipitation, due to the increased moisture-holding capacity of a warmer atmosphere.[10]
6 The sea level will rise due to the thermal expansion of seawater in a warmer ocean and the increased melting of glaciers and continental ice sheets.[11]

Except for the first result, which is difficult to explain without getting into intricate details of the theory of radiation, the other results are easy to explain using simple thermodynamics and physics arguments. More elaborate models over time have replicated these predictions and increased our confidence in them.

The IPCC also assesses the observational evidence for climate change. This has changed considerably since the first assessment, as we now have 30 more years of data. Since global warming accelerates over time (Figure 7.2), this additional data helps us better detect the human-induced warming signal against the noise of natural fluctuations of warming and cooling in global surface temperature.

The first IPCC assessment in 1990 was noncommittal on whether global warming had been unequivocally detected, but the next four assessments slowly changed the tone (see Box).

- 1990: "The size of this warming is broadly consistent with predictions of climate models, but it is also of the same magnitude as natural climate variability. Thus the observed increase could be largely due to this natural variability, alternatively this variability and other human factors could have offset a still larger human-induced greenhouse warming."[12]
- 1995: "The balance of evidence suggests a discernible human influence on global climate."[13]
- 2001: "Most of the observed warming over the last 50 years is likely to have been due to the increase in greenhouse gas concentrations."[14]
- 2007: "Most of the observed increase in global average temperatures since the mid-20th century is *very likely* due to the observed increase in anthropogenic greenhouse gas concentrations."[15]
- 2013: "It is *extremely likely* that more than half of the observed increase in global average surface temperature from 1951 to 2010 was caused by the anthropogenic increase in greenhouse gas concentrations and other anthropogenic forcings together."[16]

Notice that in the last four assessments, the detection of the global warming signal went from being "discernible" to "likely" to "very likely" to "extremely likely." The introduction of the term *discernible*, which occurred late in the drafting of the second assessment, in itself created a controversy regarding the IPCC peer-review process.[17] A small but vocal minority of scientists began to disagree with the IPCC assessments, starting with the second assessment. Some of the objections of this minority have to do with statistical issues in detecting the warming signal in observations, which are beyond the scope of this book. Other objections have to do with trust in climate models and their predictions; we shall address some of these objections in Chapter 13.

7.2 Howcatchem: The Attribution of Recent Climate Change

Commonly, global warming is described as a crime that has been committed, with climate scientists as detectives trying to figure out "whodunit." The standard detective story, featuring fictional sleuths like Sherlock Holmes or Hercule Poirot, starts with the crime and the evidence at the crime scene. After several plot twists, and a few deductions on the part of the detective, there is usually a surprise reveal of the criminal in the finale. This is perhaps a good metaphor for weather prediction, where the final result – the forecast – is difficult for humans to simply guess but can be arrived at given the initial conditions through the logic of physics and the power of computers. Weather forecasting is emblematic of hypothesis-driven deductivism, where the hypotheses fed into the computer-detective lead to the surprising, but logical, conclusion.

But there is another kind of detective fiction that turns the standard genre on its head. The most famous example of a "howcatchem" is perhaps the television series *Columbo*, which features the disheveled detective Lieutenant Columbo. At the very beginning of each episode, the audience sees exactly who committed the crime. The remainder of each episode is about how Columbo doggedly follows a trail of evidence and proves that a particular suspect is the perpetrator.

Sherlock Holmes and Columbo are both detectives, just as weather and climate models are essentially the same program – but their storylines are very different. The climate storyline is more of a "howcatchem" than a "whodunit." Detecting and attributing global warming is about the slow burn of assembling necessary evidence, rather than the surprise at the end of the story.

Most episodes of Columbo start with a death, but quite often the death appears to be natural. First, Columbo carefully examines the evidence to establish that the death was not natural – that it was in fact a homicide. He then proceeds to convince others who are initially skeptical that there is something unnatural about the death. Columbo also guesses who the perpetrator is early on, often based on an isolated piece of evidence. He collects additional evidence to rule out other suspects, and to make a convincing case for the perpetrator's arrest and prosecution.

If we consider the time series of temperature showing a slow warming of the globe since 1850 to be a "death," then we first need to establish that it is not a natural occurrence, that something unnatural is responsible for it. It is tricky to establish that the warming seen in surface temperature data since the start of the Industrial Revolution is attributable to human causes. Natural variability like El Niño events, volcanic eruptions, and irregular climate fluctuations act as background noise that can hide the signal. The human-induced signal would have been weak in the early part of the twentieth century and would only have picked up toward the end of the twentieth century, as carbon dioxide concentrations started to rise more rapidly.

The IPCC assessment uses statistical techniques on both surface temperatures, measured directly using thermometers, and so-called proxy data, where properties of ocean sediments, tree rings, ice cores, fossils, and corals are used as proxies for temperature over the last 1,000 years. The proxy data are much more uncertain than thermometer measurements, but there is no other way to obtain a longer data record. Once we have collected the data, we can compare the warming since 1950 against all warming events that have occurred over the last 1,000 years, to check if the warming we observe is unprecedented. As more data became available, the IPCC concluded that most of the warming since 1950 was due to nonnatural causes.

Once we accept that what occurred did not happen naturally, we need to find out who is responsible. Like Columbo, we start with a good hunch as to the culprit. As we have seen, the possible primary role of fossil fuel burning and the associated increase in carbon dioxide concentration in causing global warming have been known since the pioneering study of Arrhenius, followed by the work of Guy Callendar, Suki Manabe, Jim Hansen, and others.

Charles David Keeling's measurements, beginning in 1958, showed that carbon dioxide was in fact increasing rapidly. But how do we know that most of the increase in carbon dioxide came from humans burning fossil fuels, not from some natural process? We can do the equivalent of forensic analysis of the crime scene – in this case, the Earth's atmosphere – using a technique known as "carbon fingerprinting."

Living and recently dead plants have a small fraction of carbon atoms that are heavier than normal carbon atoms. These are atoms of the carbon-14 isotope. Carbon dioxide emitted into the atmosphere by rotting or burning plant material has the same fraction of carbon-14 as living plants. The heavier atom, though, decays radioactively over thousands of years. This means that fossil fuels, which consist of dead plant material from millions of years ago, contain virtually no carbon-14. The burning of fossil fuels emits carbon dioxide with no carbon-14 atoms, and this dilutes the amount of carbon-14 in the atmospheric carbon dioxide, lowering its overall fraction. This dilution is known as the Suess effect, named after Austrian-American chemist Hans Suess.[18] Keeling and others measured the Suess effect in the atmosphere to show that the observed increase in carbon dioxide is mostly associated with fossil fuel burning, essentially finding the "fingerprints" of fossil carbon in the atmosphere.

But correlation between the two positive trends in global temperature and carbon dioxide concentration does not mean anything; any stock market index will also show a long-term positive trend, but that does not necessarily mean that the stock market caused global warming. Other factors related to economic and population growth, such as changes in land use and urbanization, also have long-term positive trends that could potentially impact climate. Data cannot conclusively prove that increases in the concentration of atmospheric carbon dioxide are causing global warming. However, climate models can be used to show this, using something called an "attribution study."

Climate change attribution studies are similar to NASA's famous Twins Study,[19] which was conducted using Scott and Mark Kelly, the only pair of astronauts who are also identical twins. It is not easy to study the effect of zero gravity on aging: People also age in normal gravity, and each person ages at a slightly different rate, so it would be hard to tell if changes in the aging rate in space were due to normal variations or zero gravity. But if we had two identical bodies, one subject to zero gravity and one subject to normal gravity, then we could compare the aging rates between the two bodies. This is what NASA did. From March 2015 to March 2016, Scott Kelly spent 340 days in the space station being tested frequently, while his identical twin brother Mark stayed on Earth and underwent the same tests as Scott, at the same times. An important result from the Twins Study was that something called "gene expression" was affected by Scott's stay in space, in a way quite different from the changes Mark experienced on Earth.

The NASA Twins Study was like a nature-versus-nurture attribution study. It used identical twins in two different living conditions, zero gravity versus normal gravity, in order to attribute observed changes in gene expression to conditions in space. Since we don't have an identical twin of Earth to experiment with, we resort to climate models. We use two identical copies of a climate model, like Scott and Mark Kelly, and subject these identical copies to two different greenhouse gas boundary conditions. Then we look at the difference in the simulated global temperatures. These are referred to as "model twin" experiments in climate science.

A common attribution study involves "predicting" recent global warming starting from preindustrial conditions, say from 1850, with all observed boundary conditions, including variations in solar energy reaching the Earth; changing greenhouse gas concentrations; and concentrations of small particles, called *aerosols*, floating in the atmosphere. These aerosols can be natural – such as those produced by volcanic eruptions or dust storms – or *anthropogenic* (human-induced) – such as particles in industrial emissions. Given all boundary conditions, climate models simulate the warming that has occurred over this period with a fair degree of accuracy (Figure 7.3). Treating this experiment as the control, we can carry out another experiment in which we allow greenhouse gas concentrations to increase realistically but keep all other boundary conditions at their preindustrial values. The temperature changes seen in this experiment would be attributable to greenhouse gases, assuming the different attributions are roughly additive.[20] As we see in Figure 7.3, by themselves greenhouse gases would cause more warming than is actually observed. A similar attribution study for the effect of anthropogenic aerosols shows that these aerosols tend to cool the globe. This modest cooling effect cancels out a portion of the greenhouse gas warming.

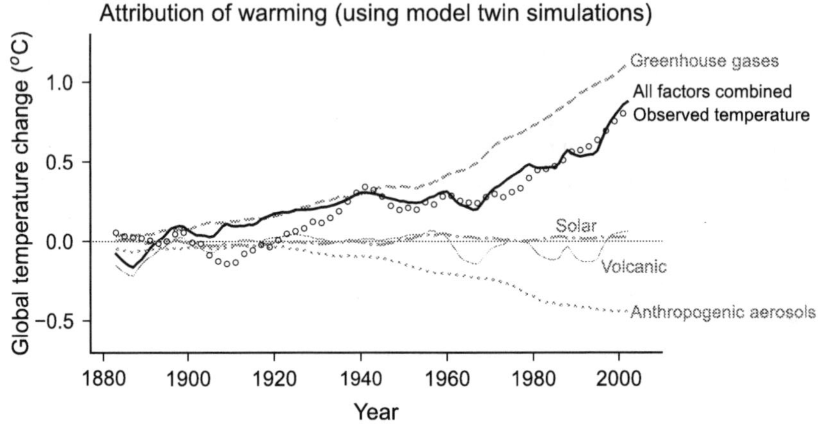

Figure 7.3 Attribution of global temperature changes to different factors that can warm or cool climate. Each line shows the temperature change associated with a specific factor, as indicated. The observed annual temperature is shown for reference (open circles), as is the sum for all factors (solid line). All temperature changes are with respect to the 1880–1910 base period and are smoothed in time using a 7-year moving average. Note that solar (dash-dot) and volcanic (thin solid) effects are natural factors, whereas greenhouse gases (dashed) and anthropogenic aerosols (dotted) are due to human influences. While natural factors have always made a steady contribution to temperature changes, their relative importance has dramatically decreased as the human-influenced factors have grown. (An ensemble of five simulations, starting from five different initial conditions, was carried out for each attribution factor. The ensemble average is shown.) (Plotted using data from NASA/GISS ModelE2 twin simulations described in Miller et al., 2014, and in Rosten and Migliozzi, 2015)

What about other possible causes of warming? Like Columbo ruling out other suspects, we need to eliminate them as primary causes of the observed warming. An obvious suspect would be the variation in solar energy output; after all, it is the sun that keeps Earth warm. If the sun were the culprit behind global warming, we would expect its energy output to have increased during the last 30–40 years,[21] that is, when the warming accelerated. But as climate scientists Andrew Dessler and Dan Cohan put it in a 2018 op-ed,[22] "the Sun [...] has an airtight alibi – we have direct measurements of the output of the Sun from satellites, and we observe that the Sun has not gotten any brighter." Using recent satellite measurements and extrapolating them back in time using the 11-year periodicity of the sunspot cycle, we can carry out an attribution study to identify the impact of solar energy variations from the start of the industrial era. We see that the solar effect is too small to explain the long-term warming signal (Figure 7.3).

We can use an alternative approach to determine if solar energy variations can explain the warming signal. It is also known as "fingerprinting," but it involves spatial patterns, rather than isotope ratios. If we look only at trends in surface temperature, all types of variability look similar, with irregular ups and down over time. But like a meticulous detective, we need to carefully examine how the warming (or cooling) varies with latitude, longitude, and height in the atmosphere.

Model calculations tell us that the signal associated with increasing greenhouse gases shows warming in the troposphere, but cooling in the stratosphere. The warming is also stronger at higher latitudes, especially in the winter. We can treat these spatial patterns as "fingerprints" of human-induced warming. Other possible causes of warming or cooling, such as changes in solar energy output or volcanic eruptions, are associated with different spatial patterns or "fingerprints," which can be calculated using climate models. Once we have collected the fingerprints associated with different suspects, we carefully examine the observational data for these fingerprints. For example, increased solar radiation would be expected to warm both the lower atmosphere and the stratosphere, which is different from the observed global warming signal.[23] This provides additional confirmation that solar output variations are not major contributors to recent global warming, countering a claim often made by climate contrarians.[24]

Recall that it was John Mauchly's desire to calculate the influence of sunspots on weather that eventually led him to build the first electronic computer. Climate models can now answer the question that motivated Mauchly. Analyzing large amounts of weather data using a computer, as Mauchly originally intended, can identify the small signal of sunspot influence on climate. But this will not lead to predictions of any practical value because this signal is much weaker than the greenhouse warming signal.

"Just one more thing," as Columbo would say: Climate change attribution is less sensitive to model errors than climate change prediction. One of the biggest uncertainties in predicting global warming is compensating errors between processes represented in climate models. But such errors typically affect the two experiments of twin attribution studies in approximately the same fashion. Computing the difference between the two experiments would mitigate the impact of these errors, just as the NASA twin study would have been less affected by errors that affected both twins the same way. This means that comparing the relative contributions of different sources to warming during the past 100 years is a more robust application of climate models than trying to predict future warming for the next 100 years. The same can be said for estimating the spatial patterns or "fingerprints" of different contributors to warming. These applications rely less on the predictive power of climate models and more on their constraining power, a topic we shall return to in Chapter 9.

7.3 Extreme Weather Attribution: The Butterfly Did It?

What distinguishes hurricanes, typhoons, nor'easters, and other storms from random variations between normal sunny and cloudy weather? These storms are bigger-than-normal weather fluctuations. They seem to appear out of nowhere and then grow rapidly, moving about in ways that are sometimes hard to predict.

Mathematically speaking, these storms are caused by an instability of the atmospheric flow. Atmospheric winds, temperature, and moisture are sometimes arranged in such a way that a small "kick" or perturbation to the system grows rapidly to form these storms. Such an atmospheric flow configuration is said to be "unstable." How fast the storm grows depends on how unstable the flow configuration is.

The atmosphere may be unstable over a large region, ripe for a storm, but we cannot easily predict exactly when and where the storm will form, because we cannot observe the small perturbations that trigger its formation. Lorenz used the flapping of a butterfly's wings as a metaphor for the initial kick or perturbation (Section 2.4). The unpredictable appearance and subsequent growth of the storm is the Butterfly Effect in action.

What causes a particularly strong storm? That's the extreme weather attribution "whodunit" question. Is it the butterfly flapping its wings or the unstable air that it is flying in? There could be many butterflies fluttering about in a large region of unstable air, but only one may flap its wings hard enough to trigger a storm. So we would blame that butterfly – or the random small perturbation it represents – for triggering the storm at that particular point in space and time.

What if global warming made the atmosphere more unstable, say, by providing more moisture to support the growth of storms? This could have two different types of consequences. The threshold for the amount of flapping needed to generate the storm may be lowered. So the butterflies flapping their wings may generate more storms. Additionally, or alternatively, storms may be able to grow faster and become much stronger after they form, due to more moisture being available.

What if we observe an extremely powerful storm – let's call it Hurricane H? Did the recent global warming cause Hurricane H, or did a butterfly cause it?

If storms as powerful as H had occurred during the "normal" climate of the past, then we would blame it on a butterfly. Without that butterfly flapping its wings, the storm would not have happened at the time and place it did. If the storm was truly unprecedented, we can attribute that storm – and all such future storms – also to climate change, which acts as an essential accomplice. Without climate change, such storms would not have happened at all.

To be truly unprecedented, though, it is not enough if such a powerful storm has not been seen in a short climate data record. We have accurate storm records for less than 100 years, but our climate has been "normal" for more than 10,000 years, after the last ice age ended. What appears to be unprecedented in a 100-year data record may not be so in a 10,000-year data record, in which a much greater number of storms would be present. Paleotempestology,[25] the study of hurricanes in the distant past using geological data, can provide important clues about past extreme events, but it is hard to be definitive because the evidence is usually circumstantial. Therefore, it is easier to argue statistically that Hurricane H is truly unprecedented if Hurricane H is far more powerful than any storm occurring in the recent past, as opposed to just slightly more powerful.

Climate models are getting better at simulating extreme weather phenomena like hurricanes, although they still need a much finer grid to capture details like the clouds along the hurricane eyewall. If models are able to accurately simulate the statistics of hurricanes, model simulations can serve as proxies for the longer data record. We can also use "model twin" experiments, like the ones described in Section 7.2, to study attribution for a variety of extreme weather events,[26] such as heat waves, cold spells, tropical cyclones, and storm surges.

We should not fully attribute an individual weather event such as a hurricane or heat wave to global warming unless it happens to be extraordinarily strong. It is better to make partial or probabilistic attributions[27] – we can attribute to warming the trends in the statistics of weather events, such as the average intensity of severe weather events, or the frequency or likelihood with which they occur. If we think of random weather events as being determined by rolls of a set of dice, then global warming can load the dice, preferring some outcomes over others.[28] Each roll still has a mostly random outcome, but, with enough rolls, we can tell if the dice are loaded.

Since we often lack enough data to reliably detect trends in many types of extreme events, we resort to "model twin" experiments. We can then compare the intensity and frequency of extreme weather events in simulations of preindustrial conditions against those occurring in current conditions. This allows us to statistically attribute fractional changes in the intensity or the frequency of extreme events to global warming. For example, attribution studies assess that rainfall due to Hurricane Harvey was 15 percent higher due to global warming, and that heat extremes over land are five times more likely to occur.[29] (Model imperfections mean that there are still uncertainties in the exact numbers we can assign to the fractional attributions, though.[30])

Figuring out how extreme weather will evolve in the future is a crucial aspect of predicting climate change. Extreme events, though infrequent, are responsible for a significant portion of the damages associated with global warming. We expect that a warmer atmosphere will lead to more heat waves and fewer cold spells, which is confirmed by the last IPCC report.[31] A warmer atmosphere can also hold more water vapor, which can lead to more intense rainfall events, likewise confirmed by observations and simulations. But there is no "blanket" linkage between climate change and the increasing likelihood of *all* extreme events, as it is sometimes assumed. For example, some media headlines were quick to blame climate change for the devastating freeze suffered by much of Texas in February 2021,[32] even though Texas had experienced extreme cold events in the past.[33] The evidence for climate change increasing the likelihood of such extreme cold events is quite mixed;[34] if anything, global warming is expected to decrease the likelihood of cold spells.

Since extreme events are rare, by definition, the statistical signal associated with their trends is harder to detect against the background of weather noise. According to the last IPCC report, there isn't yet sufficient observational evidence to either support or rule out global trends in many other types of extreme events, although there is more evidence for regional trends in phenomena such as wildfires. More than half the increase in fuel aridity in the western United States since the 1970s has been attributed to human-induced climate change,[35] resulting in devastating wildfire seasons. While careless campers or downed power lines may provide the initial sparks, it is dry fuel conditions caused by the combination of climate change and poor forest management that feed the wildfires. One of the most noticeable impacts of wildfires is the release of smoke particles into the air. These microscopic particles – which are responsible for breathing problems, burning eyes, and apocalyptic orange skies – also play an important role in climate.

7.4 The Minuscule but Mighty Aerosol

Little drops of water,
Little grains of sand,
Make the mighty rain cloud
And the haze over land.

Modified from the nineteenth-century poem by Julia Carney

One of the weakest arguments made against the existence of the greenhouse effect is that there is so little carbon dioxide in the air (only about 400 parts per million) that it cannot be responsible for all that heating. Following that line of reasoning, we could also argue that very small amounts of arsenic in drinking water or of carbon monoxide in indoor air cannot really harm us. Of course, that would be absurd. If one's aim is to cast doubt on climate models, it would be better to argue that very small things in the atmosphere matter very much, and that climate models cannot possibly get them right! It is obvious that climate models can get very big things mostly right, such as the movement of air in cold fronts or water flow in ocean currents. We can measure the big things easily, and the computational grid of the model can capture them well.

Computer models also get the properties of some very small things right, such as the absorption properties of molecules of carbon dioxide, which are governed by the laws of physics. However, there are some things neither very big nor very small that current models have trouble with, such as clouds – which are made of tiny water droplets or ice crystals – and aerosol particles – which can reflect or absorb sunlight. The properties of these droplets, crystals, and particles are hard to measure. They are microscopic but much larger than molecules, and they come in many different sizes and shapes. Since they are not easily described by the simple laws of physics, their behavior can be complicated to understand and model.

Aerosols are liquid or solid particles that are so minuscule that they can float in the air for long periods of time.[36] They are found naturally in the atmosphere. If you have been to the beach, the salt in the sea spray is an example of an aerosol particle. Minuscule particles of salt continuously floating in the air above the ocean are referred to as *marine aerosols*. Dust particles blown away from desert regions like the Sahara and emitted from volcanic eruptions, are also natural sources of aerosols. Aerosols in the lower part of the atmosphere, the troposphere, only stay around for a couple of weeks. But if they enter the upper part, the stratosphere (Figure 3.1), they can remain there for years, as in the case of aerosols generated by powerful volcanic eruptions or asteroid impacts. Attribution studies show that volcanic eruptions can have a significant short-term global cooling impact (Figure 7.3).

Human activities also generate aerosols. The burning of fossil fuels generates sulfur dioxide. Sulfur dioxide interacts with other gases to form sulfate aerosols, which include tiny droplets of sulfuric acid. These were responsible for acid rain in the United States before the 1990 amendments to the Clean Air Act curbed sulfur dioxide emissions. Sulfate aerosols, which are also generated naturally during volcanic eruptions, reflect sunlight and thus cool the atmosphere. The burning of vegetation, on the other hand, generates aerosol particles called "black carbon," or soot, which absorbs sunlight before the sunlight reaches the ground. This warms the atmosphere. The reflection or absorption of sunlight by aerosols is known as the *aerosol direct effect.*

Clouds and aerosols are closely connected. A cloud is made up of water vapor condensed or frozen into small droplets or crystals. This condensation or freezing happens around small particles in the atmosphere known as *nuclei.* Aerosols serve as nuclei for the formation of clouds. Predicting the role that clouds play in climate is a challenge. While clouds low in the atmosphere reflect sunlight and tend to cool the climate, clouds higher up can trap infrared radiation and warm the climate.[37] Depending on how cloud distribution shifts as the climate changes, clouds may either amplify the greenhouse effect or mitigate it.

Clouds can range in horizontal extent from about 100 meters to hundreds of kilometers. But a climate model grid box today has sides that are many tens of kilometers wide – much larger than an individual cloud. This means that climate models do not currently capture clouds, so approximate formulas need to be used to represent them (Figure 4.2). Even if the grid box width were to shrink to 100 meters, the effect of aerosols would still need to be represented using formulas, because aerosol particles would still be a million times smaller. There is no possibility of building a climate model with a grid box the size of a microscopic aerosol particle.

Since climate models do not directly simulate clouds, the heating (or cooling) associated with clouds and the aerosol particles that help them form is currently one of the biggest uncertainties in climate modeling. The impact of aerosols on cloud formation is known as the *aerosol indirect effect.* Some aerosol effects increase cloud formation; others decrease it. Climate models differ in their estimates of the net effect of aerosols on clouds, which has a significant impact on global warming predictions. The net effect is believed to be a cooling, but its exact strength is uncertain.

The first IPCC assessment barely mentioned the effect of human-made aerosols, whose properties were just beginning to be understood.[38] Subsequent reports started to provide estimates for the net cooling effect of aerosols, but the numbers varied. The fifth IPCC assessment report in 2013 estimated the cooling effect to be about 1 watt per square meter, which was about 40 percent higher than in the previous assessment. To compare, the warming effect of greenhouse gases is about 3 watts per square meter, which means that aerosols could potentially reduce net warming by a third.

The effect of aerosols helps explain an important feature in the global temperature record. The global temperature warmed from the early 1900s to the early 1940s, but it started to cool after that, and the warming did not resume until the 1970s (Figure 7.2). Not coincidentally, it was after this that scientists started to seriously worry about global warming. But what caused the cooling after the Second World War? The 1950s and 1960s were a period of rapid industrial growth, and air pollution was not yet regulated. This led to a large increase in aerosols in the atmosphere, especially the reflective sulfate aerosols. Attribution experiments using climate models show that aerosol-induced cooling can explain the lack of warming during that period (Figure 7.3), despite the increase in carbon dioxide concentration measured by Keeling (Figure 5.2). Controls on air pollution have slowly decreased the concentration of atmospheric aerosols since then, while greenhouse warming has gotten stronger, allowing the global warming signal to emerge in the 1980s and 1990s.

7.5 Internal Variability: Much Ado about a Hiatus

For about a decade after Jim Hansen's 1988 testimony, global temperatures continued to warm. The warming was not monotonic. There were bumps and dips from year to year, but, averaged over multiple years, there was a clear warming trend (Figure 7.2). But, after an exceptionally warm year in 1998 associated with an El Niño event, the warming appeared to pause for several years. This created a stir in the climate change community. Contrarians who did not believe in global warming seized upon this as proof that global warming was not actually occurring. Climate scientists, on the other hand, proposed different theories for the pause, treating it as an unprecedented event: Possible causes ranged from the sunspot cycle to El Niño events in the tropical Pacific.

The pause or "hiatus" in the warming lasted over a decade, at which point the warming started to tick up again – and in 2014, it resumed at a faster pace. (Recently revised estimates of global temperatures mean that the slowdown may not have been as much of a pause as originally believed.[39]) A long pause in the global warming signal had occurred before, from the mid-1940s to the mid-1970s, due to increasing aerosols in the atmosphere (Section 7.4). Aerosols could not explain the more recent hiatus, as they were no longer increasing. This leads us to a general question: What causes dips, bumps, and pauses in global temperatures over the years?

We can understand the pauses by viewing the global average temperatures for a particular year as the sum of signal and noise. The signal is greenhouse warming. If the noise has cooling that counteracts the warming, we see a pause. Such noise can arise from variations in the external boundary conditions – such as solar influence – or factors internal to the climate system – such as warm El Niño events (or their cool counterparts, La Niña events). If the noise or the external boundary conditions have strong periodic or oscillatory behavior, we can use that to predict climate. The 11-year sunspot cycle, with which John Mauchly had hoped to predict weather (Section 1.1), is oscillatory, but its climate influence is weak (Section 7.2). There are also some climate phenomena which are labeled oscillations, such as the North Atlantic Oscillation (NAO) or the Atlantic Multidecadal Oscillation (AMO), but they are not necessarily periodic. (The name oscillation sometimes just refers to their spatial dipolar structure, as in the case of the NAO.) With the exception of El Niño, many of these "oscillations" can be modeled as random phenomena on annual-to-decadal timescales.[40] El Niño does exhibit oscillatory behavior, but its period is irregular, ranging from 3 to 5 years, making it hard to predict beyond about a year.

Climate prediction involves more than predicting averaged atmospheric weather. It also requires predicting oceanic motions, including El Niño events and other natural variations. Like the atmosphere, the ocean displays the Butterfly Effect, or chaotic error growth. Unlike those of the atmosphere, the oceanic initial conditions are poorly known – three-dimensional snapshots of the ocean are hard to obtain, limiting predictability. The ocean acts like a flywheel for the climate system, owing to its large heat capacity (Section 4.6), and the oceanic Butterfly Effect occurs in slow motion, over years to decades, instead of days to weeks. Eventually, the chaotic error growth in different components of the climate system saturates, resulting in random noise, usually referred to as *internal variability* or *stochastic uncertainty*. This noise is stronger than the greenhouse warming signal over the first decade or two of a climate prediction, but it becomes relatively less important later on.[41] As in the case of ensemble weather prediction (Section 2.5), we can use an ensemble of climate predictions, corresponding to different atmospheric and oceanic initial conditions, to estimate this uncertainty. Climate scientist Clara Deser refers to this stochastic uncertainty as "certain uncertainty" because it is irreducible[42] – better climate models will not diminish it, although they can help better quantify it.[43]

Model simulations show that the "pause" between 2001 and 2014 can be explained by internal variability. An analysis of a large ensemble of short-term climate change predictions, using multiple models, indicates that it is possible for some regions to experience cooling for the one or two decades beyond 2020, even if the model's climate sensitivity is large.[44] This means that pauses in global warming cannot be ruled out in the near future, but they should be progressively less likely to occur, as the global warming signal is expected to dominate over internal variability.

7.6 Storylines, Scenarios, and Projections

All the world's a stage,
And all the men and women merely players
William Shakespeare, *As You Like It*, Act II, Scene VII

To make a weather prediction, we construct a computer model that represents all the important atmospheric processes, directly using a grid for large-scale processes and using approximate formulas for processes that occur on finer scales than the grid. We measure the initial conditions of the atmosphere as best we can, plug them into the model, and use the model equations to calculate what happens in the future, typically up to two weeks ahead. Climate prediction is similar, but we predict several centuries ahead.

Initial conditions in the atmosphere don't matter for climate prediction because we predict climate well beyond the weather predictability limit of two weeks. Initial conditions in the ocean can matter for the first 20–30 years of a climate prediction because the ocean has a much longer predictability limit. But for predictions on the scale of centuries, neither initial conditions in the atmosphere nor initial conditions in the ocean matter. (If and when slowly evolving components like continental ice sheets become part of climate models, details of their initial conditions will become important – and so will any Butterfly Effect associated with their initial conditions.)

To make climate predictions, we need to:

1 Include additional models for the ocean and sea ice to interact with the atmosphere because those interactions become important when we predict beyond a couple of weeks into the future.
2 Provide boundary conditions for greenhouse gas concentrations and long-term solar variations for the entire duration of the climate prediction (Chapter 5). (These are also needed for weather prediction, but we can simply assume them to be constant over a two-week period.)

When simulating past climate, as in the attribution experiments of Section 7.2, we can use the observed boundary conditions. But to predict future climate, we need to first predict the boundary conditions themselves. We can predict the seasonal cycle of solar variations very accurately using astronomical data about the Earth's orbit. However, we cannot really predict future emissions of greenhouse gases because they depend crucially on decisions that we make about our socioeconomic future. Instead, we use different assumptions about our future decisions and make predictions about emissions that are conditional on those decisions.[45] Such conditional predictions are referred to as *projections*. The assumptions underlying these projections are collectively referred to as a *storyline* or *scenario*.

A good storyline or scenario is not meant to be a long-term prediction;[46] it is more like a well-crafted piece of socioeconomic fiction, like a novel or a play about our future world. A sample scenario narrative, titled "regional rivalry – a rocky road," begins like this:[47]

> A resurgent nationalism, concerns about competitiveness and security, and regional conflicts push countries to increasingly focus on domestic or, at most, regional issues. Policies shift over time to become increasingly oriented toward national and regional security issues. Countries focus on achieving energy and food security goals within their own regions at the expense of broader-based development. Investments in education and technological development decline.

Scenarios provide answers to the question "What if X, Y, and Z happen?" for different plausible combinations of X, Y, and Z. The focus of a scenario is on creating a consistent plotline, in which different sources of emissions respect economic and technological constraints. Scenarios can also take into account current trends. The major assumption here is that these constraints and trends can be extrapolated centuries into the distant future. We have no choice but to operate under this assumption, since we cannot predict human behavior in the same way that we can predict the Earth's orbit. Ultimately, though, the atmosphere does not care about the socioeconomic aspects of an individual scenario plotline. All it cares about are the physical aspects, the total amount of greenhouse gases emitted and changes in land use, which determine the change in global temperatures. Therefore, the primary requirement of scenarios is that, as a group, they span the range of plausible future emission trajectories.

For his 1988 climate projections, Hansen used three very simple boundary condition scenarios, labeled A, B, and C. Scenario A (High) projected exponential growth in greenhouse gas concentrations, at the rate of 1.5 percent per year. Scenario B (Medium) projected linear growth in concentrations, extrapolating the recent trend into the future. Scenario C (Low) projected that concentrations would remain constant after the year 2000, assuming strong mitigation efforts.

In recent years, more elaborate approaches have been used to generate scenarios for greenhouse gas emissions. A variety of consistent storylines is created for different socioeconomic scenarios of future population growth, energy use, agricultural practices, economic development, technological innovation, and emission mitigation policies. The storyline for a "high emission" scenario might assume no mitigation policies and strong economic growth; the storyline for a "low emission" scenario might assume some combination of strong mitigation policies and lower economic growth. These storylines are fed into *integrated assessment models* that generate emission scenarios,[48] predicting how much carbon dioxide, methane, and other greenhouse gases will be emitted each year, and what their atmospheric concentrations will be.

In 2007, a new IPCC expert committee came up with a different set of four scenarios called Representative Concentration Pathways (RCP). These scenarios were labeled RCP2.6, RCP4.5, RCP6.0, and RCP8.5. The numbers following "RCP" – 2.6, 4.5, 6.0, and 8.5 – denote the amount of extra heating due to increased greenhouse gas concentrations, measured in watts per square meter.[49]

The high-end scenario, RCP8.5, assumes that emissions will continue to increase throughout the rest of the century, with little effort undertaken to curb them (Figure 7.4a). The low-end scenario, RCP2.6, assumes that ambitious mitigation efforts will cause carbon emissions to peak around 2020 and then decline. RCP6.0 is a middle scenario which assumes that modest mitigation efforts will lead to an emissions peak around 2080, followed by a decline. (Another scenario, RCP4.5, lies in between RCP6.0 and RCP2.6, but in order to keep this discussion simple, we will not discuss it further.)

The emission scenarios are fed into a simplified climate model to compute how carbon dioxide concentrations will change in the future (Figure 7.4b), in effect extending the Keeling Curve (Figure 5.2b). In the RCP8.5 scenario, the continued increase in emissions causes the concentration to increase at an accelerating rate and to exceed 900 ppm by the end of the century. In the RCP6.0 scenario, carbon dioxide concentrations increase more slowly, but continue to grow even after emissions peak around 2080. Since carbon dioxide persists for thousands of years in the atmosphere, emissions accumulate over time and cause concentrations to steadily rise;[50] only when emissions decline to zero (as they do in the RCP2.6 scenario) will high carbon dioxide concentrations slowly decline as the gas is absorbed by the ocean and land.

As a companion to each cycle of the IPCC assessment, a Coupled Model Intercomparison Project (CMIP) is performed. Modeling centers around the world carry out a number of long climate prediction simulations, following a uniform protocol for different scenarios of emissions of greenhouse gases. Analyses and validation of these simulations form a crucial part of the assessment report. An important task of these simulations is to predict the change in global temperatures and precipitation until the year 2100 and beyond, as well as to provide estimates of climate sensitivity.

The climate model simulations corresponding to the IPCC Fifth Assessment are known as CMIP5.[51] Many different models from around the world participated in CMIP5 to carry out climate simulations for the period 1850–2005, using observed boundary conditions for greenhouse gas concentrations, aerosols, and other factors (similar to the "all factors" attribution in Figure 7.3).

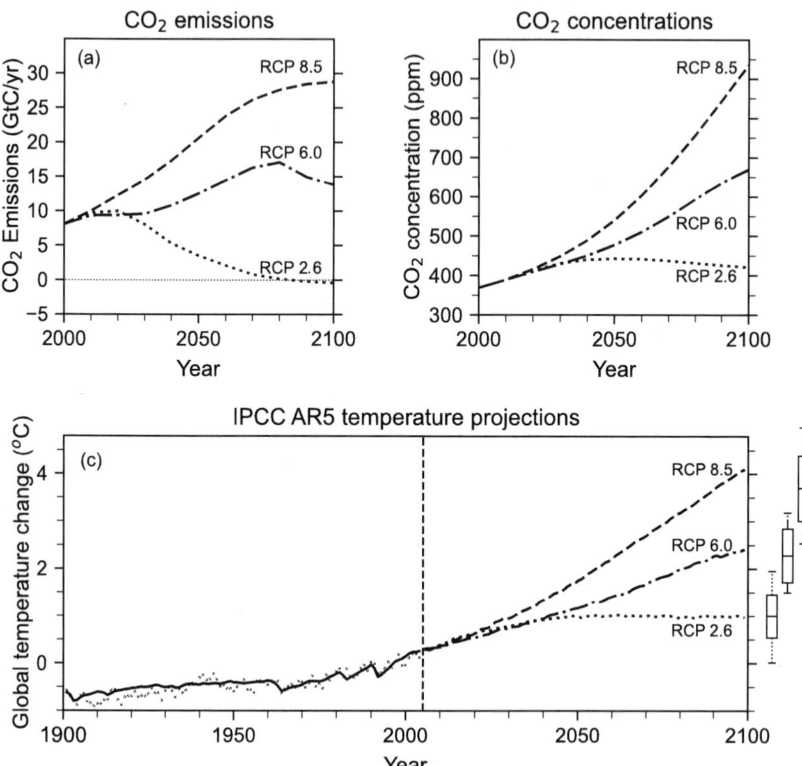

Figure 7.4 (a) Annual global emission of carbon dioxide (in gigatons of carbon per year) in three different RCP scenarios: 8.5 (dashed), 6.0 (dash-dot), and 2.6 (dotted). (b) Atmospheric concentration of carbon dioxide (in parts per million) for the RCP scenarios. (c) Global temperature change, averaged across models, relative to the 1986–2005 base period. The solid line from 1900 to 2005 represents the historical simulation, and the gray dots show the observed annual temperatures for the same period. After 2005, projected temperature changes for the different RCP scenarios are shown. The numbers of models using RCP scenarios 8.5, 6.0 and 2.6 were 30, 17, and 26 respectively. The rectangular boxes show the mean and the one standard deviation range for the spread among the different model projections averaged between the years 2080 and 2099; the "whiskers" show the minimum and maximum projected values. *Note that the spread across models is neither an error bar nor a probabilistic estimate of uncertainty.* (Based on the IPCC Fifth Assessment, Collins et al., 2013; panel (c) adapted from Knutti and Sedlaček, 2013, by permission from Springer/Nature Climate Change, © 2013)

These simulations are referred to as "historical simulations," and their purpose is to verify that models can "predict" the known past, given the right boundary conditions. As we see in Figure 7.4c, the average model "prediction" matches the observations fairly well.

Beyond 2005, simulations were carried out using greenhouse gas concentrations from the RCP scenarios as boundary conditions. These are referred to as "projections," and their purpose is to assess the range of possibilities for future climate change. The projected temperature change for the different scenarios,[52] averaged across the models, is shown in Figure 7.4c. Note that the temperature changes mostly track carbon dioxide concentration changes, not carbon dioxide emission changes. It is the cumulative effect of emissions that determines the warming.[53]

Figure 7.4c also shows the spread in projected warming among the models, averaged over the 2080–2099 period.[54] Different models project considerably different degrees of global warming, even for the same emission scenario. For RCP8.5, the high-emission scenario, the projected global warming ranges between 2.6°C and almost 5°C across the models, with an average of about 3.7°C. The average projected warming is substantially lower for the mid- and low-end scenarios, but the spread of the estimates is still significant. The reason for the spread is that different models have different parameterizations – they represent fine-scale processes like clouds and aerosols in different ways (Section 4.3). Overpredicting or underpredicting the number of clouds, for instance, can affect the degree of warming that a model predicts. This is referred to as *model uncertainty*, to distinguish it from the *scenario uncertainty* associated with different storylines of the future.

7.7 Warming Globally, Impacting Locally

Cyberpunk science fiction writer William Gibson observed: "The future is already here. It's just not evenly distributed yet." This is certainly true of global warming, which some may think is a problem of the future. One of the most unambiguous indicators of global warming is sea level rise, which is unaffected by factors such as weather noise, urban growth, and land use changes. Rising sea levels are already a serious problem for many island nations. On the island nation of Kiribati, 81 percent of households have already been affected by rising sea levels, according to a 2016 UN report. Other impacts are equally clear: Warming oceans and acidification are killing coral reefs around the tropics, heat waves are becoming more frequent around the world,[55] and warming temperatures have become a major factor contributing to worsening wildfires in California.[56]

Scientists who study regional climate change actually have a metric, called *Time of Emergence*, that encapsulates Gibson's observation.[57] Time of Emergence is an estimate, for each region of the globe, of how many years from now it will take for the climate change signal to emerge from the background noise of internal variability in that region. As you may imagine, it is harder to discern certain trends in regions with highly variable weather as opposed to regions with more monotonous weather. Climate models have shown that the global warming signal is stronger in the higher latitudes and weaker in the tropics, due to local feedbacks in the former, such as the ice–albedo feedback. But that does not mean the global warming signal is easier to detect in the higher latitudes, because these regions have more variable weather than the tropics. The global warming signal is weaker in the latter, but the noise of weather is weaker yet. This means that climate change can be detected decades earlier in tropical countries than in those outside the tropics.

Much of the discussion of global warming is focused on spatially averaged measures like global-average temperature and climate sensitivity. But just as "all politics is regional," all climate change is also regional.[58] Just as there is no "average voter" in real politics, there is no "global-average region" that we all live in. If the global-average temperature were to warm by 2°C, some regions would experience greater warming, while other regions would experience less. Extreme weather events would amplify these differences. Global warming thresholds like 1.5°C or 2°C are purely numeric constructs; nothing dramatic will happen globally at the moment that these thresholds are crossed. However, dramatic impacts may start to appear locally well before this threshold is reached, and these impacts will continue to appear after the thresholds have been crossed.

To contextualize regional and global changes, consider the peak of the last ice age, known as the Last Glacial Maximum, which occurred about 20,000 years ago (Figure 5.3). At this time, the global-average temperature was around 5°C lower than it is today. This may not sound like a huge number compared to the seasonal temperature change in some regions. However, much of northern Europe and northern North America was covered by gigantic ice sheets during the Last Glacial Maximum. The conservation of water requires the sea level to have been lower at that time, and it was, by about 120 meters (400 feet) on average. Similarly, we can expect any scenario even approaching a 5°C global warming to lead to drastically different regional climates, which could disrupt regional economies and societies in catastrophic ways. For example, the global warming projected by climate models (Figure 7.4c) would result in some dramatic regional impacts. In the high-emission RCP8.5 scenario, the Arctic becomes ice-free in the summer by mid-century.[54] Marine ecosystems are threatened in all three scenarios, as the ocean continues to become more acidic.[59]

Global warming thresholds, such as 1.5°C/2°C, are the subject of many heated debates.[60] These thresholds can serve a motivational purpose, but, scientifically speaking, there is no global "cliff's edge" for warming. If your region is expected to see a 4°C rise in daily maximum temperatures, that information should be more important to you than the expectation that global-average temperature will warm by "only" 2°C.[61] Unfortunately, though, trying to predict regional climate change is much harder than trying to predict the global average.[62] There are a variety of processes that control temperature and rainfall in each region, such as the prevailing winds, proximity to the ocean, presence of mountains, and type of vegetation. Climate models exhibit different errors in representing these local processes, but, in computing the global average, many of these model errors tend to cancel out, as does the uncertainty associated with regional internal variability. Global-average temperature is a metric that climate models can more reliably predict. Focusing on this single number also makes it easier to compare different models and scenarios. In the end, though, planning for climate change locally will require better predictions of regional climate change.[63]

Rising sea levels are among the most visible effects of climate change. The global-average sea level rose by about eight inches between 1888 and 2018, and the rate has accelerated in recent decades: Over the last three decades, it rose by about three inches. According to the IPCC, under mid-range emission scenarios, the global average may rise about half a meter (1.5 ft) by the end of the century,[64] seriously impacting coastal regions. In high-emission scenarios, sea level rise is double that amount, greatly increasing the risk for coastal cities and island nations. Even if emissions are drastically reduced, due to the inertia of the climate system, sea levels will continue to rise for centuries.[65]

As the atmosphere warms, it warms the ocean and melts ice over land and sea. The warming causes water to expand and raises the sea level. This thermal expansion explains about 40 percent of the observed rise in sea levels, and the partial melting of glaciers and ice sheets over land explains the rest. (The melting of sea ice actually has only a small impact on the sea level,[66] just as the melting of ice in a cup of water does not change the water level in the cup.) These processes explain the rise in the globally averaged sea level, but explaining local sea level changes along a coast is much more complicated. Steady winds blowing toward the shore can raise the sea level, and winds blowing away from the shore can lower it. Geological effects like the subsidence of land that occurs when groundwater or oil is extracted can also lead to changes in the relative sea level.

One of the most counterintuitive things about regional sea levels is what happens when a giant continental ice sheet melts. Say the Greenland ice sheet were to melt completely; it would lead to a catastrophic seven-meter rise in global sea levels. We might naively expect the rise in sea levels near Greenland to be about the same as, or even higher than, the rise in other regions. But actually, it turns out that the relative sea level near Greenland would fall, due to two regional effects.[67] Currently, the gravitational pull of the massive ice sheet is pulling the local sea level up, and the weight of the ice sheet is pushing the local land level down. Without the ice sheet, not only would the real sea level drop, but the land would also rebound upwards, lowering the relative sea level even further. Calculations show that this fall in sea level could reach up to 2,000 km from Greenland, returning land to coastlines from Newfoundland to Norway. Further away from Greenland, though, the sea level would compensate for this effect and rise much more than seven meters, devastating coastal communities. It is important that we rely on scientific modeling, rather than human experience or intuition, to understand long-term climate change.

Other impacts of global warming, like changes in rainfall patterns and the increased likelihood of droughts and floods, will also be very regional – sometimes in similarly counterintuitive ways. Rainfall is likely to increase on average, as warmer air can hold more moisture.[68] But the increased rainfall may not always occur in regions that need it. Certain regions will see decreases in rainfall and more droughts. In fact, despite the expected increase in global-average rainfall, land surfaces are also expected to get drier on average.[69] This counterintuitive behavior happens because land warms more quickly than the ocean, as evaporative cooling cannot always counteract the warming over dry land as it does over the ocean. If the land is warmer, but the same amount of water vapor is blowing in from the ocean, relative humidity will be lower, and the land will be drier. Discussions of climate change tend to emphasize harmful regional impacts, which are likely in most areas. A few places may actually see changes that are beneficial in the long run,[70] but rapid change in any direction will require adaptations in infrastructure and land use, with attendant costs.

Predicting regional climate change is an ongoing challenge because model projections often disagree on how climate will change in different regions.[71] This is fundamentally due to errors in models, which are amplified regionally. But internal variability or stochastic uncertainty compounds this problem if a single simulation, or a limited ensemble of simulations, is used to assess regional climate change. While one simulation may project strong warming trends over much of North America during the next 50 years, another simulation may simulate weaker warming or even some cooling in parts of North America.[72] This can happen even if both simulations use the same model and the same boundary conditions, because of slight differences in the initial conditions. This manifestation of the Butterfly Effect can be even greater for local weather extremes. For example, one simulation may predict that daily summertime heat records will be broken almost continuously in the city of Dallas, Texas, starting in the late 2060s, whereas another may predict fewer and more intermittent heat extremes, even under the high-emission RCP8.5 scenario.[73] The large regional uncertainty in projections makes it much harder to unambiguously attribute observed regional climate change to specific human or natural factors, the way we were able to attribute global climate change (Figure 7.3).

***

In the 1980s, global warming went from being a scientific curiosity to becoming a global policy issue, one that required a comprehensive assessment. By the time the IPCC was established in 1988, 38 years after the ENIAC forecasts, the young scientists and visitors connected with the Princeton Meteorology group had become powerful within the field of climate science. Von Neumann's web of influence extended far and wide. The first generation included scientists such as Jule Charney, Joe Smagorinsky, Phil Thompson, Norman Philips, George Platzman, Bert Bolin, and Harry Wexler (Figure 1.3). These scientists had either trained or hired the next generation of pioneers: Charles David Keeling, Suki Manabe, Kirk Bryan, and Akira Kasahara.

Climate predictions made by computer models play an important role in the IPCC assessments. These models have become much more complex and elaborate over time, but the most prominent features of their predictions remain the same. This raises the question: What exactly are the benefits of increased model complexity? In the next several chapters, we shift our focus from the history of climate prediction to its philosophical underpinnings. We discuss the power of models and their limitations, as well as the public perception of model predictions.

PART II

The Present

8

Occam's Razor

The Reduction to Simplicity

he devoted himself to the study of motion, and, to the great [consternation] of all the philosophers, he disproved many conclusions of Aristotle himself, until then maintained as obvious and indubitable. He did so with experiences, solid demonstrations, and discourses. Among other opinions, he rejected Aristotle's idea that the velocities of bodies made of the same material but with different weights, falling through the same medium, should be proportional to their weights. Instead, he proved that all of them fall with the same velocity. He did so by means of repeated experiences made from the top of the Leaning Tower of Pisa, with the other lecturers and philosophers and the students attending them. Similarly, he refuted Aristotle's idea that the velocities of the same body falling in different media should depend on the ratio between the resistances or densities of the media, showing the manifest absurdities that would follow against common sense.

Vincenzio Viviani, *Historical Account of the Life of Galileo Galilei*[1]

It's perhaps one of the most famous experiments in the history of science, although it is disputed whether it happened exactly as described above. Toward the end of the sixteenth century, the scientist Galileo Galilei held the chair of mathematics in the University of Pisa in Italy, home of the famous leaning tower. Since the time of the Ancient Greek philosopher Aristotle, it was believed that heavier objects fell faster than lighter objects. Legend has it that Galileo carried out a dramatic experiment atop the Leaning Tower of Pisa. He dropped two spheres from the tower, one heavy and one light. Both landed at the same time, proving the nearly 2,000-year-old Aristotelian physics wrong.

Scientists have always sought the simplest possible explanations for what they observe. Galileo's simpler theory of motion, derived from experiments with inclined planes as well as falling objects, supplanted Aristotle's more complicated theory. Later, Galileo's ideas would become part of Newton's laws of motion – a set of three simple laws that describe a bewildering range of physical phenomena. Galileo's pioneering use of a telescope to make astronomical observations allowed him to introduce another important simplification in science. It had long been believed that the planets and the sun moved around the Earth. The Ancient Greco-Roman astronomer Ptolemy had provided a convoluted explanation of the path of planets in this Earth-centric system, involving complicated geometric models called "epicycles," or circles moving on circles.

Galileo's observations of the phases of Venus and the motion of Jupiter's moons led him to proclaim that it was simpler to explain planetary motions by assuming that the Earth moved around the sun and not the other way around.[2] Instead of complicated epicycles, a much simpler system of elliptical planetary orbits around the sun, proposed by German astronomer Johannes Kepler, could be used to explain the observations.[3] There are many examples in science of simple theories beating out more complicated theories that involved superfluous concepts, like phlogiston or aether. The criterion of choosing the simplest of many competing theories is known as *Occam's Razor*, or the principle of parsimony.

8.1 Philosophy Break: The Principle of Parsimony

> *We are to admit no more causes of natural things than such as are both true and sufficient to explain their appearances.*
>
> Isaac Newton, *Principia Mathematica*

The principle of parsimony was popularized by a fourteenth-century English monk named William who lived in the English village of Ockham.[4] William of Ockham was a well-known theologian and philosopher who wrote extensively in Latin, which was the language of scholars at that time. His Latin name is *Gulielmus Occamus*, which is shortened to *Occam*. Occam made extensive references to the principle of parsimony in his writings, with statements such as *Frustra fit per plura quod potest fieri per pauciora* ("It is futile to do with more things that which can be done with fewer"). This principle could be stated alternatively: Other things being equal, simpler or more parsimonious explanations are preferable to more complex ones (Figure 8.1).

Related to Occam's Razor is the principle of reductionism, which we discussed earlier. This is a principle that works well in physics, in which the universe can be studied by reducing it to its smallest components, be they

Ockham chooses a razor

Figure 8.1 Occam's choice. (© Chris Madden; used with permission)

atoms or subatomic particles. We can use Occam's Razor to select between competing theories describing these fundamental components. But reductionism does not work well in fields like biology, in which the behavior of complex organs or organisms is studied holistically using the principle of emergentism, the opposite of reductionism. Occam's Razor could still be used to select among competing biological theories, but the theories themselves may not be that simple.

Although Occam's Razor is a conceptually appealing device, it is hard to use in practice, and easy to misuse. For example, it has been used to defend creationism against the complex theory of evolution – a theory with one big assumption competing against another with several small assumptions. As we will see in Section 13.6, physicists have implicitly used Occam's Razor to question the reliability of complex climate models. Ironically, it has also been used by atheists to argue against the need for religion, even though William of Ockham was a Franciscan monk!

Simplicity does not necessarily mean aesthetics or elegance; it means a parsimony of assumptions. The fewer assumptions required by a theory, the more broadly applicable and hence more testable or falsifiable it is. The preference for simplicity is conditional; *ceteris paribus*, or "other things being equal," is an important qualifier. Occam's Razor only comes into play when choosing between two theories that fit the available data equally well. Given only eighteenth-century data, Einstein's theory of special relativity would have been rejected in favor of Newton's laws of motion, which are simpler. If one theory explains more phenomena, or explains the same phenomena more deductively starting from simpler assumptions, it is the better choice according to Occam's Razor. That's why creationism is not preferable to the theory of evolution. The latter explains the same phenomena more deductively, starting from a set of simple assumptions with few details, rather than using a single, but extremely broad, assertive assumption with a lot of implied detail – that is, the characteristics of every living being.

8.2 Predicting versus Understanding: Complexity versus Simplicity

Simple models . . . they're very complicated

Zen koan attributed to Syukuro (Suki) Manabe

Predicting weather using a computer model is not the same as understanding how weather works. A mechanical engineer who knows the mathematical equations governing air motions could write a computer program several thousand lines in length for weather prediction, and a software engineer could run the program to make weather forecasts. The two engineers do not need to understand anything about the science of fronts, storms, hurricanes, typhoons, tornadoes, and so on. If weather forecasts produced by this model are very good, there isn't much more to do.

This brute-force approach may actually produce decent forecasts a day or two ahead, especially if there are good initial conditions for the model, because computers today are powerful enough to solve three-dimensional fluid flows on a fine spatial grid. But in the 1950s, when the Princeton Meteorology Group wanted to use the world's first computer to make a weather forecast, the situation was quite different.

By modern standards, computing power in the 1950s was minuscule. Recall that ENIAC, the first computer, had a memory of 40 bytes or characters – barely enough to store a few words from this sentence, let alone a thousand-line program (Section 1.1). The mathematical equations governing air motion were too complex to be solved directly on this computer. They had to be simplified drastically, while still retaining their essential ability to predict weather. John von Neumann, the brilliant mathematician and physicist who established the group, knew how to program the computer. But to make the right simplifications to equations, he needed someone who had an understanding of the emergent properties of weather, not just knowledge of the mathematical equations.

Jule Charney and other meteorologists recruited by von Neumann possessed that specialized knowledge. They knew that three-dimensional equations governing air motions could be simplified to just two horizontal dimensions to create a model that captured the essence of weather systems, ignoring the vertical dimension. They devised a spatial grid consisting of a few hundred points to represent the atmosphere over the United States. It is this simplified model that they used to make the world's first numerical forecasts with the ENIAC. Simplification was the key to their success, and their understanding of weather phenomena was key to that simplification. Subsequent forecasts by the group used slightly more complex models, but these were still much simpler than a full-blown three-dimensional model.

Suki Manabe at GFDL used a complementary type of simplification, retaining only the vertical dimension and ignoring the two horizontal dimensions to come up with his estimate of climate sensitivity in 1967 (Section 4.5). He made these simplifications because the fully three-dimensional climate model that he and Joe Smagorinsky had already built was quite unwieldy. With the nimbler one-dimensional radiative–convective model, he could carry out many different calculations and estimate climate sensitivity. Manabe had already simplified the representation of the process of convection for the three-dimensional model, which allowed him to incorporate it into the one-dimensional model.

Ed Lorenz also used simplification to make his discovery of weather chaos in 1961 (Chapter 2). His simplifications were much more drastic. Instead of using a few hundred grid points to represent weather, he used essentially just three points to heuristically capture the chaotic nature of weather. His three-variable model was able to demonstrate the sensitive dependence of weather on initial conditions. If he had simplified weather processes even further and used a two-variable model, he would not have encountered chaos at all! (There's a mathematical result called the Poincaré-Bendixson theorem which states that a two-dimensional system cannot exhibit chaotic behavior.)

These are three examples of scientists wielding the analytic knife to dissect a complex system and understand its innards. The trick to creating a reduced model from a complex model is knowing when to stop – when you have captured the essence of the behavior you are interested in, but before you lose it. The fact that the scientists were familiar with the emergent properties of the complex system was useful in this reductionist exercise.

Not all simplified models are successful. Manabe was known to remark that simple models were in fact quite complicated. Lewis Fry Richardson's attempt at making a weather forecast, described in his 1922 book, used a very simple model of weather that produced a terrible forecast, in part because it was unable to handle the noise in the initial conditions (Section 1.2). One of Manabe's colleagues, Fritz Möller, published a 1963 paper using a simplified climate model, but his simplified model was missing a key ingredient – the proper representation of the water vapor feedback (Section 4.5). This led to a poor estimate of climate sensitivity. Although Möller's model was unsuccessful, it paved the way for Manabe to formulate his own.

There can be two different motivations for creating simplified models – necessity and curiosity, although the two are not always separable. Charney and Manabe needed simpler and therefore faster models to get around the limitations of computational resources while still maintaining some degree of realism. Manabe could have worked with the computationally expensive three-dimensional model, but, with that model, he could not have carried out numerous calculations to address his curiosity. Lorenz's model was a heuristic model, far from reality – built with the sole purpose of satisfying curiosity and improving understanding.

Creating a complex model is like taking a photo. Constructing a simplified model is quite different from photography; it is more like drawing a caricature of a famous person, say a US president. You need to be able to capture an essential detail of that person and emphasize it, like Richard Nixon's nose or Barack Obama's ears. A good portrait photographer will not necessarily make a good caricaturist, but having multiple photos of the subject can be very useful to the caricaturist. Just as a cartoon can capture the essence of a complex idea, simplified models can be very useful in understanding and explaining the behavior of more complex models. But simplified models are often not a substitute for complex models in real-world applications, just as a caricature isn't a substitute for a photo when it comes to identification. The principle of Occam's Razor does not disqualify complex models; complex models are more realistic and explain a broader range of phenomena than simpler models.

In debates about whether we should trust complex computer models of climate, some argue that global warming can be explained by the simple physics associated with the greenhouse effect of carbon dioxide and water vapor. That is indeed true, but only in a qualitative sense. Simple physics arguments can predict that doubling carbon dioxide concentration will warm the globe, but they cannot say if the resulting warming (also known as climate sensitivity) will be very low (say, 1°C) or very high (say, 5°C) because simple arguments cannot quantify some crucial feedbacks, especially the cloud feedback. More comprehensive models that represent all important feedbacks are essential to obtain quantitative estimates of the parameters of climate change.

Another danger associated with simplified models is that they can exhibit spurious behavior, that is, behavior not exhibited by the complex model.[5] Such behavior is not of interest, unless it can be demonstrated that the complex model was wrong in not exhibiting that type of behavior. For example, a simple thermodynamic model of climate can be constructed using the relationship between temperature and the amount of water vapor the atmosphere can hold, known as the Clausius-Clapeyron relation. This means that water vapor in the atmosphere would increase by 7 percent for every 1°C of global warming, suggesting that global-average rainfall may increase at the same rate. But more complex models show a much weaker rainfall increase of about 2 percent per 1°C of warming because the circulation of air that causes rain slows down with warming.[6]

8.3 Biology Analogy: The Fruit Fly Model of Climate

One of the problems with climate science is that it is not possible to carry out controlled experiments with the system we are studying, that is, our planetary climate system. The scale of resources needed to modify the planetary climate in a controlled fashion would be impractically large – and this is not to mention the profound legal and ethical issues that such experimentation would raise. We will revisit this issue in the context of geoengineering, but climate scientists cannot carry out planetary-scale experiments simply out of scientific curiosity.

Another discipline faces similar problems in carrying out controlled experiments on the subject of interest: medicine. There are many practical and ethical issues in carrying out experiments on human subjects. To know if a particular drug is effective or if a certain genetic treatment works, scientists carry out experiments on "simpler" animals, such as mice or fruit flies. It is common in biology for papers to refer to the "mouse model" or the "fruit fly model," where the word *model* refers to a natural biological "abstraction," rather than a mathematical or computational abstraction, as in the case of climate science.

The genetic code of the common fruit fly, *Drosophila melanogaster*, contains more than 60 percent of human disease genes, allowing it to serve as a proxy for human subjects.[7] Its short generation and rearing time lend themselves to rapid and inexpensive experimentation. Due to the fruit fly's ubiquitous use in research, its properties are well known in the scientific community. That makes it easy to plan new experiments and also for other scientists to replicate experiments that use it.

Climate modeling would benefit from having an analog to the fruit fly model. It is difficult to replicate simulations between different climate models because the models are structurally different. Climate models also have a rapid development cycle, which means that by the time problems with the current generation of climate models are analyzed, a new generation is being developed, and some of the insights from the analysis may no longer be relevant. A third problem that complicates experimentation and replication is that state-of-the-art climate models are computationally very expensive to use, almost by definition, because model complexity is often increased to match the capacity of the latest supercomputer.

A standard computationally inexpensive, structurally less complex climate model, whose properties are widely analyzed and understood, could serve as the "fruit fly" of climate models. The GFDL scientist Isaac Held, a student of Manabe, has strongly advocated creating such a standard hierarchy of models with different degrees of complexity to study the climate system, analogous to the fruit fly model and the mouse model.[8]

One advantage of the fruit fly is its natural evolutionary connection with the human body, which is the real subject of interest. There is no such analog for the real climate system. Other terrestrial planets such as Venus or Mars are too different from Earth to serve as alternative models for it, and they are hard to observe in any case. A simplified climate model is typically constructed by taking a more complex climate model and replacing some of the complicated formulas, used to represent fine-scale processes, with much simpler formulas. Often these simpler formulas are empirical, derived directly from data, making the model more inductive and less deductive.

One of the challenges is to construct simplified models that have a "lasting value," like the fruit fly model.[9] The Held-Suarez model, named after Held and NASA scientist Max Suarez, is one such model.[10] It is commonly used as a standard model for the atmosphere and replaces complicated formulas for atmospheric convection and radiation with a much simpler formula inspired by observations. Such an approach works well if the simplified model is used to study phenomena that are not sensitive to the fine-scale processes that have been "oversimplified."

A common aphorism in modeling is that "models are for insight, not numbers,"[11] reflecting the caveat that models are imperfect representations of reality. This caveat very much applies to simplified models. The best candidates for the mechanistic approach of simplified modeling are qualitative emergent phenomena, such as the polar amplification of global warming.[5] For quantitative estimates of emergent properties that depend crucially on approximate formulas for fine-scale processes, such as climate sensitivity, this simplified approach does not work well because it does not address the structural uncertainty issue; the simpler formula structure in the simplified model just adds another structure to the mix, and a less realistic structure at that. This can at times cause the simplified model to exhibit some unrealistic behavior. We may not want to spend much time studying such unrealistic behaviors in models, just as we may not want to spend much time studying a disease found in the fruit fly that has no counterpart in humans.

However, if we found an interesting phenomenon in the fruit fly that could have a human counterpart, we would try to verify that using more complex models, like a mouse model, for example. In the same vein, simplified models like the Held-Suarez model can be very useful as hypothesis generators to explain climate phenomena. Since these models are computationally inexpensive, many different hypotheses can be tested out in their simulations.[5] Then the most promising of those hypotheses can be validated on more complex models and confirmed with data.

By construction, simplified models are more elegant and less elaborate than realistic models. But simplified models are not necessarily better in the sense of Occam's Razor, since they usually do not capture the same range of phenomena that more complex models do and therefore are less falsifiable. They have more explanatory power, but this usually comes at the cost of predictive power. They may also have fewer constraints if an important climate system component is omitted. Do these caveats about simplified models mean we should only work with realistic models? No, because there is a flip side to the argument. The state-of-the-art climate models Suki Manabe and Jim Hansen used in the 1970s to predict climate change would be considered simplified models by today's standards, as they neglect some important processes. The comprehensive, sophisticated, state-of-the-art climate models of today will also likely be considered simplified models in the future.[12] The caveats about today's simplified models also apply to today's realistic models. We need to be confident that the insights we gain from our models, simplified or realistic, apply to the even more complex real world. Experimenting with a hierarchy of structurally simpler and computationally inexpensive models can increase that confidence.[13]

8.4 Spherical Cows and Tipping Points

As the name suggests, a tipping point is related to the tipping or tilting of a tall object, say a cup on the table. If you pull the edge of the cup slightly to tilt it, gravity will try to bring it to its upright position, because the cup is stable to small perturbations. If you tilt it too much, beyond its tipping point, it becomes unstable and will almost instantaneously tip over. Simplified explanations in physics often start with the trope: "Consider a spherical cow . . . ." As a matter of geometry, spherical cows are easier to tip over than real cows. In this section, we argue that simplified models are more likely to exhibit tipping points than their more realistic counterparts.

In the complex climate system, a tipping point is a threshold for change that, if crossed, will lead to much more rapid and dramatic change. Tipping points attract a fair amount of attention in discussions of climate change because they portend cataclysmic events or *abrupt climate change*[14] – climate change that happens much more rapidly than indicated by the projected global warming scenarios (Figure 7.4c). They are typically high-impact, low-probability events, such as disintegration of the West Antarctic ice sheet or Amazon rainforest dieback. But will global warming make some of them more probable? Or is the climate system stable enough to resist being tipped over?

We can view stability itself as an emergent property of a complex system. A complex system involves a balance between stabilizing (negative) and amplifying, or destabilizing (positive), feedbacks. In a stable, complex system, the stabilizing feedbacks are stronger than the amplifying feedbacks. If the system is pushed in one direction, the stabilizing feedbacks will try to compensate and pull it back to equilibrium, up to a point. That point becomes a tipping point if an amplifying feedback dominates beyond it. A system with many interacting components is likely to have more stabilizing feedbacks than amplifying feedbacks because multiple components need to act in concert to sustain an amplifying feedback loop, whereas completely random responses by individual components will act to break the amplifying feedback loop.

The fruit fly is a stable complex system. Many different properties of the fruit fly must balance each other for the species to survive. Its genetic code will naturally mutate to disturb the balance, but natural selection will reestablish the balance, ensuring that only viable mutations survive and propagate. If we alter the genetic code of an individual fruit fly artificially to create a new variant, there is no guarantee that the mutation will produce a viable result.

Similarly, the complex climate system is maintained by a balance between stabilizing and amplifying feedbacks. The balance has to be reasonably robust because we know that the climate was relatively stable for many thousands of years before the Industrial Revolution (but not for hundreds of thousands of years). A realistic climate model mimics the behavior of the complex climate system. That includes mimicking the stability properties because realistic climate models are tested for their stability under preindustrial conditions and will be rejected if they are not stable.[15]

Simplifying a complex model can disturb the balance between stabilizing and amplifying feedbacks. It is akin to creating an artificial genetic mutation, but without the benefit of natural selection to ensure stability. If an important stabilizing feedback is inadvertently eliminated in the simplification process, the simplified model will be prone to instability.[16] Does this mean simplified models cannot be used to explore the possibility of tipping points? No, but any tipping points exhibited by simplified models should ultimately be confirmed with more comprehensive models.

Can we use a simple model to predict quantitative thresholds for tipping points? We can always make predictions using a simple model, but, for the prediction to be credible, a compelling case would have to be made that all the processes either omitted or crudely approximated in the simple model will not affect the prediction in any way. Climate models are complex because climate is a complicated system. It is unlikely that an amplifying feedback leading to a tipping point can operate without interacting with many other climate processes, all of which will need to be represented accurately for any quantitative prediction regarding the tipping point to be credible. Therefore, we should treat any tipping point predictions using simple models as providing insights but not necessarily accurate numbers,[17] with the insights subject to future revision as additional effects are considered.

An extreme example of a simplified model is our old friend the three-variable model used by Ed Lorenz to discover deterministic chaos. It is a highly simplified model of weather that has led to important qualitative insights. But it is also prone to tipping point behavior because it is highly nonlinear. The simulated "weather" in this model flips chaotically between two distinct states. Real weather states form much more of a continuum.

With these caveats about simplicity in mind, it is worth briefly describing some tipping points that have been proposed by climate scientists. Typically, they involve an amplifying feedback that is no longer countered by stabilizing feedbacks past the tipping point and that continues to amplify until a new equilibrium is reached. Recall that such a "runaway" water vapor feedback was invoked to explain extremely hot conditions on Venus, although we ruled this out as a possibility for the Earth (Section 3.6). There are two important amplifying feedbacks that could generate tipping points for Earth's climate: the *ice–albedo feedback* and the *carbon cycle feedback*.

The Earth has two large ice sheets, the Greenland ice sheet and the Antarctic ice sheet. Ice is very good at reflecting sunlight because it has high reflectivity or albedo, and hence it keeps the polar regions cool. As the ice melts due to global warming, at some point, there will be less ice coverage in these regions. That means more sunlight will be absorbed locally, leading to even more melting. This is the ice–albedo feedback. This amplifying feedback could accelerate the pace of global warming, and the effects could include a catastrophic rise in sea levels. If the entire Greenland ice sheet were to melt, global sea levels would rise by up to seven meters on average, over many centuries (Section 7.7). Complete melting of the Antarctic ice sheet is unlikely, but even partial melting could raise sea levels by several meters.

The second important amplifying feedback involves the cycling of carbon between the atmosphere and the land. The Earth stores carbon underneath the land surface in the form of permafrost in the Arctic region. Global warming will thaw this permafrost and release some of the stored carbon into the atmosphere in the form of methane or carbon dioxide. This will amplify greenhouse warming and in turn lead to further thawing of the permafrost. Over centuries, this too could accelerate global warming and disrupt ecosystems. A similar feedback argument could be made for carbon stored in the form of methane hydrates in the cold deep ocean.

There are other potential feedbacks that involve shifts of the atmospheric monsoon circulation or slowing down of the oceanic thermohaline circulation that transports heat from the tropical regions to the polar regions.[14] These amplifying climate feedbacks can create climate tipping points – push climate beyond that point and the amplifying feedbacks trigger catastrophic climate change. The existence of such tipping points is a real possibility, but we have a poorer scientific understanding of these "dramatic" possibilities than we do of the "boring" trends predicted for global temperature. Thankfully, some of the scariest runaway feedbacks are physically constrained to be slow. Research into runaway feedbacks has mostly relied on simple models of the climate system because even today's state-of-the-art complex climate models do not represent the relevant slow processes and long timescales very well.[18] Complex climate models have only recently begun to include carbon cycle feedbacks and representations of continental ice sheets.

An important factor to consider for very slow climate processes is the Butterfly Effect, or the sensitive dependence of predictions on initial conditions (Chapter 2). For the atmosphere, this effect saturates after a few weeks, and for the oceans it saturates in a few decades. But for ice sheet or permafrost predictions, the initial condition error could continue to grow for centuries or even for millennia, compounded by structural model errors. This reinforces the point made earlier that while simple models can help explain the mechanisms of tipping points, they are less reliable for making quantitative predictions regarding the exact timing or likelihood of such tipping points. Also, the feedbacks responsible for tipping points typically occur in specific regions of the globe. Since different regional climate change patterns can result in the same global average, it is not possible to accurately predict regional tipping point thresholds from simple metrics like globally averaged surface temperature without resorting to comprehensive climate models. But comprehensive models still exhibit large uncertainties in their regional climate predictions (Section 7.7).

There is evidence for the occurrence of abrupt climatic changes in the Earth's past, such as that of 12,000 years ago, following the end of the last ice age (Figure 5.3), and that of about 55 million years ago, at the start of the Eocene, when the Earth was considerably warmer than it is now.[19] This suggests that the Earth's climate has "tipped" from one equilibrium to another in the past. Therefore, these dramatic, but hard-to-quantify, climate tipping points should be taken seriously, yet, absent stronger evidence, not as seriously as the gradual, but quantifiable, global warming. Likewise, simple model simulations featuring these dramatic feedbacks should not be taken literally.[20] The steps we take to mitigate global warming, by keeping our climate closer to a stable equilibrium, will also reduce the likelihood that we cross these tipping points.

***

In this chapter we focused on the explanatory power of models. Often, explaining climate phenomena requires studying progressively simpler versions of a complex model to understand its behavior. But this reductionist or mechanistic approach involves a tradeoff between simplicity and accuracy. When discussing "fruit fly models" of climate, we noted that simplified models are less realistic by design, sacrificing some important elaboration for elegance. Simplified models can help explain the emergent behaviors in climate models, but they will be less accurate and less skillful in predictions. They can also be more prone to exhibiting tipping point behavior than complex models, as some of the stabilizing feedbacks may have been eliminated. In an emergent science like climate science, in contrast to a reductionist science like physics, it seems that we can have simplicity or accuracy – but not both.

Explanation and prediction are not the only benefits provided by climate models. They also impose strong physical constraints on the climate system, as will be discussed in Chapter 9.

9

Constraining Climate

A Conservative View of Modeling

Continuing our climate prediction journey, we travel to Boulder, Colorado, where Susan Solomon worked before she set off on her expedition to Antarctica to study the ozone hole. Boulder is a scenic city, nestled in the foothills of the Rocky Mountains. We drive west along the redundantly named Table Mesa Road toward the two towers of the Mesa Lab of the National Center for Atmospheric Research (NCAR). Unlike the Georgian revival style of the IAS or the utilitarian style of GFDL in Princeton, the architectural style of the Mesa Lab, built in 1967, is unabashedly modern (Figure 9.1a). It was once considered so ahead of its time that its exterior was used in the 1973 science fiction comedy *Sleeper* to represent a building in the year 2173. The Mesa Lab was designed by the famed architect I. M. Pei, who also designed the Louvre pyramid. Its elemental concrete forms evoke the ancient Anasazi cliff dwellings of southwest Colorado, and the reddish-brown concrete exterior is designed to blend into the sandstone cliffs of the Flatiron rock formations that form its backdrop.

The National Center for Atmospheric Research, like GFDL, is a laboratory funded by the US government, but it is operated by a consortium of universities. It was established in 1960 to carry out research in all areas of the atmospheric sciences. For much of its history, NCAR, like GFDL, had a state-of-the-art supercomputer located in its basement, to be used in weather and climate modeling. But supercomputers continue to grow larger and more power hungry; in 2012, the NCAR supercomputer was moved to the energy-rich neighboring state of Wyoming. Supercomputers are voracious consumers of electricity, which could be a constraining factor in the future development of climate models. We will address this issue in Chapter 16.

The first director of NCAR was an atmospheric physicist and astronomer named Walter Orr Roberts. He hired as his associate director Phil Thompson, a meteorologist who had worked with John von Neumann for two years, at the

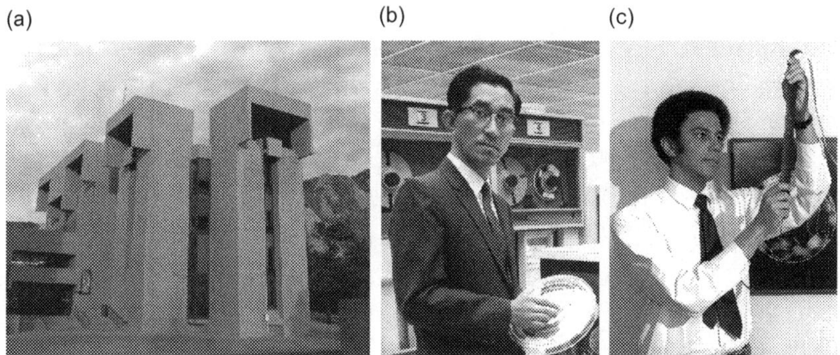

Figure 9.1 (a) Mesa Laboratory of the National Center for Atmospheric Research (NCAR) in Boulder, Colorado. (b) Akira Kasahara in front of magnetic tape drives for the CDC 6600 supercomputer at NCAR (1970). (c) Warren Washington examines microfilm showing model simulation results (1973). (Photos: NCAR/UCAR)

start of the numerical weather prediction project (Chapter 1). Thompson joined NCAR after he retired from the Air Force, helping to manage research and recruit new talent to this young institution.

Among the first climate scientists Thompson hired was Akira Kasahara (Figure 9.1b), formerly of Professor Syono's group at the University of Tokyo.[1] Recall that Kasahara had previously worked with two other members of the Princeton group – John Freeman and George Platzman – and had played a role in the hiring of Suki Manabe by Joe Smagorinsky. The National Center for Atmospheric Research also hired another young scientist named Warren Washington (Figure 9.1c), who had just completed his Ph.D. at Pennsylvania State University.

Unlike research at GFDL, which was very mission oriented, research at NCAR was characterized at the time by a freewheeling spirit. When Kasahara asked Thompson what he should work on, he was told that he could work on whatever he wanted. So Kasahara decided to continue modeling tropical cyclones. Washington asked Thompson the same question and received a similar answer. Being more junior and dissatisfied with this lack of direction, he approached Thompson's boss, Roberts, who told him, "I really want you to work about half the time on helping us get started on modeling. The other half, work on what you think is important."[2]

Washington then told Kasahara that he wanted to build a climate model. Kasahara was immediately receptive to the idea, familiar as he was with the pioneering work on the "infinite forecast" by Norman Phillips of the Princeton Meteorology Group. Kasahara and Washington started designing their climate model in 1964,[3] using Lewis Fry Richardson's 1922 book on weather forecasting for inspiration.[4] Each NCAR scientist had roughly one programmer to assist in creating and managing the model software, which ran on NCAR's first supercomputer – a CDC 3600. Kasahara and Washington's model had horizontal grid boxes 500 × 500 km in size over a fully global domain; this allowed the use of the model to study phenomena, like the Indian monsoon, involving winds flowing across the equator.[5]

***

Thus far we have discussed early climate model development at GFDL and NCAR, but there were at least two other climate models developed in the 1960s. The four earliest climate models were developed at institutions widely known by their four-letter acronyms: GFDL, NCAR, UCLA, and LLNL (Lawrence Livermore National Laboratory). The scientists who developed them were a varied bunch: three Japanese and four American scientists, including one African-American scientist – Warren Washington – who would go on to have a distinguished career at NCAR. The very first three-dimensional climate model was developed by Chuck Leith at LLNL. The other three climate models grew out of direct collaborations between Japanese and American scientists: Akio Arakawa and Yale Mintz at UCLA; Manabe and Smagorinsky at GFDL; Kasahara and Washington at NCAR. Coincidentally, all three of these Japanese scientists had come to the United States from Professor Syono's research group in Tokyo!

The LLNL model was not developed further after Leith moved to NCAR in 1965 and shifted his focus to turbulence research. The UCLA effort was influential, but it ended after several decades, when it became difficult for a university to sustain the increasingly complex model-development effort. The GFDL and NCAR climate models continue to be actively developed. In Section 9.1, we discuss an important attribute of climate modeling before returning to the story of climate modeling at NCAR.

9.1 Conservation of Mass and Energy

Follow the money . . .

whispered by Deep Throat to Bob Woodward
in the movie *All the President's Men*

To figure out why people behave the way they do, you can ask them what they think, or you can study their actual spending habits. Good intentions are an unlimited resource, but money is conserved. People often say that they support increased spending on public education and public transportation, but when new taxes to pay for these services are on the ballot, they may vote against them; if they were to pay the new taxes, they would have less spending money. Following the money is also a good way to study political corruption or the financing of terrorism; it helps us uncover a chain of causal linkages. If someone is spending more money than we would normally expect them to have, then they must be getting the money from somewhere else.

Similar principles of conservation also apply to the scientific study of weather and climate. When we observe the warm Gulf Stream ocean current off the coast of the United States, we can track warm water masses back to the tropical regions where they originated because water temperature is conserved. Weather and climate models carefully track conserved properties of the atmosphere and the ocean – such as air and water mass, momentum, and energy (or temperature) – to make predictions. These conservation properties are inviolable physical constraints,[6] more stringent than limits on the supply of money (after all, governments can print money, and economic growth can create wealth).

Computer models of weather and climate incorporate the laws of thermodynamics and Newton's laws of motion, which can be stated in their *predictive* form as: (1) A body in motion stays in motion unless acted upon by a force, (2) the rate of change of momentum of a body (defined as the product of its mass and velocity) is proportional to the force applied, and (3) every action of force has an equal and opposite reaction. Alternatively, these laws can be expressed in a *conservative* form: The first law states that momentum is conserved when there is no force; the second law states that a force acting on a body can change its momentum; and the third law states that the agent exerting the force must change its momentum by the exact negative amount, so as to keep the total momentum of the system – agent plus body – constant.

In the absence of friction, Newton's laws imply the conservation of both momentum and energy. These are powerful constraints that every model of weather and climate must obey. Of all possible directions in which air can move, only those directions which satisfy the conservation requirements are allowed. This narrowing of possibilities about the future provides useful information, even though it is not an actual prediction. (In the presence of friction, momentum is no longer conserved, but energy is still conserved; friction simply converts energy from one form to another.)

In addition to the conservation properties for air, the conservation of the mass of water and carbon are also important constraints on climate. Water exists in solid, liquid, and gaseous states. It can transform between different states, but its total mass must always be conserved. Carbon exists in gaseous form as carbon dioxide and methane, as well as in solid form within living organisms and rocks, and buried underground as fuel deposits. The total mass of carbon must also be conserved.

Energy can also transform from one form into another, rather as money can be transformed from one currency into another. For instance, the force of friction converts kinetic energy into heat energy, as when you rub your hands together. The Earth receives a lot of energy from the sun, mostly in the form of visible light, and almost all of this energy is radiated back into space as heat or infrared radiation. When water and carbon change from one state to another, they can store energy or release the stored energy.

Predicting weather is like managing money in the short term, such as planning a monthly budget. Small imbalances between the inflow and the outflow can be tolerated: If our monthly income is $1,000, and our monthly spending is $1,030, the $30 difference can be covered in the short term by dipping into our savings. If an atmospheric model overpredicts air temperature by 1°C, it can still be a pretty good weather forecast model because the air temperature changes that need to be predicted are usually much larger than a degree.

Predicting climate is like managing money in the long term, such as planning for retirement. Small but systematic discrepancies between the monthly inflow and the monthly outflow can add up to large amounts over decades. A spending plan that only roughly balances the budget each month may not work well for retirement planning. Putting $100 per month versus putting $130 per month into your retirement account may not make a big difference on a month-to-month basis, but if the imbalance is sustained, your retirement income could change by 30 percent. An atmospheric model that overpredicts temperature changes by 1°C may not make a good climate prediction model, because the error is comparable to the global warming signal. Another difference between weather and climate prediction is that unpredictable external factors like economic growth play a much bigger role in climate prediction, as they do in retirement planning.

Although weather models and climate models contain essentially the same computer code, they are subject to different error tolerances and have different values for adjustable coefficients in their parameterizations. Climate models must conserve mass, energy, and carbon with much greater accuracy than weather models. They need to carefully track very small differences between very large energy inputs and outputs. Weather models are highly validated for short-term weather predictions, but there is a greater error tolerance allowed in their energy budgets. This means that a good weather model isn't necessarily suited for long-term climate prediction, even if the same computer code can be used to make both types of predictions.

9.2 Flux Adjustment and Non-Conservation

The ocean, because it is able to store about 1,000 times as much energy as the atmosphere, acts as a flywheel or energy reservoir for the climate system (Section 4.6). When the greenhouse effect warms the atmosphere, some of that heat energy is transferred to the ocean, mitigating the atmospheric warming trend. The ocean warms, but this does not happen uniformly across the globe; different areas of the ocean absorb energy at different rates, depending on local circulation patterns, through a process known as ocean *heat uptake*. The spatial pattern of the heat uptake can affect the rate of warming simulated by climate models.

The atmospheric and oceanic components of a climate model have errors in them. This means that climate simulations look different from the real world in some respects. In one region of the simulation, say the North Atlantic Ocean, it may not be as cloudy as it is in the real world, leading to too much solar heating in the simulation; in another region of the simulation, say the South Pacific Ocean, it could be windier than it is in the real world, leading to too much evaporative cooling in the simulation. This means that the simulated atmosphere over the North Atlantic may be warmer than it is in the real world, and, over the South Pacific, it may be cooler than it is in the real world. If the simulated North Atlantic Ocean is prepared for a certain amount of heat uptake, the warmer simulated atmosphere would then give it more heat than would be expected in the real world. In a long-term climate prediction calculation, the imbalance in this climate model would make the sea surface in the North Atlantic warmer and warmer each year, and the sea surface in the South Pacific colder and colder each year. This error is known as *climate drift*.[7] Climate drift can happen in a model simulation without any change in the greenhouse gas boundary conditions.

At the time of the first IPCC assessment in 1990, coupled climate models that included both the atmosphere and the ocean were in the early stages of development. Manabe's GFDL climate model was one of them. Like other models, Manabe's model faced the climate drift problem, due to the mismatch between the heat uptake expected by the ocean model and that expected by the atmosphere model. To reduce the drift, Manabe used a technique called *flux adjustment*. He added a correction term in the energy equation that adjusted the heat provided by the atmospheric model, so that the ocean model would receive the right amount of heat. The correction term explicitly compensates for errors in the atmospheric model and the ocean model; it varies from region to region, but, for each region, it does not change from year to year. This technique solved the climate drift that the model displayed, but it violated the law of conservation of energy in each region, because the correction term represented the creation (or destruction) of energy.

To use the analogy of Laplace's Demon, climate simulation requires the Atmosphere Demon and the Ocean Demon to regularly exchange gifts of energy and water with one another to maintain equilibrium. When one demon expects a certain gift and receives another, unhappiness and model drift result. Flux adjustment is an intermediary that takes the gift offered by one demon and replaces it with the gift expected by the other demon, keeping them both happy and at equilibrium.

One way to avoid the need for flux adjustment is to reduce the errors in the atmospheric model and the ocean model – by using better formulas for different processes and also by adjusting the unknown coefficients in the formulas, through a process called *model tuning*. We shall discuss this process in more detail in Chapter 10. In the late 1980s and early 1990s, climate models were not that elaborate and were hard to tune. It was also argued then that model tuning was merely an implicit compensation for errors, while flux adjustment made the compensation more explicit and therefore more acceptable. Nevertheless, the large correction terms involved in flux adjustment were hard to justify for a model that was supposedly derived from the basic laws of physics.[8] So modeling centers worked hard in the early 1990s to improve their models to reduce the climate drift problem and avoid the need for flux adjustment.

9.3 Community Climate Modeling

Meteorologists have shared data since the very beginning; it is a mutually beneficial practice. To create a weather map for Europe, for example, we need to collect pressure data from each country in Europe. If countries didn't share data with one another, no single country would be able to draw a weather map, and we would have no initial conditions with which to compute weather forecasts. The WWW program coordinates this process of sharing weather data internationally (Section 2.2).

Unlike weather and climate data, models do not necessarily need to be shared to be useful. Initially, climate modelers were protective of their models, which they had invested a great deal of time in developing. The GFDL climate model, for example, was only shared with close collaborators in the early days. Over time, climate modeling became more open. In the 1980s, a new climate model was developed at NCAR, known as the Community Climate Model or CCM. The CCM was created to address the broader research community's need for climate models.[9] Developing a new climate model requires a sustained effort over many years, so it is hard for small university departments to create an elaborate model from scratch and maintain it. The code for CCM was made freely available to anyone who wished to use it, enabling universities to access a state-of-the-art climate model.

The CCM, despite the word "climate" in its name, was just an atmospheric model. But to predict climate over many decades, the role of the ocean had to be taken into account. That required a coupled ocean–atmosphere model. At GFDL, Manabe and Bryan had built a simplified barebones coupled model in 1969; by 1975, they had built a model with realistic geography.[10] The National Center for Atmospheric Research was a relatively late entrant into the coupled climate modeling game, and its coupled model faced climate drift problems similar to those described in Section 9.2, such as an overly cold tropical ocean or an overly warm Antarctica.[8] In the 1990s, NCAR decided to develop a new community model that included all the main components of the climate system: atmosphere, ocean, land, and sea ice. This model was initially called the Community Climate System Model (CCSM), but, as more components were added to make it more comprehensive, its name was eventually changed to the Community Earth System Model (CESM).

One of the explicit goals of the CESM project was to minimize the climate drift problem and avoid using flux adjustment, enforcing local energy conservation whenever and wherever the atmosphere and ocean exchanged heat. The atmospheric component of the CESM was an improved version of the CCM. The oceanic component was derived from an ocean model developed by the US Department of Energy. One of the deficiencies of ocean models at the time was that their grid boxes were too large to capture the fine-scale swirling motions known as *eddies* that move energy around in the ocean. National Center for Atmospheric Research scientists added new formulas to their ocean model to represent the effects of eddies, which brought the heat uptake expected by the ocean much closer to the amount of heat the atmosphere provided.[11]

With these improvements, the CESM could be used for calculations extending more than 300 years into the future, with much reduced climate drift and without the need for a correction term for the heat uptake. Solar energy input averaged over the Earth's surface is about 340 watts per square meter. In a long-term climate calculation, the CESM can conserve energy with errors of less than 0.3 watts per square meter,[12] or better than 0.1 percent accuracy compared to the energy input. This level of accuracy is important because the extra heating associated with doubling carbon dioxide concentrations is only about 4 watts per square meter.[13] The CESM was among the first models to address the climate drift problem without using flux adjustment. Since then, all major climate models have followed suit and have become better at conserving energy.

9.4 Earth System Models: Carbon Cycles and More

Early climate models were essentially just models of the atmosphere, along with highly simplified representations of the land (such as a bucket) and the ocean (such as a slab of water). Then, more realistic ocean models were coupled to the atmospheric model (Section 4.6), with the realization that the ocean plays a crucial role in absorbing the extra heating associated with the increased greenhouse effect and therefore slows down the pace of global warming. More complex models of land processes, as well as models of the Arctic and Antarctic sea ice, were also added to climate models. But these were all still physical components of the climate system.

Chemistry and biology also play a very important role in climate.[14] For example, the ocean removes some of the excess carbon dioxide in the atmosphere by absorbing it, which mitigates global warming. Chemical and biological processes on land can also store and release carbon in rocks and plant material. Since carbon dioxide and other greenhouse gases play a crucial role in climate, models need to incorporate chemical and biological processes that emit those gases, in addition to physical processes. The more complex climate models that do this are commonly referred to as *Earth System Models*, to reflect their more comprehensive nature.[15] Although they are more realistic than mere coupled ocean–atmosphere models, these models have some downsides. They are a lot more complex, and the chemical and biological processes they represent are not as easily quantifiable as physical processes[16] – Newton's laws govern the motion of air and water, but there is no simple equation for the growth of trees!

The crucial ingredient of Earth System Models is carbon, which, like energy, is conserved. Carbon exists in four main forms: rocks, fossil carbon (long-dead plant material), living plants, and gas (carbon dioxide and methane). When we burn fossil carbon (oil or gas extracted from the ground), we generate carbon dioxide (and energy). Different forms of carbon flow through the climate system in a process known as the carbon cycle. Similar cycles are also associated with other chemical ingredients of the climate system; nitrogen, for instance, undergoes a nitrogen cycle.

Photosynthesis captures carbon dioxide from the atmosphere and stores the carbon in plants, which causes carbon dioxide concentrations to drop during the growing seasons. The carbon in plants eventually makes its way into animals. In the fall, decaying plants and animals release carbon dioxide, causing the atmospheric concentration of carbon dioxide to increase. The famous Keeling curve shows the sinusoidal seasonal cycle of carbon dioxide superimposed on the steady year-to-year increase (Figure 5.2a). Land is therefore a carbon reservoir, acting both as a source and a sink. The ocean is also a carbon reservoir, one that can likewise absorb and release carbon dioxide.

We can attribute the steady increase in carbon dioxide emissions to human activities, primarily the burning of fossil fuels, through techniques such as carbon fingerprinting (Section 7.2). But if we do an actual budget of human carbon emissions, the increase in atmospheric carbon dioxide is considerably slower than what we would expect. Human activities emit almost 10 gigatons of carbon every year, but the atmospheric increase in carbon dioxide can only account for about half that amount. Where did the other half go?[17] About a quarter of the carbon is absorbed by the upper ocean and the rest is absorbed by the land, and both of these effects mitigate the pace of global warming. But this comes with undesirable side effects: When carbon dioxide dissolves in the upper ocean, it forms carbonic acid. This makes the ocean less hospitable to certain forms of marine life because the increased acidity weakens the shells of marine organisms.

When we burn fossil fuels such as coal, petroleum, or natural gas, we are burning fossil carbon, which is dead organic material from millions of years ago. Fossil carbon is like an inheritance: Once it is spent, it is gone forever. Carbon in living plants is like money in a savings account. Carbon dioxide gas is like money in a checking account, with a lot of deposits and withdrawals. Sustainable living, in this analogy, would mean living on a monthly earned income, not running through an inheritance.

An important property of money is the rate at which it accumulates, or the rate at which it is spent. A balance in a checking account might last for a few days to weeks. In a savings account, a balance may last for years. Money invested in a home or other property may last for decades. A similar property for a greenhouse gas is its *atmospheric lifetime*, a measure of how quickly it accumulates or decays in the atmosphere. Carbon dioxide, by far the most important greenhouse gas, has an atmospheric lifetime of centuries to millennia.[18] This means that if we stop emitting carbon dioxide today, over thousands of years, the concentration of carbon dioxide in the atmosphere will slowly decrease as more of it is transferred from the upper ocean into the deep ocean, and as it is absorbed even more slowly by rocks on land through the process of chemical weathering. In contrast, the second most important greenhouse gas, methane, has a much shorter atmospheric lifetime, of about 10 years; unlike carbon dioxide, it can be removed from the atmosphere by chemical reactions. If we stop emitting methane, the excess methane in the atmosphere will be gone in a matter of decades.

The long atmospheric lifetime of carbon dioxide in the atmosphere means that, even if our carbon dioxide emissions were to drop to zero tomorrow, the Earth would not start to cool rapidly (e.g., see RCP2.6 in Figure 7.4). Carbon dioxide would continue to warm the globe over the course of several centuries, but its concentration would slowly decrease due to absorption by the ocean and land. At the same time, the ocean heat uptake would slowly reduce as the deep ocean warmed. Models suggest that the two effects would roughly cancel out,[19] leading to continued global warming at a much-reduced rate, or even roughly constant global temperatures.

Adding the carbon cycle to a climate model increases its complexity, but it puts an additional constraint on the system: the conservation of carbon. This constraint allows us to track the inventory of carbon in different components of the climate system and better predict how it will change in the future. It also makes the model more realistic by permitting additional feedbacks, known as *carbon cycle feedbacks*, which are possible mechanisms for tipping points in the climate system (Section 8.4).

The possibility of amplifying carbon cycle feedbacks, such as the permafrost feedback, is concerning, but these feedbacks are rather slow. Perhaps the most worrisome aspect of climate change is carbon dioxide's long lifetime in the atmosphere. As noted, even if we reduce carbon emissions to zero now, it will take many centuries for carbon dioxide concentrations to drop significantly. Delaying reductions in emissions makes it that much harder for us to mitigate global warming. Trying to slow down climate change is like trying to stop the motion of a massive ship, like that of the *Titanic*. If there is an iceberg ahead and the ship doesn't start turning or slowing down early enough, the ship will be unable to avoid disaster. The climate system, like the *Titanic*, cannot turn on a dime.

9.5 Better Models, More Uncertainty?

Without a carbon cycle in a climate model, climate prediction requires future concentrations of carbon dioxide to be specified as the boundary condition, as part of a scenario. A simplified climate model is used to compute concentration scenarios from the emission scenarios produced by integrated assessment models (Section 7.6). In an Earth System Model, we can use the future emission scenarios directly as a boundary condition. The model will itself calculate the concentration of carbon dioxide over time, enforcing the appropriate constraints and conservation properties.

To understand how adding the carbon cycle to a climate model affects that model's predictions of global warming, we consider different sources of errors. The squared amplitudes of independent random errors in different parts of a system tend to add up. (An important caveat, though: There are other types of error that do not simply add up and that can actually be amplified or reduced through feedbacks.) This property of squared random error can be used to make some important points because it is "conserved" in a heuristic sense. We can write simple heuristic equations for the total squared random error. (For convenience, we drop the qualifier "squared random" below.)

The total error of long-term climate predictions can be expressed as

Prediction Error = Model Error + Scenario Error + Stochastic Error

When using a noncarbon-cycle model, we can express Scenario Error and Model Error as:

Scenario Error = Carbon Error + Other Errors (economics, energy use, ...) [No carbon cycle]
Model Error = Atmosphere Error + Ocean Error + ...

where "Carbon Error" refers to the error in the simple model that converts carbon emissions to carbon dioxide concentrations.

When we switch to a model using a carbon cycle, the Carbon Error term moves from the Scenario Error equation to the Model Error equation:

Scenario Error = Other Errors (economics, energy use, ...) [With carbon cycle]
Model Error = Atmosphere Error + Ocean Error + ... + Carbon Error

There is another wrinkle. In the noncarbon-cycle case, all models use the same scenario for carbon dioxide concentrations and have the same Carbon Error contributing to the total. When the carbon cycle is introduced, each model has a different Carbon Error, depending upon its formulation of the carbon cycle, which appears to increase the total Prediction Error.

This heuristic argument explains why the Model Error may appear to grow as more processes, such as the carbon cycle or continental ice sheets, are added to make a model more comprehensive. Calculations using climate models with carbon cycles indicate that the spread in the values of predicted warming by 2100 for different models will widen, primarily at the higher end, due to carbon cycle feedbacks.[20] Does this mean that making the model more comprehensive actually increases the uncertainty in predictions? Not really – it just quantifies a previously unquantified uncertainty.[21] As NCAR scientist Kevin

Trenberth writes, more knowledge can sometimes lead to less certainty, as we increase "our understanding of factors we previously did not account for or even recognize."[22]

9.6 Predictive, Constraining, and Explanatory Power

When you have eliminated the impossible, whatever remains, however improbable, must be the truth?

Arthur Conan Doyle, Sherlock Holmes in *The Sign of the Four*

In the 2004 movie *The Day after Tomorrow*, which is ostensibly about climate change, New York City freezes over in a matter of days because cold air from the stratosphere descends at the center of a mammoth snow hurricane, cooling the atmosphere at the rate of "ten degrees per second."[23] Is this scenario remotely plausible, or even physically possible? Climate models may not always predict climate change precisely, but, because they impose physical laws such as the conservation of mass and energy, they are good at ruling out impossible or outlandish scenarios. The rapid freezing of New York City by descending stratospheric air is physically impossible in such a short time frame due to the law of conservation of energy. (Northern North America did freeze during the last ice age, but that occurred over thousands of years.)

Scientific models using the hypothesis-driven approach emphasize the *predictive power* of models, or how models can tell us what will happen in the future. The data-driven approach focuses on the *explanatory power* of models, or how models can explain what has already happened. But the existence of constraints on the climate system suggests a third type of power associated with models. We can call it the *constraining power*, or how models can rule out what cannot happen in the future, by using inviolable physical laws. Purely inductive or data-driven models lack this power because they cannot rule out future possibilities absolutely.

To illustrate the difference between these different powers, we consider a hierarchy of climate models with increasing levels of complexity, to predict the globally averaged temperatures:

1 *Extrapol*: A spreadsheet model that fits a line or exponential curve to the observed global average temperature data (with no physical information)
2 *NoCarb*: A climate model with no explicit carbon cycle (but with a prescribed carbon dioxide concentration in the atmosphere)
3 *SimCarb*: A climate model with simple carbon cycle constraints
4 *ComCarb*: A climate model with complex carbon cycle constraints

We use the carbon cycle as the distinguishing constraint purely for the purposes of illustration. A similar argument could be made for any other conserved property of the climate system.

The addition of a carbon cycle to a model highlights some important issues regarding model validation. *NoCarb* uses the evolution of carbon dioxide concentrations as the boundary condition. *SimCarb* and *ComCarb* use the evolution of carbon emissions from all human and natural activities as the boundary condition; internally, the carbon cycle model will compute the carbon dioxide concentration, taking into account various atmospheric, marine, and terrestrial processes.

A commonly used criterion in testing a model is to check how well the model simulates the last 150 years of climate change, given the history of carbon dioxide concentrations or carbon emissions. The (unproven) expectation is that the better a model's ability to simulate the past, the better its ability to predict the future. Other important criteria used to evaluate a scientific model include the model's structural simplicity and the falsifiability of its predictions.

Which of the above models satisfies these criteria best? In terms of raw simplicity, the spreadsheet model *Extrapol* is the clear winner: Climate models can have more than a million lines of code. But due to that very simplicity, the linear or exponential curve fitting in *Extrapol* is unable to simulate the ups and downs of global temperature during the last 150 years. It will perform poorly in the test of simulating the past, and we shall not discuss it any further.

Model *NoCarb* is more amenable to testing on past temperature measurements than *Extrapol*, but it cannot be tested on any measurements relating to the carbon cycle because it does not include the carbon cycle. The two models that include the carbon cycle, *SimCarb* and *ComCarb*, are the most testable, as they can be validated for carbon measurements as well. Therefore, they have the most constraining and explanatory power. If all three models fit the past data roughly the same, we should pick one of the two models that includes the carbon cycle. How do we choose between those two? Unless *ComCarb* predicts additional variables that can be verified by data, or unless it fits past data much better, *SimCarb* is the one that satisfies the Occam's Razor criterion of simplicity and should be our choice. Complexity without additional verifiability is not a virtue.

Suppose that *NoCarb* fits the past temperature data much better than either of the two models with a carbon cycle. Does that mean *NoCarb* is better than *SimCarb* and *ComCarb*? No, because the playing field is not level. We would need to provide carbon dioxide concentrations as inputs to *NoCarb*; *SimCarb* and *ComCarb* only need carbon emissions as inputs.

Let us assume for the sake of argument that the reason for the good fit of *NoCarb* was that the permafrost carbon cycle feedback did not play an important role in the past. Recall that when permafrost thaws due to warming, it releases methane, which can further amplify the warming. Even if the carbon cycle models were capable of capturing a relatively small permafrost feedback well, errors in other aspects of the carbon cycle may have degraded their performance relative to *NoCarb*.

In the future, permafrost in the Arctic may thaw and trigger an amplifying feedback. If we used one of the carbon-cycle models to make a prediction using a carbon emission scenario as the boundary condition (Section 7.6), it would be able to predict this amplifying permafrost feedback, which may well dominate over smaller errors in the carbon cycle. However, *NoCarb*, which uses a carbon dioxide concentration scenario from a simplified climate model as its boundary condition, would in this case make a poor prediction – it does not include the carbon cycle and would thus not predict the amplifying permafrost feedback. Note that the carbon cycle is not constrained in the simpler *NoCarb* model: It is simply specified. When evaluating and comparing climate models, constraining power and comprehensiveness may be as important to consider as the power to "predict" the known past.

***

Today's climate models are very good at constraining energy and mass budgets, even if their predictions cannot be easily verified. Reaching equilibrium in these climate models requires that they conserve energy with very high accuracy. When we evaluate the utility of models, we must take their constraining power into account, in addition to their predictive power. The more comprehensive Earth System Models provide additional constraints on important climate processes like the carbon cycle, albeit at the cost of increased model complexity.

The reductionist approach to climate modeling decomposes a large system into smaller parts, but if some of those parts cannot be modeled accurately, the accuracy of the larger system is limited. In the late 1990s, the NCAR CESM and other climate models of the next generation discarded the crutch of flux adjustment and started to conserve energy. But there were still model errors hidden in the adjustable coefficients of the parameterizations used to represent certain smaller-scale processes, such as clouds that were too fine to be captured by the models' large grid boxes. The compensations that occur between different errors have troubled climate models from the very beginning, starting with the simple model of Arrhenius. In Chapters 10 and 11, we discuss this persistent problem in some detail.

10

Tuning Climate

A Comedy of Compensating Errors

Our discussion of Occam's Razor in Chapter 8 began with the Leaning Tower of Pisa, where Galileo is reputed to have performed his famous experiment demonstrating that two spheres of different weights fall at the same rate. The tower, the very platform Galileo used to demonstrate the simplicity of mechanics, also turns out to be a testament to the complexity of mechanics. In 1990, engineers discovered that the lean of the famous tower was starting to increase at an alarming rate and immediately initiated measures to prevent its collapse. It turned out that stabilizing the tower was an extremely difficult geoengineering problem, with an 800-year history that vividly illustrates the concept of compensating errors in complex systems.

In the twelfth century, the wealthy Italian city-state of Pisa began to construct a majestic cathedral, with an eight-story freestanding bell tower as its crowning glory (Figure 10.1).[1] The tower's design was a hollow cylinder with a spiral staircase inside and a marble surface outside. Construction of the tower began in 1173 with a shallow foundation in the alluvial soil, which was quite soft around Pisa – made of sand, clay, and shells.[2] Even as the lower floors were being constructed, the tower began to tilt northward by about 0.2 degrees from the vertical. Due to the softness of the soil in the region, this was not considered unusual. Construction continued, with columns built taller on the sinking side to compensate for the tilt. This resulted in the shape of the tower deviating from a straight cylinder, ending up banana-like due to the differing column heights.

Construction stopped again in 1278, when the tower was built up to the seventh story. The tower now tilted the other way, to the south, by about a degree: the soil must have been softer on the south side. When construction resumed in 1360, extra weight was added on the northern side of the bell chamber to try to counter the southward tilt. This effort was unsuccessful; by 1838, the tilt had increased to over five degrees, and the base of the tower had

Figure 10.1 Old nineteenth-century engraving of the Leaning Tower of Pisa. (Getty Images)

sunk below the ground. Attempts to excavate the walkway around the base of the tower breached the shallow water table on the south side and caused the tower to tilt further to the south by about half a degree.

In the 1930s, the dictator of Italy, Benito Mussolini, decreed that the tower needed to be straightened. Grout was injected into holes drilled in the tower's foundation. Unfortunately, this caused the tower to lean even more to the south. By 1990, the tilt had increased so much that it was feared that the tower would collapse, as had happened with other towers in Italy.[3] The tower was closed to visitors, and the government formed an international committee composed of engineers, mineralogists, architects, archaeologists, and historians to address the problem. As a temporary measure to counterbalance the tilt, the committee ordered 600 tons of lead ingots to be placed on the north side of the tower. This worked, and the tower's tilt slowly started to reverse. However, the lead ingots were an eyesore, not aesthetically acceptable at this national treasure. A more permanent solution needed to be found.[4]

The committee came up with an ambitious plan to inject liquid nitrogen into the soil around the tower to freeze it, then excavate to install anchored cables to stabilize the tower.[5] But as soon as the soil was frozen, the southward tilt started to increase at an alarming rate. The plan was immediately suspended, and an additional 300 tons of lead ingots were placed on the north side to correct the damage. It was then decided that soil should slowly be removed from below the tower's north side. Gravity would then cause the southward tilting to be arrested and then reversed. Slanted holes were drilled to begin extracting the soil in 1996, and, when the project was completed almost three years later, it had successfully reduced the tilt to around four degrees. The tower reopened to the public in 2001, and the tilt is expected to be stable for several centuries to come.

The story of Galileo demonstrating the simplicity of basic mechanics atop the Tower of Pisa continues to inspire young scientists. But it is the 800-year history of the complexity of structural mechanics that likely rings a bell for seasoned climate modelers. Errors in climate models are often referred to as biases, analogous to the lean of the tower. For example, a climate model that runs too hot, simulating a climate that is warmer than reality, is said to have a "warm bias," and a climate model that runs too cold is said to have a "cold bias." Any climate model will initially exhibit a certain type of bias; as the model evolves and attempts are made to fix the old bias, the model will often come to exhibit a new type of bias. Like the columns of varying lengths on different floors of the Tower of Pisa, the computer program for every mature climate model includes within it a collection of fixes to compensate for biases in the model's simulated climate.[6]

10.1 Model Tuning: Orchestra or Whack-a-Mole?

The lurching history of the Leaning Tower shows how compensating errors can maintain the equilibrium of a complex system. According to Newton's laws of motion, the forces acting on the tower must balance for it to be stable. The weight of the tilting tower must be balanced by soil resistance. When these forces were not in balance, adding lead counterweights on the appropriate side compensated for the error and restored stability.

A climate model maintains the global thermal equilibrium by balancing different sources of energy, including solar heating, infrared cooling, the reflection of sunlight by clouds, warming due to the greenhouse effect, and cooling due to industrial haze. If the climate model is accurate, all the heating terms should balance out to a net heating of approximately zero under pre-industrial conditions[7] because there was no global warming due to greenhouse gases back then.[8] But in practice, models have errors,[9] and this net heating will not be zero when a model is first constructed.

When we run a new climate model for periods of decades to centuries, the global temperature will slowly drift. Errors too small to be detected during the testing of the model over short-term periods will slowly accumulate and degrade the long-term climate simulation. For example, the cooling effect of haze may be represented too weakly in the model, resulting in a positive heating error. The model simulation will then exhibit a systematic warm error in global temperature that grows in time, even with no change in such boundary conditions as greenhouse gases.

To reduce the systematic model error, a procedure known as *model tuning* is carried out. Parameters in a model, especially those associated with uncertain terms like the effects of clouds,[7] are adjusted to compensate for the largest errors in order to bring the net global heating as close to zero as possible for preindustrial conditions. The goal is to reduce the systematic error in the model's global temperature simulation to an acceptable level, rather like controlling the tilt of the Tower of Pisa. Once a model is tuned, it is ready to be used for climate predictions using different scenarios. After choosing a prediction scenario, the concentration of carbon dioxide and other greenhouse gases is increased. This will lead to increased greenhouse effect, and the net heating will become positive, warming the global temperature for the right reasons.

The term "tuning" used for models is borrowed from music.[7] If you go to an orchestra performance, you first hear a cacophonous noise while each instrument is tuned, while knobs are adjusted, and strings are tightened. Slowly, the noise dies down, and soon a beautiful sound emerges. The instruments in an orchestra tune to the concert pitch, typically the A produced by the oboe. Instruments can also be tuned on their own, to a pitch produced by a tuning fork or an electronic tuning device. If an instrument is not tuned to the same pitch as the rest of the instruments in the orchestra, the mismatch will produce unpleasant sounds called "beats" when the entire orchestra performs.

Unfortunately for climate modeling, there really is no concert pitch for the components of the model to tune to, and there is usually no exact analog to a tuning fork that can be used for individual components. Many deficiencies in the components only become apparent when they are all used together, and small errors in the standalone tuning of an individual component slowly add up over the course of a long-term climate simulation. This means that climate model tuning has to be done "on the fly." It's like tuning instruments in an orchestra without a standard pitch: Start the performance playing musical pieces while adjusting each instrument and continue to play new pieces until the orchestra slowly settles into a reasonably harmonious state. The situation is easier when we replace an instrument during the performance because we can assume that the other instruments are already in tune and try to adjust the replacement instrument to match their pitch. Sometimes the replacement

Figure 10.2 The whack-a-mole game of model tuning, reflecting the global interconnected nature of errors in climate models. Altering the temperature in the Pacific Ocean, for example, can lead to changes in the rainfall over California and Brazil.

instrument can be stubborn and fail to match. Finally, the tuned orchestra may end up playing the musical piece in a different key because we don't know what the true key should be.

The purpose of belaboring this broken orchestra analogy is to convince the reader that model tuning, unlike musical tuning, is neither a benign nor desirable thing. Nor is it a dark and sinister thing, as critics of models seek to portray it. It is only an unavoidable necessity that makes models useful. It would be more accurate to refer to model tuning as *model calibration*, or even more explicitly, as *error compensation*, thus removing any subliminal positive connotations.

A more apt analogy for error compensation may be the arcade game *whack-a-mole*. The game consists of a soft mallet and a rectangular box with several circular holes at the top (Figure 10.2). A plastic mole can appear or disappear in each hole, controlled by the mechanism within the box. When the game starts, moles will start to randomly pop up in holes. If you hit one on the head with the mallet, it will pop back in and you will move on to another mole. You hit that one next, and so on. The faster you "whack" the moles in, the more points you win.

When climate modelers fix an error in one place in the model and rerun it, soon a new error will pop up in a different place.[10] For example, fixing a rainfall error in Brazil may result in making Texas more prone to droughts. Fixing that problem in Brazil can result in a different error – say, overly warm Atlantic temperatures. That could lead to more landfalling hurricanes in Florida and so on. This has nothing to do with the Butterfly Effect; it is simply a consequence of the air flow patterns that connect different regions of the globe.

The mathematical equations that form the foundation of a climate model impose strict conservation laws and other constraints. When you alter one piece of a climate model, other pieces must adjust to some extent to maintain the constraints. John von Neumann reportedly said of models, "with four parameters I can fit an elephant, and with five I can make him wiggle his trunk." This may be true of less physically constrained models, but, for a climate model, when you tune it so that you can wiggle the "trunk," you may end up losing a "leg" due to the constraints.

Mathematically, we can think of model tuning as a nonlinear optimization problem. We can associate model error with a multidimensional cost function that depends on many different adjustable parameters in the model. For example, if we visualize the nonlinear cost function as a surface in a three-dimensional space, tuning involves finding a local minimum or "valley" of the function surface to minimize the cost. The adjustable parameter values corresponding to this valley are the optimal "tuned" values for a particular model configuration. If we alter a model component, we will change the shape of the cost function surface, and we may no longer be in a valley.[7] Further tuning of other components will be needed to compensate for new errors and locate a new valley of the cost function surface.

Students new to climate modeling are often excited when a climate model simulation closely fits a particular set of historical data after tuning. It needs to be pointed out to them that the closeness may be coincidental. If a model known to be imperfect fits a particular set of observed data perfectly, there must be a latent problem in the model. When the model is improved – that is, when the grid is made finer or when processes are represented better – it will likely no longer fit that data perfectly. The model will change over time, but the historical data will not change.

While model tuning mitigates the symptoms of the disease of model errors, it is not clear how well it addresses the underlying disease itself; it is a palliative rather than a curative treatment. Model tuning is more like adjusting the column heights of the Tower of Pisa to address the weight imbalance – the proximate cause of the tower's tilt – rather than fixing the tower's weak foundation, which turned out to be the ultimate cause of its tilt.

10.2 Correctness, Confirmability, and Usefulness of Models

The models we use are incomplete and imperfect representations of reality. A familiar quotation in modeling, by the statistician George Box, encapsulates this: "All models are wrong, but some models are useful." This leads us to ask: Does tuning a climate model make it more correct? Logically speaking, the tunability of a model is a necessary condition to establish the correctness of a model, but it is not a sufficient condition.[11] Tunability is also necessary for an orchestra, but it is far from sufficient to ensure a good performance. Confusing the necessary with the sufficient is a logical fallacy.

For example, as discussed in the context of the carbon cycle, a climate feedback that played only a minor role in the past could become important in the future (Section 9.6). In this case, tuning the model to the past is not sufficient to ensure that this feedback is correctly simulated in the future, if the future is expected to be different in unprecedented ways.[12] Recall the 1985 discovery of the ozone hole (Chapter 6). Even the most optimal tuning of the coefficients of formulas in the pre-1985 models could not have correctly predicted the ozone hole because the surface chemistry of polar stratospheric clouds, crucial to the process of ozone depletion, was not even included in the models. (This is reminiscent of the disclaimer that appears at the bottom of brochures for financial investment products:[13] "past performance is not an indicator of future outcomes.")

A standard question that is asked of climate models is: How does a model simulate past variations in climate since the preindustrial period? We know the right answer to this question from data, and most climate models are explicitly or implicitly tuned to reproduce the global warming signal of about a degree Celsius seen thus far.[14] An "untuned" climate model will typically simulate the past poorly due to model deficiencies, giving the wrong answer for the wrong reasons. The ultimate goal of tuning is to progress from getting the wrong answer for some wrong reasons to getting the right answer for all the right reasons. But in practice, tuning often gives the right answer for a combination of some right reasons and some wrong reasons.[15] While it is easy to check the answer, it is much harder to verify that all the reasons for the answer are also correct. Tuning for the correctness of a single metric like the global-average temperature compounds this problem because different regional warming patterns can average out and appear to have the same global warming.

Sometimes, tuning the model simulation to agree with the observed climate can take primacy over physical constraints during model development.[16] Climate scientist and hurricane expert Kerry Emanuel describes the case of a researcher who decided to ignore the physically important effect of frictional heating due to strong winds because it made his hurricane model worse.[17] For this researcher, as Emanuel says, "getting the 'right answer' was the goal, even if it is obtained for the wrong reasons."

Rather than a model's correctness, we can consider a different property, *confirmability*.[18] Tuning a model confirms that it can reproduce known observations.[19] The more tunable a model is, the more it will be able to reproduce multiple aspects of observations, say both rainfall and temperature at multiple locations, rather than only temperature at a single location. As we noted previously, it is hard to tune a model in all respects due to all the physical constraints that must be satisfied. It therefore seems plausible that the more comprehensively tunable a model is, the closer it is to being constrained in the correct way, and the more usable it becomes.[20]

There is a less familiar variation of the Box quote that elaborates on the usability issue: "Remember that all models are wrong: the practical question is how wrong do they have to be to not be useful." We need to shift our focus from the correctness of models to the usability of models. Tunability, or confirmability, definitely makes a model more usable. It is common for climate models from a particular country or region to be tuned to reproduce the local climate well.[7] That makes predictions from that model better suited for use in local climate change impact assessment and adaptation planning.[21] A more confirmed model would therefore find much broader usability.

***

Tuning is a form of inductivism in apparently deductive models. The mathematical equations underlying climate models are hypothesis-driven, but the uncertain coefficient values in some of these equations are essentially determined by data as part of the tuning process. The resulting models are hybrid products of deductive and inductive reasoning. Weather models have far fewer coefficients that need to be carefully tuned, making them much more deductive than climate models.

One of the primary goals of climate model tuning is to reduce the drift in simulations for preindustrial conditions and keep the climate static. In this context, it is worth recalling the cautionary tale of one of the most famous, or infamous, model-tuning efforts in the history of science.[22] When Albert Einstein formulated the equations of general relativity in 1917, his solution predicted a universe that was not static. Since observations at the time indicated that the universe was static, Einstein introduced a new term in his equations in order to keep the universe they predicted static. This new term, called the cosmological constant, was essentially a "fudge factor" that produced a desirable result. However, observations by the American astronomer Edwin Hubble in 1929 showed that the universe was actually expanding, which was consistent with Einstein's original equations. Einstein soon dropped the cosmological constant, reportedly calling its introduction his "greatest mistake," because his equations could otherwise have predicted the expanding universe. (With the recent discovery of dark energy, the cosmological constant has been revived, with its sign reversed, as a possible explanation for the accelerated expansion of the universe – highlighting the complex interplay between hypothesis-driven science and data-driven science.)

The birth and demise of the cosmological constant remind us that it is not possible to prove a model correct. Confirmed predictions of a model could be due to the cancellation of compensating errors, which could happen fortuitously or explicitly through model tuning. In particular, a model's "prediction" of what is already known has far less value than its prediction of unexpected future events. Compensating errors are the bane of climate modeling, but we cannot simply wish them away. They are a symptom of the limitations of the reductionist approach to climate modeling, in which we focus our attention on smaller and smaller parts of an individual climate model. The cosmological constant was an explicit compensating term in a relatively simple and transparent model. Climate models are much more complex and opaque;[23] there are numerous compensating terms, and often they are hidden deep inside complicated mathematical formulas and computer programs. In Chapter 11, we will discuss some workarounds to mitigate the impact of these hidden compensating errors on climate prediction.

11

Occam's Beard

The Emergence of Complexity

Physics is the purest of the natural sciences, shorn of complexity by Occam's Razor; chemistry has complex stubble; and biology has a veritable beard of complexity. So goes the popular stereotype, which holds that the other natural sciences want to be like physics and that even the social sciences suffer from physics envy and try to emulate its precision.[1] On the xkcd purity scale (Figure 11.1), weather would be quite close to physics, but climate would be somewhere between biology and chemistry. Biogeochemical processes, especially the carbon cycle, are important for climate but not for weather. The more empirical disciplines of epidemiology and economics would be to the left of biology.

To better understand the distinction between the sciences, imagine that our universe is a giant computer simulation.[2] The code in a computer program consists of pieces called functions. A *simple function* may add or multiply two numbers. A *compound function* carries out more complicated actions by using a combination of different simple functions. For example, to calculate our taxes, we might use a compound function that adds and multiplies many numbers, subject to different conditions. Computing Earth's climate would be the job of a vastly more complex compound function. In science, we are trying to reverse engineer the program for the universe. Physics reverse engineers the simple functions, which are the fundamental building blocks of the universe. Other sciences are attempting to reverse engineer the compound functions. Physics can tell us how one molecule can bounce off another; biology analyzes how millions of molecules in the brain work together to think. Of course, this is an overgeneralization – subfields of physics, such as condensed matter physics, clearly deal with emergent phenomena[3] – but it is broadly true.

In philosophical terms, physics is very amenable to reductionism, to studying the universe as a collection of molecules, atoms, and subatomic particles. Chemistry and biology are less amenable to reductionism. They emphasize

Figure 11.1 Purity. (xkcd.com)

emergent properties that arise from interactions between different parts of a complex system, such as the behavior of complex organisms. Comprehensive models of climate now include chemical and biological processes. Therefore, it seems only natural to extend the biology analogy used to motivate the "fruit fly models" of climate mentioned previously, and to thus embrace emergent complexity in climate models.

One of the biggest deficiencies of climate models – and other complex emergent models – is structural uncertainty. To use the code reverse engineering analogy, there may be ten different compound functions that compute taxes in the real world, but the reverse-engineered model may be trying to represent all of them using just a single compound function. This will lead to errors that cannot be fixed simply by adjusting coefficients in that function. As discussed in Chapter 10, this structural uncertainty results in the problem of compensating errors, leading to uncertainty in estimating important emergent properties like climate sensitivity. Although this problem is fundamentally intractable, climate scientists have developed workarounds to mitigate it.[4] To explain an important workaround for the comedy of compensating errors, we turn to a literary tragedy as a metaphor, a story about families and unhappiness.

11.1 The Anna Karenina Principle

The novel *Anna Karenina* by Leo Tolstoy opens with one of the most famous lines in literature:

Happy families are all alike; every unhappy family is unhappy in its own way.

This line has inspired the Anna Karenina principle, which holds that all well-adjusted systems are alike, and that each poorly adjusted system is different in a different way. This principle, in various forms, has been applied in the physical and social sciences.[5] We invoke a modified version of this principle to describe the collective behavior of climate models, focusing in particular on compensating errors.

Suppose we are studying the behavior of dozens of families. Let us assume that some of them are happy, but we don't know which ones. If we numerically average the behavioral aspects of all the families, the behaviors of the unhappy families will tend to cancel out, and we will be left with a numerical description of the behavior of happy families. For example, say that unhappy families converse either too much or too little, and happy families converse just the right amount. If we average the time that all of the families under consideration spend conversing, it will be close to the right amount. We can call this the strong version of the Anna Karenina principle.

What if there are no truly happy families? Then we can also consider a weaker version of the Anna Karenina principle, where we assume that there are several families that have some (but not all) behavioral aspects of a truly happy family, such as the right amount of conversation – even if we don't know beforehand what the right amount of conversation is. We still require that the unhappy behavioral aspects of each family be different from the unhappy behavioral aspects of the other families. Provided this is the case, if we numerically average the behavior of all families, then the common happy aspects will survive the averaging process, but the opposing unhappy aspects, such as too much or too little conversation, will tend to cancel each other out.

A similar averaging argument is used in public opinion polls before elections. Polls tend to have hidden sampling biases that cannot be known until after the election takes place. But different polls are likely to use different sampling and analysis methodologies. Therefore, the average of many polls, or an aggregated poll, can cancel out some of these unknown biases and provide better predictions.[6] The underlying assumption here is that the element of truth common to all polls will be amplified by the averaging process, provided the polling methodologies are sufficiently different.

11.2 An Ensemble of Climate Demons

In Section 2.5, we considered an ensemble of identical Weather Demons (or weather models). Each Weather Demon in the ensemble received a slightly different set of initial conditions with which to make a weather forecast. By averaging the forecasts from all the Weather Demons in the ensemble, the errors associated with uncertainty in the initial conditions are reduced, but the errors associated with the weather model itself are not reduced.

For the climate problem, we consider an ensemble of Climate Demons (or climate models), rather than Weather Demons – but each Climate Demon in the ensemble is different from every other Climate Demon in the ensemble. We know that climate models are imperfect and do not usually simulate all aspects of climate correctly at the same time;[7] they get some aspects of climate right and others wrong. Compensating errors are also typically different in each model. Model tuning can reduce some of these errors, but it does not eliminate them.

Since every climate model has some skill in "predicting" climate in the present and recent past, we assume that every climate model has some elements of the "perfect" model, such as good representations of certain fine-scale processes. However, we may not be able to identify these good elements simply by looking at model simulations because simulations represent the emergent outcomes of complex interactions in the climate system.[8] A tuned model often gets the right answer for a combination of right reasons and wrong reasons, and we can't always distinguish between the two (Chapter 10). Since there are no perfect models, we assume the weaker version of the Anna Karenina principle, restated as: *All good climate model features are alike; every bad model feature is bad in a different way.*

We can make a climate prediction with each model in our ensemble and then average the predictions of all the models. Such a prediction is called a *multi-model average* prediction. To better understand how it works, we write a schematic equation for the heat error in a climate model:

$$\text{NetHeatingError} = \text{SolarHeatingError} - \text{InfraredCoolingError} - \text{CloudCoolingError} + \ldots$$

We focus on the largest contributors to the errors, which are commonly related to solar heating, infrared cooling, and clouds. The net error on the left side is the sum of the error contributions from each term on the right side. For each climate model, the error contributions and the net error will be different. So we replicate the above equation for each of our models, using subscripts to differentiate between the models:

$$\text{NetHeatingError}_1 = \text{SolarHeatingError}_1 - \text{InfraredCoolingError}_1 - \text{CloudCoolingError}_1 + \ldots$$
$$\text{NetHeatingError}_2 = \text{SolarHeatingError}_2 - \text{InfraredCoolingError}_2 - \text{CloudCoolingError}_2 + \ldots$$
$$\text{NetHeatingError}_3 = \text{SolarHeatingError}_3 - \text{InfraredCoolingError}_3 - \text{CloudCoolingError}_3 + \ldots$$
$$\text{NetHeatingError}_4 = \text{SolarHeatingError}_4 - \text{InfraredCoolingError}_4 - \text{CloudCoolingError}_4 + \ldots$$

When we calculate the multimodel average, we simply average the corresponding terms down each column across all the equations, to obtain a single equation:

$$\text{NetHeatingError}_{MA} = \text{SolarHeatingError}_{MA} - \text{InfraredCoolingError}_{MA} - \text{CloudCoolingError}_{MA} + \ldots$$

The subscript MA denotes the multimodel average. For each type of error, some models will have positive errors and some models will have negative errors – distributed randomly, according to the Anna Karenina principle that we have assumed. When we average many such random error numbers, the average will typically be smaller, because the positives and negatives will tend to cancel out.[9] This reduction in error magnitudes will also be reflected in the net error on the left side.

What we are arguing heuristically is that even if every one of our many models is imperfect, we can cancel out some (but not all) of the errors by averaging the predictions of all of these imperfect models. We don't need to know which models are better and which are worse for the argument to work, which is an important feature of this approach. Climate models are very complex: Each model has some good and some bad features. Therefore, the models are often weighted equally in computing the average, assuming some sort of "model democracy."

Perhaps we could do better by assigning higher weights in the model average to models that provide a better simulation of the current climate, that is, models that are better tuned. The process of tuning can be thought of as persuading an unhappy family to "put on a happy face," or at least to appear less unhappy. Should happier-looking families be given greater weight? Does putting on a happier face mean that you are truly happier? How do we assign weights to aspects of happiness displayed in different parts of the face, such as the mouth, the cheeks, or the eyes?

Happiness is hard to characterize with a single number; it means different things to different people. We face the same problem when we try to characterize model performance with a single number.[10] How do we weight better simulations of temperature, rainfall, or sea ice in one region or another? It may be better to settle for a modified version of model democracy, in which we treat all models roughly the same, but with some restrictions.[11] There should be some minimum criteria that a model must meet to participate in our modified model democracy.[12] For example, we could choose to include only models that reasonably simulate past climate variations and that are also sufficiently comprehensive, respecting all the constraints of the climate system. Also, we need to consider the structure of each model, that is, the details of how its different parameterizations are formulated. "One model structure, one vote" may be better than "one model, one vote" because we want to sample the structural diversity among models; we do not want to allow a situation in which multiple variations of a single model structure dominate the average and reduce the cancellation between errors.

We have argued that, theoretically, the multimodel average prediction will have a smaller error than individual model predictions. In practice, multimodel averages for simulations of the current climate do show reduced errors, compared to the errors of each of the individual models.[13] But what about predictions of the future?

We cannot verify whether multimodel averaging works for century-scale climate predictions because we will not have that data for a while. However, climate models are also used to predict phenomena like El Niño, a warming of the tropical Pacific Ocean that happens roughly every three to five years. El Niño prediction falls somewhere in between weather and climate prediction: Initial conditions in the ocean are important, but compensating model errors also play a major role in determining the quality of the forecast. Multimodel ensembles are routinely used for predicting El Niño operationally, and the multimodel average performs as well as or better than any single model in terms of El Niño forecast skill.[14]

Although our argument for the value of multimodel averages is not rigorous, and is by no means a proof, the fact that it does work for improving El Niño prediction provides some support for the Anna Karenina principle. We can extrapolate the same approach to century-scale climate predictions as well. That is why climate assessments, such as those produced by the IPCC, typically present multimodel averages for climate predictions. Given that every individual model has compensating errors in its simulations, this seems a reasonable compromise.

11.3 Herd Mentality and Model Diversity

Multimodel averaging works best when there are roughly equal numbers of positive and negative compensating errors in the ensemble, which requires that the compensating errors are symmetrically distributed around the value zero. When there are too many models in the ensemble that have positive errors associated with clouds, as opposed to negative errors, averaging the predictions across the ensemble will not cancel out the cloud errors. If the models in the ensemble are simply constructed randomly, there is no guarantee that their errors will be distributed symmetrically around zero. But if the models are tuned to mimic reality, as is typically the case, such a symmetric distribution around zero becomes more likely for the current climate, because the very purpose of tuning is to bring the errors closer to zero in predictions of the current climate.[15] However, when the same models are used to simulate the future climate, the symmetry may be lost because it is not possible to tune models to the unknown climate of the future – we don't know where the true "zero error" of the future is, and, as we have seen, the models could be missing or misrepresenting a feedback that is unimportant in the current climate but becomes important in a future climate.

Another important factor to consider in model ensembles is "herding" or "herd mentality," which is also a problem that afflicts opinion polling. One pollster might copy the techniques of another, or, alternatively, refrain from publishing a polling result that contradicts the majority of polls to avoid being called an "outlier." This would compromise the independence of polls and defeat the purpose of aggregating polls.[16] The field of climate modeling is also affected by herd mentality, partly due to individual modelers' desires not to be outliers, but also due to the inherently collaborative nature of climate research, which encourages the exchange of ideas between different modeling centers.[17] For example, if a particular technique to represent clouds looks promising, many different climate models may then start to use that particular technique. Even if each model uses somewhat different values for the adjustable parameters, this cross-fertilization may result in more cloud errors of one sign than of the opposite sign.

To avoid the herding problem, we need *model diversity*: Climate models need to be diverse, exploring uncertainty along different dimensions and considering different methodologies for simulating processes (such as clouds) that are poorly represented. Modelers use the terms *structural uncertainty* and *parametric uncertainty* when discussing similarities and differences between models. In plain language, we can refer to these as qualitative (structural) and quantitative (parametric) similarities and differences. To illustrate this distinction, we return to the family analogy.

One of the biggest costs for a homeowning family is their mortgage – the interest on the large loan they took out from a bank to buy the home. In the United States, there are two main types of mortgages: fixed rate and adjustable rate. As the name suggests, the fixed-rate mortgage means that the family pays a fixed interest rate for the life of the loan. Adjustable rate means that the interest rate varies from year to year; it is usually determined by adding a certain percentage to a variable benchmark interest rate, like the rate offered by US Treasury bonds. There are also other types of mortgages, with different payment structures. Based on family creditworthiness, a bank may offer different families slightly different interest rates for the same mortgage type.

The difference in the type of mortgage taken out by a family is a qualitative, or *structural*, difference. Different formulas are used to calculate the payments for the different types of mortgages. The difference in interest rates between mortgages of the same type is a quantitative, or *parametric*, difference; only the numbers are different, not the formulas. For climate models, structural differences are differences in the type of formula used to represent the effect of clouds.[18] Parametric differences are differences in the numeric values for the adjustable coefficients in the chosen cloud formula. To ensure model diversity, it is essential to have both structural and parametric differences among models.

11.4 Catch-22: Reproducibility and Replicability

One of the central tenets of science is the reproducibility of results. If one scientist performs an experiment and produces a result, other scientists should be able to repeat the experiment and obtain the same result. This is an especially critical issue in fields like medicine, where incorrect conclusions can affect human lives. Is drinking wine good or bad for you? Is it better to eat less fat and more carbohydrates, or the other way around? Over the decades, popular studies have vacillated between these diametrically opposed dietary recommendations.

In 2005, the medical researcher John Ioannidis published a paper with the provocative title "Why Most Published Research Findings Are False." Although it has been argued that some findings of the paper fall victim to its own title, the paper does make the very important point that the bias toward publishing only "positive" results means that published findings are not collectively as reliable or reproducible as they claim to be individually. Consider the following extreme scenario. Let us suppose that drinking wine has no impact on your health. In science, it is common to publish if it is 95 percent likely that your result is correct, meaning there is only a 5 percent chance that it is wrong. But if 100 scientists around the world carry out independent studies to determine if drinking wine improves your health, five of their studies will pass this publication threshold purely by chance because many other factors also affect the participants' health. Those five scientists will publish their results believing that they have passed the threshold, but they would all be wrong! Of course, this is an extreme example, and not all papers are wrong, but it highlights a serious problem in many areas of science that use statistics. A related issue is when scientists unwittingly cherry-pick their data to find publishable results, which is known as *p-hacking* when done wittingly.[19]

Any field of science in which the signal-to-noise ratio is a problem needs to worry about the reproducibility issue. Long-term climate prediction falls very much into this category, but reproducibility itself becomes harder to define for models. In scientific measurements of the real world, there is a single underlying reality – a single signal – even if it is obscured by noise. In long-term climate prediction, there are multiple realizations of reality, one for each model, and each one may have a different signal. Since we don't have measurements of the future of the real world, we do not have a good idea what the true signal is. This means that we need to revisit and redefine the concept of reproducibility itself.

In the strictest sense of the word, climate predictions are reproducible – but only for the same model. If we repeatedly run the same climate model with identical initial and boundary conditions on the same working computer hardware and software, we will get the exact same simulation every time.[20] (There are rare occasions in which a computer hardware fault leads to non-reproducibility, but then the computer can no longer be considered a working computer. There could also be issues with bugs in the software, but modern climate models are no buggier than other comparable complex software products.[21]) This type of reproducibility in models is called bit-for-bit reproducibility, where "bit" refers to the binary digits in the numbers.

As Ed Lorenz found out when he discovered the Butterfly Effect, changing even one digit in the initial conditions can change a weather forecast – and also a climate forecast, which is essentially just a very long weather forecast. Changing software versions or using a computer with a different number of processors can have an effect similar to changing the initial conditions slightly. But these are not really reproducibility issues, because climate predictions are inherently statistical, not deterministic: Climate predictions typically average an ensemble of forecasts rather than relying upon on a single forecast.

Related to reproducibility is transparency. A complex climate model cannot be developed by a solitary scientist or even by a small group of scientists where opaqueness can be maintained. The level of collaboration among large numbers of scientists that is required to make a climate prediction necessitates transparency and ease of access. Some, but not all, climate models are open source and freely downloadable. International collaborative agreements ensure that the data used for boundary conditions are freely accessible, and so are the model simulation results. This means that anyone with a powerful enough computer can check for themselves that climate predictions are reproducible.

Unlike in many other fields in science, there are no hidden problems with strict reproducibility in climate modeling. There is, however, a closely related problem in climate modeling hiding in plain sight: *replicability*.

Reproducibility is about repeating a prediction using the same model with the same initial and boundary conditions. Replicability is about replicating an important result using a different climate model but the same boundary conditions. Say that one model predicts that all the Arctic sea ice will be gone in 15 years. That is an interesting result, the kind of result that generates press releases and news articles. Will a different climate model predict the same? Not necessarily. Another model could well predict a more delayed loss of sea ice, because of uncertain model parameters and differences in the balance of compensating errors between the two models.[22]

In climate modeling, we don't expect strict replicability. Different models predict different degrees of warming for the same emissions scenario, as is evident from the model spread shown in Figure 7.4c. If all models agreed exactly on their predictions, they would have to have the same representations of fine-scale processes, and they would end up being the same model. As anyone familiar with the competitive sociology of climate modeling knows, this doesn't happen! Scientists who develop parameterizations of fine-scale processes like clouds are typically proud and protective of their creations. Even within a climate modeling center, there are at times bitter battles over which parameterizations should be "blessed" as the official choice for the latest version of the model released by the center. At a conference, a speaker may tout the virtues of one parameterization over all the others. All of this happens despite the fact that there is considerable cross-fertilization between climate modeling centers, often leading to broad structural similarities in parameterizations.

Essentially, we have a Catch-22, that is, the need to satisfy two contradictory requirements. Say we want climate predictions to be exactly replicable. Models will replicate each other's results only if they have the exact same parameterizations. But if they do, the Anna Karenina principle no longer applies: We lose model diversity and the ability to use multimodel ensembles to average out some of the compensating errors. So, we need models to remain sufficiently different from each other, and thus not replicate each other.[23]

Instead of exact replicability, we settle for model replicability in a Goldilocksian sense – not too similar, not too different, just enough for model diversity. For example, it is expected that models will not exhibit outlandishly low or high values of climate sensitivity,[24] with data-derived estimates of climate sensitivity serving as a rough guide.

This necessity for diversity in models highlights an important but often overlooked point. The spread in model estimates of future global warming or climate sensitivity is not necessarily an estimate of the true uncertainty in that parameter: It is not an error bar in the conventional sense used for experimental measurements.[25] For example, the spread in the projected warming for each RCP scenario (Figure 7.4c) is based on the ensemble of models that happened to be available for the IPCC assessment, that is, an "ensemble of opportunity."[26] It is not a representative sample of model diversity; it could be oversampling certain types of models and missing certain other types of models altogether.[27]

There is a conflict between the requirements for the replicability of predictions and the requirement for model diversity. Replicability requires models to be as similar as possible; but, for their compensating errors to cancel out and produce a more "skillful" multimodel average, models need to be as different as possible. In other words, the broader the spread across several individual model estimates of parameters like climate sensitivity, the better their multimodel average is likely to be in estimating the true value of the parameter.

If we view the spread in model estimates as a measure of uncertainty, this sounds paradoxical: The greater the uncertainty among the individual estimates, the less the uncertainty in their average. The resolution of this apparent uncertainty paradox lies in recognizing that the spread among the individual estimates is a rough measure of model diversity.[28] Statistically, when we pick a larger sample from a diverse population, we are likely to get a better estimate of the average properties of the population.

To estimate the true error bar for climate parameters, we need to sample model diversity as uniformly as possible; we need to poll enough unhappy families to characterize all the different ways in which they can be unhappy. The problem of not sampling enough parameter values – undersampling of parametric uncertainty – can potentially be addressed by using more powerful computers to perform simulations for a wide range of plausible parameter values.[29] The problem of not sampling enough model structures – undersampling of structural uncertainty – has no easy solution: There are essentially unlimited possibilities for formulas that can be used to parameterize fine-scale processes in models.

One issue we have not addressed is the replicability of statistical analyses of climate observations. This is important for identifying the global warming trend in noisy data, but it is beyond the scope of this book, as it is not directly related to climate modeling. Suffice it to say that statistical analysis with noisy data is difficult, for present and past climates, but scientists have confidence that global warming is real and attributable to human activities, as summarized by the IPCC.

***

Climate models have evolved from their origins in the simple, reductionist principles of atmospheric physics to an emergent complexity encompassing many disciplines, including chemistry and biology. But these models are imperfect in many different ways. Rather than trying to find the "one true model," the Anna Karenina principle provides a philosophical framework for analyzing imperfect climate models as a group, using emergentism rather than reductionism. In this framework, diversity among models is a desirable feature.[30]

Aggregating predictions from an ensemble of diverse models is a work-around for the problem of compensating errors. This problem arises from computational limitations and structural uncertainties in representing processes, which limit the replicability of climate predictions among different models. The weakness of the ensemble approach is that we have to assume that the set of available models is sufficient to span the range of possible parametric and structural uncertainties. But that may not be the case if the available models are not sufficiently diverse.

We need these heuristic workarounds because model complexity has grown steadily over the years. But has there been a corresponding increase in the skill of climate predictions?

12

The Hansen Paradox

The Red Queen's Race of Climate Modeling

> *"Well, in our country," said Alice, still panting a little, "you'd generally get to somewhere else – if you run very fast for a long time, as we've been doing."*
>
> *"A slow sort of country!" said the Queen. "Now, here, you see, it takes all the running you can do, to keep in the same place. If you want to get somewhere else, you must run at least twice as fast as that!"*
>
> Lewis Carroll, *Through the Looking Glass*

One of the most iconic moments in the history of climate change was the testimony of NASA scientist Jim Hansen at a US Senate hearing in the summer of 1988. His forceful statements about the human influence on the Earth's climate and its dire consequences were instrumental in raising public awareness of what had hitherto been regarded as an academic curiosity (Chapter 7). During the hearing, Hansen presented simulations from the GISS climate model, which predicted a global warming of 0.6, 1.1, or 1.5°C within 30 years – for Low, Medium, and High scenarios, respectively, of future greenhouse gas emissions (Figure 7.1a).

Was Hansen right? In 2018, the 30th anniversary of Hansen's Senate testimony, mainstream climate scientists revisited these predictions, as did contrarians who do not believe in climate models. The actual global warming from 1988 to 2018 was around 0.8°C, between the warming figures predicted by Hansen for the Low and Medium scenarios (Figure 12.1). With hindsight, we do not need to use a predicted scenario for greenhouse gas emissions over the last 30 years, because we know exactly what happened. Gavin Schmidt, who succeeded Jim Hansen as director of the NASA GISS lab, estimates that actual greenhouse gas emissions were also between the assumed emissions in the Low and Medium scenarios, about 40 percent higher than those in the Low scenario and 20 percent lower than those in the Medium scenario.[1] If we adjust

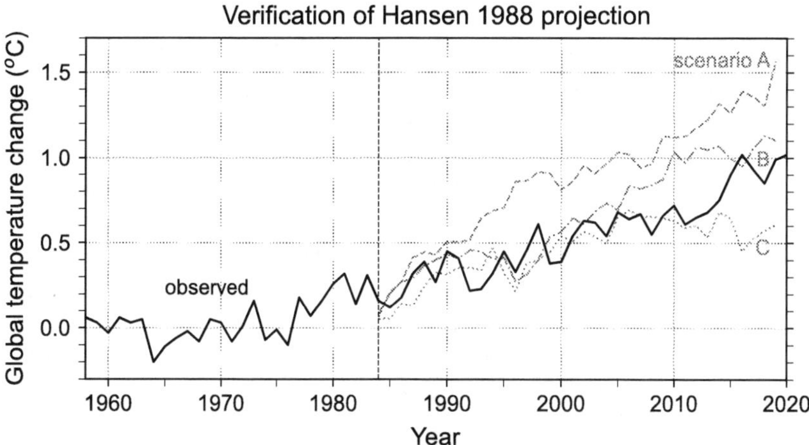

Figure 12.1 Verifying the projections made by Hansen in 1988 (Figure 7.1a) using observed surface-temperature change (solid line) over the period 1984–2020. The actual temperature evolution appears to track scenario C until 2010, but, afterwards, it appears to be closer to scenario B. (Plotted using projection data provided by Schmidt, 2007a, and surface temperature data from NASA GISTEMP, v4; GISTEMP Team, 2021)

Hansen's prediction for the Medium scenario downward by about 20 percent, the predicted warming matches the observed warming rather well. Some contrarians dispute this interpretation, saying that the observed warming is closer to the Low scenario[2] – but, even then, it seems that Hansen wasn't too far off the mark.

There were significant discrepancies in the details between the emissions assumed in the Medium scenario and the actual emissions over the 30-year period. This should not be surprising, because scenarios aren't meant to be predictions (Section 7.6). The actual emissions of greenhouse gases other than carbon dioxide, such as methane and CFCs, were substantially lower than assumed in the Medium scenario. Emissions of CFCs dropped due to the success of the Montreal Protocol in combating ozone depletion by curtailing the use of CFCs – none of which was anticipated when the scenario was created.

The biggest uncertainties in Hansen's predictions came from scenario errors. Correcting for scenario errors and adjusting old model predictions can be considered an exercise in retro-forecasting. A recent study carried out such an exercise for many published climate model predictions during the period 1970–2007 and concluded that most model predictions for globally averaged surface temperature were generally accurate when adjusted for scenario error.[3]

The climate models Hansen and his contemporaries used in the 1980s would be considered quite primitive and incomplete by today's standards. Thirty years after his famous prediction, models have become much more comprehensive and have finer spatial grids. Today's models are capable of representing a number of important processes that were not even considered by Hansen's model, including the complex properties of aerosols.[4] If it is argued that Hansen's less comprehensive model made correct predictions 30 years ago, then, because of the neglected processes, at least some of that success will need to be attributed to the fortuitous compensation of errors.

Arguing about the "correctness" of Hansen's 1988 predictions, though, creates a paradox for climate modeling: If it is argued that the climate prediction made using Hansen's old jalopy of a model was "correct enough," then why have we put so much effort into building the elaborate Cadillac climate models of today? If Hansen's model was only fortuitously correct due to some compensating errors, then should we expect that our current models will continue to have the benefit of such error compensations, albeit in a more elaborate setting? And if Hansen's model was not correct enough, then when was, is, or will be the "aha" moment at which we should declare that from this point in time, climate models can be considered "correct enough"?

The issue of whether climate models have become "correct enough" is not just of academic interest. In 2016, the Australian government used this issue to rationalize drastic cuts to its climate research program when Larry Marshall, a former Silicon Valley venture capitalist, was appointed director of the Commonwealth Scientific and Industrial Research Organization (CSIRO). Justifying the cuts as essential to foster a "startup" culture within CSIRO, Marshall stated in his announcement that

> [CSIRO's] climate models are among the best in the world, and our measurements honed those models to prove global climate change. That question has been answered, and the new question is what do we do about it, and how can we find solutions for the climate we will be living with?[5]

These cuts were proposed in part because climate models were considered by Marshall to be good enough. Normally, cuts to climate research are proposed by climate contrarians who say that the models will never be good enough to be trusted! The CSIRO cuts were reversed somewhat after a worldwide outcry, when it was pointed out that climate monitoring is still needed, and that climate models do more than just predict climate change.[6] Climate models' constraining power is key to efforts to plan for mitigation and adaptation.[7]

When predicting a singular event like global warming, it is not sufficient for a model just to get the 30-year warming trend right; we need to be confident that it is getting it right for the right reasons, not for the wrong reasons (such as fortuitous compensating errors). It is not always easy to establish this for a complex model whose behavior we may not fully understand. A weather model, which is verified for a large number of cases, can be judged primarily on its prediction skill. To judge a climate model, we need to also consider the structure and scientific foundations of the model.

A more elaborate climate model, with more adjustable parameters, making the same warming prediction as a simpler model does not by itself merit more confidence than the simpler model. But "predictions" of the past made using a more emergent model that incorporates more scientific constraints (e.g., a model with a carbon cycle) inspire a greater degree of confidence than predictions made using a simpler model with fewer constraints (e.g., a model without a carbon cycle), even if the apparent prediction skill is the same (Section 9.6). To the extent that current models are more emergent than the earlier models used by Hansen and Manabe, it can be argued that their predictions inspire more confidence, even if they have not narrowed the uncertainty. Finer spatial grids also help, by providing a more detailed constraint on climate without introducing additional parameters.

Peer pressure is an important factor that drives modeling centers to make their models ever more elaborate. Each center seeks the prestige of hosting the model that is most comprehensive in representing processes, one that simulates climate at the highest possible spatial resolution. Climate scientists Tim Palmer and Bjorn Stevens write that

> [t]he enterprise of making models more and more comprehensive through the incorporation of computationally expensive but poorly understood additional processes has not so much sharpened our ability to anticipate climate change as left the blurry picture established by physical reasoning and much simpler models intact.[7]

And climate scientist Kerry Emanuel argues that focusing on simulation rather than understanding may actually be impeding scientific progress, asking: "[in] our quest for accurate simulations, are we computing too much and thinking too little?"[8] We need to carefully consider what is worth improving in models and assess the return on investment. Many of our most fundamental insights into the controls on global temperature have come from older and simpler models.[9]

These comments allude to climate models suffering from the "Red Queen Effect," where "it takes all the running you can do, to keep in the same place." As climate models become more elaborate, some basic problems change shape but remain unsolved. For example, important climate phenomena like El Niño are still not simulated very well, and errors in simulating rainfall patterns persist.[10] Success in predicting multidecadal trends in globally averaged temperature, as in the case of Hansen's model and other climate models, does not translate to similar success in predicting regional climate trends.[11] Assessing the risks of climate change and adapting to its socioeconomic impacts require better predictions of changes in the local temperature, not the globally averaged temperature. More elaborate models, with better representation of topographic features like mountains, will be able to make more detailed simulations of regional climate and extreme weather phenomena like hurricanes and typhoons.[12] But their long-term climate prediction uncertainty may not necessarily decrease, due to lingering uncertainties in the representations of much finer-scale processes like aerosols.

12.1 Falsifiability of Climate Models

Models cannot be proved right, but they can be proved wrong. Karl Popper argued that an important requirement of scientific theories or models is that they be falsifiable – they should make predictions that can be confirmed or refuted by data. We have adopted this approach with weather predictions for thousands of events, but we cannot use the same approach with long-term climate change predictions for two reasons: We only have one global warming event to work with, and predictions of that event are highly conditional on the socioeconomic scenario. By the time we are in a position to validate a model's prediction, decades into the future, the model itself will be long obsolete.

All is not lost. We can verify climate models by using them to make forecasts a few years ahead, such as forecasts for the many El Niño events that have occurred in the past. This does not fully test climate models because, like weather predictions, they are sensitive to initial conditions (in the ocean) and will be subject to a predictability limit. Nevertheless, this kind of forecast can increase our confidence in the models. The reason such validation works is because we have numerous El Niño events against which to test the model.

We might also consider validating climate models using singular, or one-time, events from the past. The warming that has already occurred over the past 150 years is one such event. The problem is that we have known about this event for some time, and many climate models have been either consciously or subconsciously tuned over the years to reproduce this event as best as possible. New versions of climate models would have (at least some of) this institutional memory baked into them. This was less true in the early 1980s; even early climate predictions like the one made by Hansen in 1988 were pretty good, when adjusted for the observed greenhouse gas concentrations. But Hansen's 30-year prediction is still a limited test due to the short validation period and small sample size (of one).

Another type of singular event is a volcanic eruption. On June 15, 1991, Mount Pinatubo erupted in the Philippines. The eruption was the most powerful in a century, and it injected a large amount of sulfate aerosols into the stratosphere. Sulfate aerosols cool climate by reflecting sunlight, and they can persist for years if they enter the stratosphere (Section 7.4). Soon after the eruption, Hansen used the GISS climate model to predict its impact on global surface temperatures.[13] The predicted cooling lasted two to three years and matched the observations fairly well. This was not a perfect test of the model, since it was more a test of the cooling mechanism specifically associated with stratospheric aerosols, not of the general mechanism of global warming. It was also a test of short-term climate trends; smaller model errors would not have had the time to have a cumulative impact. As a test of longer-term climate "prediction," climate models are routinely used to simulate climate events in the distant past, such as the peak of the last ice age; these simulations are then confirmed against paleoclimatic data.

We can also examine more fundamental predictions of climate models, especially if they are counterintuitive. An important property of global warming is how it changes as we move upward through the atmosphere. The troposphere – that part of the atmosphere closest to the Earth's surface (Figure 3.1) – is warming. Manabe's one-dimensional climate model showed in 1967 that the stratosphere, the part of the atmosphere above the troposphere, should actually be cooling. This is indeed confirmed by the data. Another property is the polar amplification of global warming simulated by models, which is also confirmed by the data.

Climate models have made predictions that have turned out to be wrong. A notable example is from the early 1970s, when global temperatures were getting cooler, and aerosol pollution in the atmosphere was increasing. There was concern among a few scientists that continued global cooling could become a problem and even lead to an ice age, like the ones that had happened in the past (Figure 5.3). Two scientists from the GISS published a climate modeling study in 1971 that extrapolated the cooling trend associated with increasing aerosols and raised the possibility of a rapid global cooling within several years, which could in fact trigger an ice age. That prediction has obviously turned out to be wrong. The model overestimated the cooling effect of aerosols and underestimated the warming effect of carbon dioxide.[14] Does this mean that all climate models are wrong? Not really, because this failed prediction is an isolated example, not a pattern. But it is a reason to be cautious in interpreting climate predictions based on a single model.

The term "consilience of evidence" is used to describe the case for climate change.[15] Consilience can serve as an alternative to the replicability criterion for scientific disciplines in which strict replicability is not a practical option (Section 11.4). Many independent pieces of evidence, none definitive on its own, have come together to make a compelling case that climate change is occurring. The warming trend, in addition to being observed, has been inferred from the rise in sea levels, the loss of glaciers, and the melting of ice sheets around the world. Accurate measurements of carbon dioxide and other greenhouse gases show that their concentrations are increasing in the atmosphere. Simple physics, as well as more complex climate models, link the warming trend to greenhouse gas increases. Paleoclimatic evidence for past climate changes provides a broad context for these trends, but the pace of the recent warming and greenhouse gas increases appears to be unprecedented.

Complex climate models are an important part of the consilience of evidence for climate change.[7] Could all long-term climate predictions be wrong? Arguably there is an unquantifiable risk that our models are totally missing an important feedback, which could be stabilizing or amplifying, that will be triggered in a warmer climate (Section 9.6)[16] – an unknown unknown, as the history of the ozone hole demonstrates.[17] Why, then, should we use models to plan for the future? We make decisions based on incomplete information every day in our lives, even though there are many unknown unknowns. And the alternatives – such as not thinking about the future at all or basing our decisions on intuition without any scientific constraints – are far worse. As this book argues, qualitative results from models are likely to be more robust than the quantitative results. And models' predictions themselves should be treated as consilient: We should trust climate models as a group, not relying solely upon predictions made by any single climate model.

12.2 Harder than Rocket Science: Numerical Convergence and Climate Modeling

Computers are commonly used to simulate the flow of fluids such as air or water by solving fluid flow equations known as the Navier-Stokes equations on a three-dimensional grid. This branch of science is known as *Computational Fluid Dynamics*, or CFD. One of the first uses of digital computers, in which John von Neumann again played a key role, was to simulate atmospheric shock waves associated with nuclear explosions. The numerical forecast described by Lewis Fry Richardson in 1922 is often considered the very first

CFD calculation, even though it was carried out by hand and failed badly. In 1963, climate modeling pioneer Joseph Smagorinsky developed one of the commonly used turbulence models for computing fluid flow.[18]

Computational Fluid Dynamics is an important part of "rocket science." When a company like Airbus or Boeing designs the wing for a new aircraft, it typically uses a CFD model to compute how air flows around new wing prototypes, in order to estimate air resistance and other properties of the wing. When NASA scientists design the spacecraft for a Mars mission, they need to use a CFD model to accurately simulate its landing trajectory upon entering the Martian atmosphere (Section 2.1). Computational Fluid Dynamics models include data-derived approximations for fine-scale turbulent fluid flows, calibrated using experiments with physical prototypes of wings or spacecraft placed in wind tunnels. After calibration, CFD models are cheaper to use than wind tunnels for testing new prototypes – not to mention much, much cheaper than expensive flight testing.

A CFD model typically has two properties:

1 As the computational grid becomes finer, the estimate of fluid flow properties becomes more accurate. This is known as *convergence*; the estimated answer "converges" to the final answer. We can halve the spacing of grid points to make the grid finer until the grid is so fine that the answer no longer changes appreciably. To illustrate this, consider numerical simulations of the aerodynamic design of a winglet, a device that is added to an aircraft wingtip to reduce drag (or friction). Figure 12.2a shows how the drag coefficient varies as the number of computational grid elements ("cells") is increased from 1.3 million to 36 million.[19] We see that the simulated drag slides down the slope of convergence as the cell count increases, and flattens out as the final answer is reached. Once convergence is assured, choosing the optimal grid size involves a trade-off between accuracy and computational cost. In this case, a grid resolution of 4.5 million cells was chosen to carry out further design simulations.

2 Every time grid spacing is halved, the total number of points in the model will increase by a factor of 8 (2^3) because the spacing is halved along each of the three spatial dimensions. Computational memory requirement will increase by a factor of 8 (2^3), but the computational time will typically increase by a factor of 16 (2^4) because the time interval between successive calculations of the flow (the time step) will also need to be halved to maintain computational stability, doubling the number of calculations.[20]

Convergence can improve our confidence in a model, even if we cannot prove the model correct. It is easier to defend a single, replicable answer than

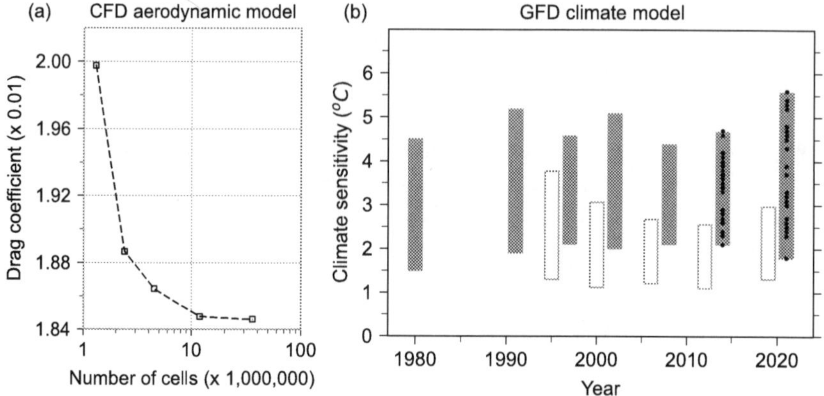

Figure 12.2 CFD versus GFD convergence. (a) A CFD model of a winglet showing the drag coefficient value converging as the number of cells in the computational grid increases; x-axis is logarithmic (Adapted from Guerrero et al., 2020, fig. 10; cc-by-4.0). (b) Climate model estimates of the range (minimum–maximum) of climate sensitivity (filled rectangles) and the transient climate response (TCR; open rectangles) in °C from the 1979 Charney report, five IPCC Assessments (AR1–AR5; 1990–2013), and the simulations available in 2020 for IPCC AR6. Small circles within the filled rectangles denote estimates from individual models. (Adapted from Meehl et al. (2020), Fig. 3; cc-by-nc)

to defend multiple answers. But reaching convergence, or getting very close to the final answer, can be a computationally expensive process since costs increase by a factor of 16 for each halving of the grid spacing. For a small problem like the flow of air around an airplane wing or the flow of water around a boat, it is possible to assure convergence and be reasonably sure that we have the correct answer.

The science of the motion of air and sea water, which is at the core of weather and climate models, is often referred to as Geophysical Fluid Dynamics, or GFD. We cannot expect to see the same phenomenon of convergence here, due to two problems:

1 GFD requires a much, much bigger domain, spanning the entire globe. We cannot use a grid of the same fineness that we would use for an airplane for the whole globe; the total number of grid points would make even the world's most powerful computer choke.
2 There is more to weather and climate than turbulent fluid flow. Poorly understood physical processes like clouds, rain, and aerosols also affect fluid flows. For climate (but not for weather), we need to additionally consider the chemical and biological processes, such as the carbon cycle, that control emissions of greenhouse gases.

Some aspects of weather models, such as the prediction of large-scale temperature patterns, are close to convergence because they are controlled by fluid flow; other aspects, like local rainfall, are starting to converge with finer grids.[21] This is evident from the steady improvement in the quality of weather forecasts (Figure 2.2). Three-day forecasts have essentially reached convergence in skill for predicting large-scale atmospheric flow patterns, and 5- to 10-day forecasts are approaching convergence. But for long-term climate prediction, the nonfluid effects such as radiation, clouds, aerosols, and greenhouse gas emissions become more important. Some of these, like aerosol processes and the carbon cycle, are much harder to model quantitatively – unlike fluid flows, they are not governed by well-known mathematical equations.

Computational Fluid Dynamics models of small-scale fluid flows also have another advantage over climate models. Their predictions can be validated using experiments. We can test spacecraft models in a wind tunnel in order to verify that the final convergent answer is also the correct answer. This is true as well for predictions from weather models, which are validated using data. But even if a climate model with a finer grid were to reach convergence, we cannot be confident that it has converged to the correct answer because we cannot experimentally verify that its long-term climate prediction is correct. (The typical "shelf life" of a climate model is less than a decade, while it would take many decades to verify its long-term climate prediction.[22])

What happens as the grid spacing of climate models gets finer, as it has over the years? Climate models in the 1960s and 1970s had very coarse grids and could barely capture the weather. As the grids became fine enough to start capturing weather, simulation errors decreased, but, in recent years, simulation properties have not converged, despite yet finer grids.[23] Part of the reason it is difficult to see convergence in climate models is that compensating errors are readjusted as the grids become finer, during model tuning.[24] If a poorly known Process X was compensating for errors in representing a better-known Process W (say, weather), then improving the representation of W may simply entail readjusting X to compensate less. Thus, the overall climate simulation quality may not necessarily improve, obscuring the benefits of convergence in Process W (in this case, improvements in representing weather phenomena).

To quantitatively assess the convergence of climate models, we can look at estimates of climate sensitivity. Climate sensitivity is an important emergent property of climate models that tells us how rapidly the planet is likely to warm (Section 3.2). Each new IPCC assessment releases a range of estimates of climate sensitivity from different models (Figure 12.2b). The models' average estimates of climate sensitivity haven't changed much since the first IPCC assessment, or even since the estimate of 3.0°C in the Charney Report of 1979.[25] Climate models are not sliding down the slope of convergence; they appear to be stuck in *an uncertainty plateau*. If anything, the spread in climate sensitivity values appears to be increasing (Figure 12.2b).[26] Other metrics used to assess the fidelity of model simulations also exhibit a lack of convergence.[27] The increased spread in sensitivity will also increase the spread in regional climate predictions, negating to some extent the increased accuracy provided by a finer spatial grid that captures more geographic detail.

Does the uncertainty plateau mean that there has been no progress in modeling over the last 30–40 years? Here's one opinion (from a 2018 *New York Times* feature):[28]

> Ken Caldeira, a climate scientist at the Carnegie Institution for Science in Stanford, Calif., has a habit of asking new graduate students to name the largest fundamental breakthrough in climate physics since 1979. It's a trick question. There has been no breakthrough. As with any mature scientific discipline, there is only refinement.

In other words, climate physics is in the puzzle-solving phase of incremental scientific progress; no paradigm shift has occurred in recent years to significantly narrow the uncertainty for emergent properties like climate sensitivity. There has been progress in adding new processes, like aerosol effects and the carbon cycle, but these have typically increased the spread in climate sensitivity[23] by quantifying previously hidden uncertainties (Section 9.5).

Since climate models accurately reproduce important features of the current and past climates, they can be considered "correct enough" to justify action on climate change. Admittedly, they are imperfect and incomplete, and there is no indication that uncertainty in their long-term predictions will decrease dramatically in the near future. The newer models are more sophisticated in that they incorporate more processes and feedbacks, and have more constraining and explanatory power, than the early climate models. We have more confidence in the newer models' ability to simulate the real climate system.[23] But the fact that the uncertainty in climate sensitivity estimates has not decreased means that we may have a long way to go before all of the processes that play a role in determining climate sensitivity are represented well in our models.[29]

We could perhaps expect the purely fluid dynamic components of a climate model to eventually converge to a unique solution with finer grids, because we know the fluid equations well. But what about the less well-known, but more complex, chemical and biological components of the climate system? Should we expect convergence for those components? Does that concept even apply?

12.3 Carroll, Borges, and the Reducibility Limit

"And then came the grandest idea of all! We actually made a map of the country, on the scale of a mile to the mile!"

"Have you used it much?" I enquired.

"It has never been spread out, yet," said Mein Herr: "the farmers objected: they said it would cover the whole country, and shut out the sunlight! So we now use the country itself, as its own map, and I assure you it does nearly as well."

Lewis Carroll, *Sylvie and Bruno Concluded*

Climate predictions have reached an uncertainty plateau: Even as the model grid has become finer and finer, estimates of metrics such as climate sensitivity have not improved. What if we were to keep making the model grid even finer, until we explicitly captured every process that could ever be important in climate prediction? This is a somewhat silly supposition, as it would effectively take us to the scale at which the validity of the fluid equations themselves would break down. But it is worth discussing briefly in order to clarify a few points, and we therefore give it an impressive-sounding name. We call it the Carroll–Borges (or the C-B) Limit of modeling, after two writers who actually described such a limit: Lewis Carroll and Jorge Luis Borges.[30] Both of them imagined a map so accurate that it was the same size as the country it was mapping. Needless to say, such a map would not be very practical to use. The important point, though, is that at the hypothetical C-B Limit, the model is no longer an abstraction of the real world. It becomes an exact copy or a clone of the real world. (For climate, this would be the ideal Climate Demon that makes perfect predictions.)

We can combine the C-B Limit with the concepts of convergence and reductionism. A reducible complex system can be broken down into smaller and smaller parts, each of which can be modeled accurately. If we do not understand the structure of some of the parts, we cannot break them down, or reduce them, any further to model accurately. The system then faces a *reducibility limit.*[31]

We can consider two classes of complex systems:

1 Without a reducibility limit (model exhibits monotonic convergent behavior well before the C-B Limit)
2 With a reducibility limit (model does not exhibit monotonic convergent behavior well before the C-B Limit)

For a system without a reducibility limit, all processes that affect the final result can either be explicitly simulated or are parameterizable in a convergent sense (meaning that the parameterizations of processes become more accurate as the grid becomes finer and converge to a unique answer, as illustrated in Figure 12.2a). For a system with a reducibility limit, the overall simulation accuracy improves initially as the grid becomes finer, but it stops improving at some point because crucial parameterizations hold it back. Uncertainties in estimates of emergent properties reach a plateau, and the models fail to converge to a unique answer. The only way to model such a system with very high accuracy is to clone the system, in which case the model would no longer be an abstraction – like a map that occupies the same area as the region it is mapping.

We know that CFD models, like those simulating the flow of air over an airplane's wing, do not suffer from a reducibility limit; they exhibit monotonic convergent behavior well before the C-B Limit. What about weather and climate models? Are they subject to a reducibility limit?

Before we address that question, we consider another limit. The work of Ed Lorenz and others in chaos theory showed that weather has a predictability limit. We cannot really ask whether climate has a predictability limit because climate prediction is fundamentally different from weather prediction: Uncertainty in initial conditions is unimportant, and boundary conditions are prescribed (Section 5.1). For a given boundary condition, the skill of a climate prediction is limited by the structural uncertainties inherent in climate models – or by our lack of knowledge of exactly how small-scale processes such as clouds or aerosols work. This suggests that there are limits to the reductionist approach that underpins climate modeling.

Fluid dynamicist James McWilliams hypothesized in 2007 that models suffer from "irreducible imprecision":[32]

> models have important levels of irreducible imprecision due to structural instability resulting from choices among a set of modeling options that cannot be clearly excluded. The level of irreducible imprecision will depend on the context, and this level is likely to be greater the more chaotic and multiply coupled the targeted flow regime is.

The uncertainty plateau seen in climate sensitivity estimates suggests that there is irreducible imprecision associated with a few critical but poorly known processes in climate models.[33] To avoid a double negative, this is equivalent to saying that climate models may be subject to a reducibility limit.

As Laplace's Weather Demon was shackled by the predictability limit, the Climate Demon may be shackled by the reducibility limit. The Climate Demon farms out prediction work to lesser demons that represent processes, like the Cloud Demon or the Aerosol Demon. If a few of these lesser demons are error-prone, the overall accuracy of the predictions made by the Climate Demon is limited, even if the predictions of the other lesser demons become more accurate. As they say, you are only as happy as your unhappiest child.

The predictability limit of weather is amenable to mathematical analysis, as is the parametric uncertainty in climate models. However, an open-ended problem like the structural uncertainty in climate models is not possible to handle mathematically. That means that we cannot prove the existence or absence of a reducibility limit.[34] Instead, we can ask a more practical question:

> Is the uncertainty plateau exhibited by climate models indicative of a reducibility *barrier*, or a reducibility *limit*?

If the uncertainty plateau is merely a barrier, we may cross an inflection point and move to a lower plateau of uncertainty as models continue to improve. But at the moment, we cannot be sure where such an inflection point may be located. For the purely fluid dynamic component of a climate model, we may cross such an inflection point when the model grid becomes fine enough to capture the turbulent fluid processes of clouds.[35] What about other crucial components of the climate model, such as even finer-scale physical processes like aerosols, as well as less quantifiable chemical and biological processes? For such processes, an inflection point in the uncertainty may lie further away, or it may not exist at all. In that case, there would be a reducibility limit for climate modeling at some level of model complexity; model improvements beyond that point would not reduce the uncertainty.

Before any inflection point is reached, it is also possible that the uncertainty may slowly increase for some time, as more processes like the carbon cycle are incorporated into climate models (Section 9.5). This is what appears to be happening currently (Figure 12.2b). There is even the possibility, albeit remote, of an abrupt change in uncertainty due to unknown unknowns that lurk on the path to the C-B Limit. If a previously unknown or neglected process is discovered to play an important role in a warmer climate,[36] as was the case in the formation of the ozone hole, this would constitute a major scientific advance – but it could temporarily increase the uncertainty of climate predictions (Section 9.5).

***

Since the development of the first computer model of the atmosphere in the 1950s, the reductionist approach has resulted in climate models becoming more and more elaborate, like an increasingly complex clockwork mechanism. But global climate predictions made in 1988 appear to be as good as the ones being made in 2018, at least superficially, despite 2018's models being much more complex. Does it mean that our models have converged to the "right answer"? Model estimates for important metrics like climate sensitivity do not exhibit increasing certainty; rather, the spread in the estimates of such metrics appears to have flattened, leaving models stuck in an uncertainty plateau. Large uncertainties continue to persist in predictions of regional climate. This suggests that there may be a practical limit to the reducibility of the climate system, although there is no way for us to theoretically prove the existence (or absence) of such a limit.

Weather models exhibit convergent behavior as they become more elaborate and approach the predictability limit. Climate models do not exhibit this behavior, due to the uncertainty plateau. Despite their imperfections, these models simulate many properties of the observed climate rather well. To address the question posed in the introduction – should we trust climate models? – we say that we should *trust, but verify*.[37] Climate models are the best tools we have to plan for the future, but we should trust them as a group, not individually, due to the lack of convergence. We should also verify and understand the complex models as much as possible using both data and simplified versions of the models themselves.

Climate science is "settled enough,"[38] as demonstrated by the consilience of evidence from a variety of models and data sources, to motivate urgent action to mitigate carbon emissions. Since we know that climate models are imperfect, we should try to continue to improve them. But we need to do so with realistic expectations: We may have to settle for increased knowledge and understanding, rather than increased certainty.[23] In Chapter 13, we discuss the process of model improvement, in which we incorporate hitherto unknown processes into a model as we come to know the processes better.

13

The Rumsfeld Matrix

Degrees of Knowledge

There are things known and there are things unknown, and in between are the Doors.

Attributed to Jim Morrison

Climate models inhabit the twilight zone between the known and the unknown. The difference between knowledge and ignorance can be crucial in responding to a crisis situation. However, everyday language does not always permit the nuances needed to carefully discuss the different degrees of knowledge and the different ways in which we know things. While it may be a cliché that the Inuit have more than 50 words for snow, it would certainly be convenient to have a multiplicity of words to describe different flavors of knowledge in common English. Philosophers, statisticians, and economists have developed different terms to describe types of uncertainty, but they are not easy to explain to the nonspecialist.[1]

Enter Donald Rumsfeld. While von Neumann and his colleagues were working on numerical weather predictions at the IAS in Princeton, Rumsfeld was studying political science on the nearby Princeton University campus. He served twice as US Secretary of Defense, appointed to the post in 1975 by President Gerald Ford and again in 2001 by President George W. Bush, at which point he became both the oldest and youngest person ever to hold the post. At a press conference in February 2002, in the aftermath of the 9/11 terrorist attacks, Rumsfeld was asked about reports of Iraq providing weapons of mass destruction to terrorist organizations. He replied:

> Reports that say that something hasn't happened are always interesting to me, because as we know, there are known knowns; there are things we know we know. We also know there are known unknowns; that is to say we know there are some things we do not know. But there are also unknown unknowns – the ones we don't know we don't know.

Rumsfeld used the phrase "unknown unknowns" to describe threats that we are completely unaware of, as distinguished from threats we are aware of but do not know the full details of, that is, the "known unknowns."[2] Rumsfeld continues to be a divisive figure in US politics because of his role in the 2003 Iraq War. In this book, however, we focus on his somewhat mangled but lasting contribution to the English lexicon and the categorization of knowledge – especially the lack of it.

13.1 The Arc of Model Improvement

We can parse the term "unknown unknowns" as follows. The first word refers to our awareness of the threat. The second word refers to the accuracy with which we can characterize the threat. With this interpretation, we can construct a 2 × 2 threat matrix, or uncertainty matrix, using the phrases "unknown unknowns," "known unknowns," and "known knowns." In effect, we are using the analytic knife to distinguish between different types of knowledge.

"Unknown unknowns" occupy the bottom right corner of the *Awareness–Accuracy* matrix (Figure 13.1a). Accuracy increases as we move left along the horizontal axis, and awareness increases as we move up along the vertical axis.[3]

We can use the cells of the uncertainty matrix to describe the development of weather and climate models. At the core of these models is the motion of fluids in the atmosphere and the ocean, and the propagation of radiation through the atmosphere. At large spatial scales, we are aware of the fluid

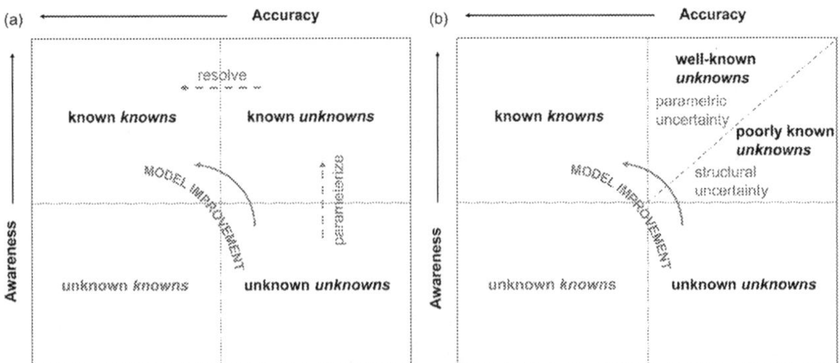

Figure 13.1 Rumsfeld uncertainty matrix illustrating the model improvement pathway. (a) Original 2 × 2 matrix. (b) Modified matrix with two subcategories of known unknowns: well-known and poorly known.

motions and radiation through clear skies, and we can characterize them accurately. These processes are said to be *resolved*. These are the "known knowns" at the top left of the matrix: We are aware and accurate. If these are the only processes that are important and if the initial conditions are known, then we can make very confident predictions using the model.

As we consider fine scales of motion, finer than the spatial grid of the model, it gets harder. We know that turbulent flows at fine scales in the boundary layer near the ground are important, we know that cloud-generating microscopic processes are important, and so on. The model grid is too coarse to represent these processes. So we use approximate formulas with adjustable parameters or coefficients to represent their effects. These processes are said to be *parameterized*. These are the "known unknowns," in the top right of the matrix: We are aware, but less accurate.

Then we go to the bottom right, the "unknown unknowns." These are processes we are completely unaware of and hence cannot represent with any accuracy. An example of this would be the surface chemistry occurring on polar stratospheric clouds that caused the ozone hole (Section 6.3). Before 1985, no one knew about this process. Within a year, scientists had identified surface chemistry as a likely mechanism to explain the rapid ozone loss during the Antarctic spring. This was soon verified by observational expeditions. Once scientists became aware of this process, it moved into the known unknowns category.

These three cells of the Rumsfeld matrix serve as a starting point for describing the categories of knowledge with which we model processes in the climate system, but they aren't quite enough to capture some important distinctions. Even something as mundane as the doneness of steak has six categories: blue rare, rare, medium rare, medium, medium well, and well done. (Seven, if you count burnt.) Surely something as profound as the knownness of knowledge deserves more than three categories. Perhaps the secret behind the enduring popularity of Rumsfeld's knowledge categories is that they combine known words, rather than relying on more obscure terminology. We resort to our analytic knife to carve out additional subcategories of knowledge, dividing "known unknowns" into two subcategories:[4] "well-known unknowns" and "poorly known unknowns" (Figure 13.1b).

Some parameterizations in weather and climate models, such as those for turbulence in the boundary layer, are much more accurately known than other parameterizations, such as those used to represent clouds. We refer to the former as *well-known unknowns* because we have confidence in the structure of the approximate formulas, even if there is some uncertainty regarding the values of coefficients or parameters in the formulas. These coefficients can be determined through calibration against data, and their uncertainty assessed.

Therefore, for predictions which involve only known knowns and well-known unknowns, we can be confident of their probability distributions.

Less well understood are fine-scale processes such as clouds and aerosols, which also play an important role in climate predictions. We have different formulas for them, but we are not sure of the correctness of their structure, let alone the values of the unknown coefficients. We shall refer to these structurally uncertain parametrizations as *poorly known unknowns*. We cannot assign specific probabilities to their predictions being correct because we don't know the probability of the formula itself being correct.

The arc of model improvement is a process of moving from unknown unknowns to poorly known unknowns and then to well-known unknowns and finally to known knowns, as shown in Figure 13.1b. Early climate models ignored the effect of microscopic aerosol particles. This process was lurking in the category of unknown unknowns until scientists recognized the importance of aerosols. The aerosol effect is currently a known unknown, but it is in the poorly known unknowns category, because we are unsure of the structure of the formulas. What can we say about processes *currently* in the unknown unknowns category? Nothing, of course!

There are some processes that we can predict will switch categories at some point, simply due to spatial grids becoming finer. Climate models' representations of the ocean are able to capture larger-scale ocean currents but not the smaller-scale ocean flows called "ocean eddies," which are the oceanic analog of atmospheric weather. Spatial grids are currently too coarse to represent these eddies accurately, so approximate formulas are used to capture their effect. As computer power increases, it is expected that climate models will switch to finer grids that will capture these eddies, moving them from known unknowns to known knowns. Clouds in the atmosphere are currently in the poorly known unknowns category, but finer grids will also capture their properties better. It is unlikely that the same will happen anytime soon for the effects of aerosols (Section 7.4), though, given that they are complex processes involving many different types of interactions at microscopic scales.

13.2 Benefits of Parameterizations: Downstream versus Upstream

Since all models are wrong the scientist must be alert to what is importantly wrong. It is inappropriate to be concerned about mice when there are tigers abroad.

George E. P. Box,[5] Statistician

We all strive for perfection. And we know that models are imperfect. It is natural that we would like to keep improving them. Even if we don't want to improve our model, someone else will improve theirs, and then we end up doing it ourselves just to keep up, leading to the Red Queen Effect. But our resources are limited, and it behooves us to ask the question: What is the value added by increasing model complexity, especially if we are stuck in an uncertainty plateau that prevents more precise estimates of emergent metrics like climate sensitivity?

When we become aware of a new process important for climate, it moves from the category of unknown unknowns to the category of known unknowns. An approximate parameterization is developed for the process and implemented in the model. Over time, a finer model grid can eliminate the need for the parameterization and move it to the category of known knowns. Making model grids finer is routinely done as more computational power becomes available. But increasing the complexity of a parameterization,[6] or parameterizing a new process, may not automatically be the right thing to do.

Suki Manabe, famous for developing moist convective adjustment and the one-dimensional radiative–convective model (Section 4.5), was against overly complicated parameterizations. His papers using fairly simple parameterizations are still considered classics in climate science, producing insights that have stood the test of time. Manabe has said that it is not worth expending a great deal of effort on parameterizing a process that we don't fully understand if other aspects of the model are treated poorly.[7] Making the analogy between a model and a stereo set, he added, "when you have a hundred dollar speaker, [there is] no point putting [in a] ten thousand dollar amplifier." His sentiments can be expressed by repurposing a quote usually attributed to Einstein: *Parameterizations should be made as complex as needed, but no more complex than that.* The excessive use of the analytic knife to dissect a complex process may conflict with the simplicity requirement of Occam's Razor.

In general, adding a new process to a model increases its complexity and computational cost. Whether the new process is worth it depends on several criteria, including its impact on the predictive, constraining, and explanatory powers of the model. (Since we know there is no way to truly quantify a climate model's long-term predictive power, we may have to settle for looking at the impact on the fidelity of simulations of the current and past climate, even if that does not guarantee predictive skill.) Not all model improvements are created equal; it makes sense to prioritize and distinguish between the different types of benefits model improvements can provide.

To better understand the impact of adding a new component to a climate model, we consider the introduction of a hypothetical new process X, which is discovered to play a role in the global heat budget. The heating term associated with process X may be large or small. If the new term is small, the effect on climate simulations is likely to be negligible, because the impact of the new term will be masked by errors in the larger terms. For a climate model in the uncertainty plateau, if the new term is large, we usually hope that the addition of process X maintains the current fidelity of simulation. Often, simulation fidelity will decline initially because the balance between compensating errors has been disrupted. Readjusting model parameters ("retuning") to recompensate can usually restore most or all of the simulation fidelity. Adding wholly new components can increase the testability and the explanatory power of the model, and may also increase its constraining power. An improved replacement of an existing component may increase the model's explanatory power, if the component itself becomes more understandable, but it may not increase the constraining power.

What are the costs and benefits of including process X in the climate model? To answer this, we wield the analytic knife to distinguish between *emergent* or *downstream benefits,* which improve the simulation fidelity and/or the constraining power of the climate model for societal impacts, and *reductionist* or *upstream benefits*, which help the scientific subdiscipline that studies process X, even if there is little or no improvement in the simulation's emergent behavior. An example of an emergent downstream benefit would be an improvement in the skill of regional rainfall simulation, needed by hydrologists for planning. This has to be balanced against the downstream costs of including the new process X: added software complexity and computational expense.

We should also consider the reductionist or upstream benefits. Incorporating a new process X, such as the carbon cycle, in a climate model would impose constraints on a component model that might have been previously absent. The conservation of carbon would need to be strictly imposed locally and globally for the carbon cycle component model. This would benefit the scientific community studying the carbon cycle because it could improve how carbon sources and sinks are characterized. Field and lab measurements of carbon processes can be used to develop and validate new parameterizations. Complex models present a constraint ecosystem in which simpler component models can "grow up," learning to follow the rules and interact with other components. For example, a marine ecosystem model that is introduced into a climate model might find that the food supply becomes limited due to the effect of ocean currents on nutrient availability. This might lead to a reevaluation of the assumptions behind the ecosystem model. Such upstream benefits have to be balanced against the upstream cost of adding process X to the model, which would be the effort put into developing and improving the component models and parameterizations.

Not all new processes need to be included in climate models. The decision should be made by considering the strength of the process relative to other terms in the budget equation, including errors in the estimates of the largest terms, which could end up masking small improvements. It should also involve carefully weighing the benefits and the costs, both downstream and upstream. Some atmospheric chemistry processes, for instance, are so computationally expensive, involving numerous chemical constituents and reactions, that it may take a long time before they can even be considered for inclusion.

The argument about the benefits of new parameterizations can also be extended to new and more accurate algorithms for solving the fluid equations. We need to consider whether any extra accuracy provided by the algorithms will be masked by the larger errors already present in the parameterizations.[8] If that is the case, there will be no downstream benefits to users of climate predictions, although there could be upstream benefits to the field of applied mathematics that develops new algorithms.

While climate modeling is stuck in the uncertainty plateau, improving model components will seldom produce downstream benefits that narrow the overall uncertainty, but the improvements may lead to greater constraining power. Adding a carbon cycle to a climate model will likely increase the spread among climate predictions, because it moves an uncertainty that was previously part of the emission scenario into the climate model (Section 9.5). But it will also add the constraining power of carbon budgets and make the model more useful for downstream applications in mitigation and adaptation planning. The carbon cycle is crucial to climate prediction. Some other biogeochemical cycles may turn out to be less important: Their addition may not provide many downstream benefits, even if there are upstream benefits for the study of those cycles and for science in general.

Another important factor to consider when parameterizing slow climate processes like ice sheet growth and decay is the predictability limit, or the Butterfly Effect. We argued earlier that this effect is not relevant for climate prediction because the decadal and longer-term evolution of the atmosphere and oceans depends on boundary conditions, not on initial conditions, due to the inherent timescales of atmospheric and oceanic processes (Section 5.1). But this argument does not apply to ice sheet processes, which have centennial to millennial timescales. Their evolution will be sensitive to both the ice sheet initial conditions and emission scenario boundary conditions, and they will be subject to the predictability limit of the Butterfly Effect as well as to the limitations on reducibility arising from the structural uncertainties of models.[9]

Although we do not know what they are, unknown unknowns are also an important reason to keep models reasonably up to date with our knowledge of the climate system, even if this effort does not lead to better prediction skill. Models did not predict the ozone hole in 1985, but they were quickly updated to incorporate new processes that could explain the newly discovered phenomenon (Chapter 6). If we had not already had "state-of-the-art" models of atmospheric chemistry, it would have taken a lot longer for us to understand the ozone hole and its implications for the rest of the planet. In the event that global warming triggers an unexpectedly strong amplifying feedback, we will need a model that can help us understand its causes and impacts. For example, if reflective clouds were to start disappearing in a warmer world, having a good cloud parameterization would be very useful in figuring out how to deal with the situation.[10] Models are like flashlights that allow us to be better prepared to deal with any unknown unknowns we may encounter as we explore our somewhat murky climate future, and we should carry the best flashlights we can afford.

13.3 Contrarians at the Gate

He who knows not,
and knows not that he knows not,
is a fool; shun him.

He who knows not,
and knows that he knows not,
is a student; teach him.

Part of Arabian Proverb

Until now, we have talked about views on global warming that are mostly mainstream among scientists. But, even among scientists, there are some who don't accept the mainstream views, who question the causes or even the existence of global warming. Climate scientists are often asked by their nonscientist colleagues whether they *believe* in human-induced global warming. The best answer is that science is not about belief, but about evidence. An overwhelming majority of climate scientists think there is strong evidence for global warming being caused by humans. Scientists who do not hold this consensus view have sometimes been called climate skeptics, but that is a misleading name, because all scientists are skeptics at heart (or should be). The name "climate deniers" has also been applied to them, but the term "denier" has other, stronger connotations. In this book, we shall use the term "contrarians" to refer to those outside the self-defined collective of mainstream scientists.

Contrarians are not a monolithic group, and contrarianism is also a somewhat relative term. Those who believe that climate predictions from models should be taken literally may view this book itself as being contrarian, because it highlights uncertainties which they may regard as contradictory to their arguments.

Although contrarians are far fewer in number than mainstream climate scientists, their arguments cannot be dismissed solely for that reason. Science is not a democracy, although it may seem like one when you view a snapshot of it operating. As philosopher Thomas Kuhn pointed out, science undergoes revolutions from time to time, but these often happen slowly. In fact, Jim Hansen was being quite contrarian during his Senate testimony in June 1988. His statement about being 99 percent certain that the signal of global warming had been detected was not the consensus view of scientists at the time. Just before Hansen's testimony, a NASA Chief Scientist warned that "no respectable scientist" would say that.[11] Hansen replied, "I don't know if he's respectable or not, but I know a scientist who is about to make that assertion."

The flipside of this argument regarding contrarianism is that revolutionary ideas appear very infrequently. Most ideas rejected by the majority of scientists are dismissed for cause: They are just poor ideas. Scientists filter ideas through a process called *peer review*, in which new ideas are vetted by other scientists before they can be published. This is far from a perfect framework, but, like democracy, it is preferable to the alternatives.

The debate between mainstream scientists and contrarians extends well beyond science. Each side has accused the other of being motivated by money or ideology, and arguments often end with ad hominem attacks. In this book, we shall take contrarian arguments at face value; we shall refrain from analyzing the personal motivations behind them. This is not to say that factors such as money and ideology are not important. They are merely beyond the scope of this book.

In the rest of this chapter, we address some of the common contrarian claims relating to climate modeling and prediction. We do not address those relating to the statistical interpretation of climate data, except to say this: Although the global warming signal was relatively weak when it was raised as a public concern in 1988, it has steadily increased and emerged from the noise over the past three decades (Figure 7.2). Most contrarians agree that the planet is warming, even as they disagree on the causes of the warming and the magnitude of its projected future trends.

13.4 Unknown Knowns: The Threat of Willful Ignorance

What gets us into trouble is not what we don't know.

It's what we know for sure that just ain't so.

Misattributed to Mark Twain

Donald Rumsfeld popularized the phrase "unknown unknowns" in the run-up to the Iraq War. What role did "unknown knowns" play in the onset of the war itself? The intelligence community knew very well that Saddam Hussein had nothing to do with Al Qaeda or the 9/11 attacks. In fact, the secularism of Hussein's Ba'athist regime was completely antithetical to the religiosity of Osama Bin Laden. Despite this known fact, unsubstantiated allusions to Hussein's role in the 9/11 attack were made by the US establishment as part of its justification for the war.[12] Adding this danger of willful ignorance to the fourth cell of Rumsfeld's uncertainty matrix completes the matrix, as noted by the Slovenian philosopher Slavoj Žižek.[13] If unknown unknowns are zero knowledge, we can think of unknown knowns as "negative knowledge." In the context of war, known unknowns and unknown unknowns are external threats; unknown knowns are internal threats. An army can protect against an external threat, but not against this internal threat.

Healthy contrarianism in science focuses on genuine scientific disagreement, where two sides whose arguments lead to two different conclusions are each convinced that their respective arguments, and conclusions, are correct. But if one side dismisses the other side's valid arguments simply because it does not like the conclusions they lead to, then this is not true contrarianism. It is a willful ignorance of inconvenient facts.[14] Climate science can help protect us from the physical threat of global warming, but it cannot protect us from the political threat of disinformation.

Climate scientists have been battling disinformation since the issue of climate change first attracted widespread public attention in 1988, after Jim Hansen's Senate testimony. At that time, the oil industry banded together to mount a campaign to discredit findings by climate scientists. The campaign sowed doubt in the minds of the public, using tactics borrowed from those used by the tobacco industry to refute studies that linked cigarette smoking to cancer.[15] ExxonMobil, the oil industry giant, is an egregious example. Beginning in the 1980s, the company funded internal climate research that acknowledged, with reasonable uncertainties in attribution, that global warming was caused by human activities.[16] The company's public communications, on the other hand, overwhelmingly emphasized the uncertainties, contradicting its own internal findings. ExxonMobil chose to publicly "unknow" what it privately knew.

Currently, much of the oil industry has publicly accepted the reality of climate change, and some in the industry are even starting to think about renewable energy alternatives, although many environmentalists remain skeptical of them. But other climate change disinformation efforts continue. Contrarians commonly question the science behind detection and attribution, the confidence in climate model predictions, and the motives of climate scientists themselves.

Climate scientists had a moment of "fast-forward" déjà vu during the Covid-19 pandemic because many of these same issues were brought to the fore, except that 30 years of debates were compressed into about 30 days! The detection of Covid-19 and the attribution of increased death rates to Covid-19 became political issues, rather than public health issues. Predictions from epidemiological disease-spread models were dismissed by contrarians as alarmist and unreliable – and it didn't help that the model predictions often disagreed with each other quantitatively, even if they were qualitatively similar. Epidemiological models are more inductive and thus more data-dependent than climate models, so the lack of data in the initial stages of the Covid-19 outbreak severely hampered their predictive power. Contrarians also questioned the motives of the epidemiological modelers, often insinuating that the results were tweaked in order to favor certain outcomes.

Two crucial aspects of the spread of Covid-19 are (1) the time lag between the initial infection and the onset of severe symptoms and (2) the exponential growth of the outbreak. In linear growth, things worsen at the same rate throughout the crisis. In exponential growth, things become more and more dire, more and more quickly, as the crisis continues. In New York City, there was a single confirmed Covid-19 infection on March 1, 2020. By the end of March, there were more than 30,000 cases.

Climate change shares the same lag problem between cause and effect as the Covid-19 spread, except that, instead of being several days, the lag is more like several decades – almost a human generation. Climate change impacts are weak in the beginning but can get much worse over time; the growth in impacts may not become exponential (unless we encounter a tipping point), but it will be faster than linear if we do not mitigate emissions sufficiently. The time lag and the initial weakness of the impacts make it easier for climate contrarians to ignore the scientific evidence and convince the public that global warming is not a real threat.

13.5 Means, Motive, and Opportunity – Or the Lack Thereof

One of the claims made by contrarians is that mainstream climate modelers overemphasize extreme scenarios of global warming and the role of human influence because doing so brings them fame and fortune – more press coverage, more papers published, more research funding, and so on.[17] When the US National Climate Assessment Report was released in 2018, a former congressman told CNN that the report was "nothing more than a rehash of age-old 10 to 20 year assumptions made by scientists that get paid to further the politics of global warming." Commenting on the same report, a fellow of the Heritage Foundation said that the "tsunami of government money distorts science in hidden ways that even the scientists who are corrupted often don't appreciate."

Let us treat these contrarian claims as formal accusations and examine how successful modelers could have been in committing this "corrupt" act of exaggerating the threat of climate change. An investigation of such an act normally considers whether the suspect had the means, the motive, and the opportunity to commit it. Climate modelers build the climate models, so they certainly have the *opportunity* to alter their models to exaggerate the predictions. But remember that a climate model has to be built more or less in plain sight; hundreds of people have access to the code of a climate model, which makes it hard to tweak a model secretly.[18] Also, climate modeling is an international activity: Modelers from different countries confirm (or refute) each other's results, which makes a coordinated conspiracy difficult to pull off.

For scientists to have the *means* to commit these "corrupt acts," they would have to actually be able to tweak the model to get the desired results. It is possible to do that in principle, but it isn't as easy as one may think, as illustrated by the whack-a-mole effect (Section 10.1). Unlike statistical models that are used in many other fields, climate models are highly constrained by the laws of physics and other sciences. Tuning a model to get a particular result, while simultaneously keeping it realistic in all other aspects, is very hard.

What about *motive*? Is the model-based evidence for climate change tainted by motives of fame and fortune? The inherent uncertainties in climate modeling allow for lower climate sensitivity values as well as higher climate sensitivity values. Are scientists consciously or subconsciously emphasizing the higher sensitivity values? Since this is not really a question about science, it is hard to answer it based on scientific arguments alone. The scientific consensus argument is often invoked in climate change debates, but we should consider history as well.

The discovery that increases in carbon dioxide could lead to global warming was indeed made by someone who sought fame, if not fortune. Like most scientists, Svante Arrhenius was seeking scientific recognition when he made that discovery in 1896 using hand calculations. But he was not trying to predict global warming in the near future; he was trying to explain global cooling that had occurred during ice ages in the distant past (Section 3.4). His discovery of greenhouse-gas-induced warming was presented as more of an afterthought. Far from calling it a threat, he actually underplayed it, based on his knowledge of emissions at that time. Arrhenius' conclusions on global warming were poorly received by the scientific community at that time[19] (although he went on to find fame in chemistry research instead, receiving a Nobel prize for his efforts).

When Suki Manabe made more refined calculations of greenhouse-gas-induced warming in 1967 using a computer, there was still no motive to exaggerate global warming. He was merely studying the factors responsible for keeping the Earth warm and habitable. Early climate modeling pioneers like Manabe, Kasahara, and Washington did not write specific grant proposals for funding. Their research institutions provided them the freedom to carry out fundamental research for curiosity's sake. At that time, global warming was considered more of a niche academic issue than a problem with immediate relevance. Until the Charney report of 1979, concerns about global warming did not even reach mainstream academia, and it was only after Hansen's landmark Senate testimony in 1988 that global warming began to attract the attention of the public.

History indicates that the discovery of greenhouse-gas-induced warming was not tainted by the temptations of fame and fortune. Contrarians could argue that, even if this was true in the past, it is not true today, when climate change research attracts media attention and government funding. But the estimates of perhaps the most fundamental metric of climate change, climate sensitivity, have barely budged since 1979, when there was far less motive to exaggerate climate change. Our current estimate of about 3°C warming for a doubling of carbon dioxide is not really that different from Manabe's 1967 estimate of 2.3°C. We can conclude that the best estimate for climate sensitivity also remains untainted by material motives.

13.6 Contrarianism and Unconstrained Ideation

We now return to Jones Hall in Princeton. Recall that this was once the home of Princeton University's Department of Mathematics, as well as temporary quarters for the IAS. Next to Jones Hall was the Palmer Physical Laboratory, which housed the Department of Physics until 1969. The closeness was by design; it was expected that mathematicians and physicists would frequently collaborate. We have traced the origin of weather and climate modeling to John von Neumann, who worked in Jones Hall and later moved to Fuld Hall, the current home of the IAS. We can also trace a major strain of climate contrarianism back to Princeton – to physicists connected with Fuld Hall and the Palmer Physical Laboratory.

An ethnographic study analyzing contrarianism in climate science noted that contrarian scientists "tend to be empiricists and physicists (i.e., theoreticians)," not modelers.[20] Among prominent climate contrarians are four physicists with connections to Princeton: William Happer, the late Frederick Seitz, the late Fred Singer, and the late Freeman Dyson. Happer, Seitz, and Singer received their doctorates in physics from Princeton University. Singer was active in space research and was a pioneer in the use of satellites to observe the Earth in the early 1960s. Seitz was the president of the National Academy of Sciences in the 1960s and received the National Medal of Science in 1973. Happer returned to Princeton later as a professor in the physics department and also served on the US National Security Council as a (contrarian) climate expert during the years 2018–2019.[21] Dyson was a physicist at the IAS and a colleague of von Neumann and Einstein. He made fundamental contributions to physics, especially quantum electrodynamics, and is well known in popular science and science fiction for developing the simple but powerful concept of the "Dyson Sphere" – the premise that technologically advanced civilizations could trap all the energy emitted from their sun and reradiate it at lower temperatures.

Why did these physicists, all accomplished in their own fields, hold contrarian views on climate change? Some of their objections relate to the statistical interpretation of observed temperature trends. Other objections have to do with the trust placed in predictions from climate models, which is the subject of this book. It is worth speculating on whether there is something specific to physics that militates against the complexity and uncertainty inherent to climate modeling. (To clarify, most physicists do not hold contrarian views on climate change,[22] but physicists are overrepresented in the ranks of contrarians.)

In an essay titled "Heretical Thoughts about Science and Society,"[23] Dyson criticizes climate models, saying that

> models solve the equations of fluid dynamics, and they do a very good job of describing the fluid motions of the atmosphere and the oceans. They do a very poor job of describing the clouds, the dust, the chemistry and the biology of fields and farms and forests.

This is not very different from some of the arguments made in this book. However, Dyson does not propose an alternative to using models to plan for the future; he merely implies that there isn't much to worry about. The general tenor of Dyson's essay, and many other contrarian statements, is that climate is

Figure 13.2 Physicists. (xkcd.com)

a complex system with great uncertainty in the observed and predicted trends, and, because of this complexity and uncertainty, there is no problem to be addressed and no action to be taken. Contrarian arguments typically exaggerate the scientific uncertainty while downplaying the impacts of climate change. Later in this book, we will address some of the broader philosophical issues regarding uncertainty raised by contrarians.

Contrarian arguments also frequently refer to mechanisms that they claim are either missing or misrepresented in climate models (Figure 13.2), such as the causes of sea level rise, the role of carbon dioxide, or certain types of cloud feedbacks. For example, Dyson writes, "To stop the carbon in the atmosphere from increasing, we only need to grow the biomass in the soil by a hundredth of an inch per year." Happer opined in 2013, "Contrary to what some would have us believe, increased carbon dioxide in the atmosphere will benefit the increasing population on the planet by increasing agricultural productivity."[24] In their writings, Dyson and Happer propose biological mechanisms that can only be simulated by adding carbon cycle and agricultural modeling components to a climate model. But the mechanisms are proposed in a qualitative and piecemeal fashion, without applying all the physical and mathematical constraints consistent with the data.

Like good intentions, seemingly plausible qualitative ideas about the causes and mitigation of global warming are an unlimited resource. But when these ideas are expressed mathematically and subject to constraints and tests within a climate model, not all of those ideas survive. This applies not just to contrarian ideas, but also to some ideas proposed by climate activists. For example, replacing all fossil fuels with crop-based biofuels may sound good, but we need to take into account relevant constraints like the amount of new agricultural land that would be needed and carefully assess whether or not the idea is viable.

Unconstrained ideation can also lead to exaggeration of the threat of climate change. When the United States pulled out of the Paris climate agreement in 2017, renowned theoretical physicist Stephen Hawking made headlines when he told BBC News that this "action could push the Earth over the brink, to become like Venus, with a temperature of two hundred and fifty degrees, and raining sulphuric acid."[25] This was unnecessary hyperbole, because even five degrees (Celsius) of global warming would be catastrophic. It also overlooks an important physical constraint – the Komabayashi-Ingersoll limit, which precludes such a runaway greenhouse scenario on the Earth (Section 3.6).

When it comes to predicting the future climate, the devil is in the details. Human scientific intuition is good at imagining a few large forces acting against each other, but it is not so good at keeping track of a large number of interacting small forces. To handle the latter, more complex, situation, we need mathematical models to impose the physical constraints. The whack-a-mole effect seen in model tuning is an expression of this; when we fix a problem with a climate simulation – like a rainfall error over one continent – another new problem – like a temperature error over a different continent – can crop up due to constraints and compensating errors.

Qualitative and piecemeal discussion of different climate mechanisms – unconstrained ideation – is like playing with a broken whack-a-mole toy. We can quickly "whack" all the moles in to win, but that's not how the real whack-a-mole toy works. Using the reductionist approach common to physics, we can study each component of the real world separately, ignoring other components. But when we combine these components in a climate model using a finite grid, and try to compute emergent properties like climate sensitivity, it gets a whole lot harder due to the complex interactions between components.

The relationship between theoretical physicists and scientists studying complex nonlinear phenomena such as the motion of air or water has always been somewhat, well, turbulent. There is a well-known (although likely apocryphal) story about the famous physicist Werner Heisenberg, one of the founders of quantum mechanics. On his deathbed, he is supposed to have said, "When I meet God, I am going to ask him two questions: Why relativity? And why turbulence? I really believe he will have an answer for the first."[26] Heisenberg died in 1976. His doctoral thesis, completed in 1923, had nothing to do with quantum mechanics; it was on turbulence.

13.7 Howaboutism: Getting down in the Trenches

Contrarianism about theories in a particular field of science from those not intimately familiar with the field is not new. In 1931, a book was published in Germany, titled *Hundert Autoren gegen Einstein* ("A hundred authors against Einstein"). It was a collected criticism of Einstein's theories of relativity by 100 scientists – but these scientists were overwhelmingly outside the domain of physics and mathematics.[27] Upon hearing of this book, Einstein reportedly responded by saying, "Why one hundred? If they were right, only one would suffice!"[28]

Whether Einstein actually said it or not, the remark highlights an important feature of science. Within its own domain, a field of science cannot make major advances by relying on consensus. However, consensus among relevant experts within the field can be a very useful in inspiring broader confidence outside the field, provided the field is sufficiently self-critical and open to new ideas.[29] If 9 out of 10 scientists within a field support an idea, and one does not, it is reasonable for those outside that field to go with the opinion of the clear majority. This does not necessarily mean that the lone dissenter is wrong, but they may first need to convince more of their scientific colleagues to take them seriously. That is why scientific revolutions tend to happen slowly.

Progress in science occurs when old models are replaced by new and better models. Old theories are replaced by new theories that explain all the old phenomena, as well as some new phenomena. Einstein proposed his theories of relativity essentially as a superset of Newton's theories. The most powerful way to criticize a model is to offer a clearly better model. It is not hard to find flaws in climate models; this book discusses many of those flaws. But contrarians who dismiss all climate models because of such flaws, without providing a constructive alternative, are seeking to replace an imperfect model with essentially nothing.

Contrarians also claim that climate models overestimate climate sensitivity. Is it even possible for us to construct a scientifically sound climate model with a very low, or even zero, climate sensitivity that can simulate current and past climate phenomena "better than" models with higher climate sensitivity? There is no evidence that we can do so, but it is a valid line of scientific inquiry.[30] It would be similar to the curiosity-driven research pursued by Manabe and others in the 1960s and 1970s, back when global warming was just a curiosity and research funding came with few strings attached. There may be some truth to the argument that the current incentive structure for research funding and publications makes it harder to carry out such "high-risk" research. But state-of-the-art climate models are available as open source, making such research easier.

Another common contrarian claim is that recent global warming is not caused by human activities, that it is merely a manifestation of natural oscillations or stochastic variability of climate. Although we can construct simple statistical models to support this claim, comprehensive climate models that incorporate all the physical constraints do not support it.[31]

Mainstream climate scientists actively explore the issue of model uncertainty. One research project, ClimatePrediction.net, has tried to address the model uncertainty question using a brute force approach. Based at the University of Oxford, ClimatePrediction.net uses crowdsourcing to carry out hundreds of thousands of simulations using a climate model that anyone can download and run using spare cycles on their personal computers. This has allowed the project to explore different values for adjustable coefficients, which has resulted in a range of climate sensitivities – a few quite low, but also many quite high.[32] There are additionally several current efforts – usually referred to as Perturbed Physics Ensembles or Perturbed Parameter Ensembles – that use similar approaches,[33] but on a much smaller scale. A fundamental limitation of all such projects, though, is that they cannot comprehensively address the structural uncertainty issue – they can adjust coefficients for a limited set of formulas, but there is an "infinite" set of formulas to choose from.

***

We need comprehensive climate models to be able to plan for the future using known unknowns, but we also need them to be able to deal with unknown unknowns as they arise. The Rumsfeld uncertainty matrix, with the added distinction between well-known and poorly known unknowns, is a useful way to characterize how climate models are built and improved. Climate is an emergent system, and complexity is essential for modeling it comprehensively, but not all complexity is essential. The challenge is to distinguish between essential and nonessential complexity.[34] For the purpose of long-term climate prediction, a basic parameterization may turn out to be sufficient, without the need for a complicated parameterization. Why offer cake to those who ask for bread?

Also included in the uncertainty matrix is the notion of unknown knowns, or "unawareness" of accurate knowledge, which relates to disinformation and contrarianism. True contrarianism is an essential element of progress in science, but willful ignorance of scientific evidence is not. Contrarians are quick to point out the limitations in the predictive power of climate models, but they do not offer comprehensive alternatives that respect the constraining power of models. Without the discipline of a model, it is easy to indulge in unconstrained ideation.

Disinformation and misinformation regarding climate change persist in part because it is not easy to communicate the technical nuances of climate prediction to policymakers and the general public. We address this communication challenge in Chapters 14 and 15.

14

Lost in Translation

"When I use a word," Humpty Dumpty said in rather a scornful tone, "it means just what I choose it to mean – neither more nor less."

"The question is," said Alice, "whether you can make words mean so many different things."

"The question is," said Humpty Dumpty, "which is to be master – that's all."
Lewis Carroll, *Through the Looking Glass*

What's in a name? According to Shakespeare, that which we call climate change, by any other name, should sound as scary. Not so, opined Swedish climate activist Greta Thunberg, who tweeted in 2019:[1] "Can we all now please stop saying 'climate change' and instead call it what it is: climate breakdown, climate crisis, climate emergency, ecological breakdown, ecological crisis and ecological emergency?"

What do we call a problem like climate change? Actually, why do we call it that? This problem originally had a rather clunky name – "inadvertent climate modification."[2] The catchier term "global warming" was introduced by US geochemist Wallace Broecker in 1975 to draw broader attention to the problem. Among scientists, this temperature-focused name coexists with the broader term "climate change," which encompasses climate impacts beyond just warming temperature, such as rainfall changes. Jim Hansen's 1988 Senate testimony played a key role in bringing the term "global warming" into general use. As the evocative term captured the public imagination, in 2001, the conservative political strategist Frank Luntz aggressively began to promote "climate change" as a less-scary alternative to "global warming," based on feedback from a focus group.

Do euphemisms, or linguistic rebrandings, like "climate change" work? Linguists use the term "euphemism treadmill" to describe how the introduction of polite alternatives for pejorative words is doomed to fail:[3] Over time the polite alternative slowly reacquires its original connotation, like "sanitation worker" for "garbage man." The term "climate change" has slowly acquired the same status as "global warming": In 2015, the Florida Department of Environmental Protection tried to ban the use of the terms "climate change" or "global warming" in any official communication because they were not "true facts."[4] On the other hand, Thunberg feels that "climate change" does not convey the sense of urgency that is needed: Due to the long lifetime of carbon dioxide in the atmosphere, every year of delay in reducing emissions makes it vastly more difficult to limit global warming to acceptable levels.[5]

The euphemism treadmill argument suggests that concepts ultimately win over labels, but it is an equilibrium argument. By relabeling climate change, we may not be able to motivate all the people all the time, but we may at least be able to motivate some of the people (the activist base) all of the time, or else all of the people some of the time (through an election cycle, perhaps). Using a more urgent name for climate change may help climate activism if the name's effects last long enough to motivate stronger climate action. (But there is also the danger that overuse of scary imagery to convey the necessary urgency may trigger feelings of helplessness and panic in some people.[6])

However, there is a practical problem in using the term "climate crisis" in scientific discussions because it describes more than the physical phenomenon of climate change – it also describes its social and environmental effects. In this book, we shall continue to mostly use "climate change" or "global warming" to describe the physical phenomenon, but we shall also use "climate crisis" where appropriate. We also avoid using the term "climate alarmism" in this book, because there is too much conflation between being alarmed by climate change and being alarmist about it, which has a different connotation.[7] If we think of the euphemism treadmill as a special case of the more general terminology treadmill, it may be time to move "climate alarmism" off that treadmill.

When communicating science to a general audience, names matter. The Antarctic ozone hole phenomenon is an example of a climate phenomenon whose particular label may have played a role in shaping the public's reaction. When the large springtime Antarctic ozone loss was first discovered using local measurements, it was called just that, "large ozone loss" or "ozone depletion." Soon it was confirmed using global satellite data, which made apparent the spatial extent of the ozone loss. The metaphor of a hole in the ozone shield,[8] accompanied by the dramatic satellite imagery, drew worldwide attention and helped spur the rapid international response to address the problem.

The potential influence of relabeling, at least temporarily, raises an important question. Does language affect how we think? A famous hypothesis in linguistics, called *linguistic determinism* or the Sapir-Whorf hypothesis, asserts that it does. Linguists have basically discredited the strong version of this hypothesis, that language completely determines our thinking. The fact that writing can be translated from one language to another suggests the primacy of concepts, rather than words. But a weaker version of the hypothesis, also known as *linguistic relativism*, survives. It says that language influences our thinking, even if it doesn't determine it.

It has been said that poetry is what gets lost in translation.[9] A certain nuance can be lost in the translation of a text between languages; the same loss of nuance can occur when an idea is translated from the precise mathematical language of science into the imprecise natural language of the media. When scientists discuss the broader aspects of how science works, or when they address policy issues, they can no longer use the precise scientific vocabulary of their field; they must use everyday language in order to facilitate general understanding. In this kind of translation, there is much scope for ambiguity, and this ambiguity can persist over time. Loss in translation can happen even between different subfields of science. It is not uncommon for non-climate-scientists who use a climate prediction to treat it as being more definitive than the community of climate scientists would themselves treat it, because the myriad uncertainties are not easy to communicate.

When scientists communicate climate issues to nonscientists, the loose use of language can obfuscate logic. The French philosopher Voltaire said: "If you wish to converse with me, define your terms." In this chapter, we analyze the definition of two terms – "model" and "data" – that are commonly used in the context of climate prediction. These two terms are often misunderstood by the media and the public, and sometimes even by scientists. We also discuss the term "business as usual," which has been the subject of some controversy and whose definition carries important implications for climate prediction. The final term we analyze is "prediction" itself.

We end this linguistic discussion by noting an amusing twist. Frank Luntz, who had originally promoted the term "climate change" as a less-threatening alternative to "global warming," had a change of heart after an uncomfortably close encounter with a California wildfire in 2017. He now believes, as he told a Senate committee two years later, that "rising sea levels, melting ice caps, tornadoes, and hurricanes are more ferocious than ever. It is happening. Just stop using something that I wrote 18 years ago, because it's not accurate today."

14.1 What Is a Model?

Whether or not you realize it, every statement you make and every position you take about the future is based on a model. A model is just an abstraction. The abstraction can be verbal, mathematical, or computational. Every quantifiable statement about the future can be expressed as the result of a mathematical formula, which becomes a model. As climate scientist Gavin Schmidt told the Associated Press in 2020, "The key thing is that you want to know what's happening in the future. Absent a time machine you're going to have to use a model."[10]

In this book, we have used the term "climate model" to refer to a set of mathematical equations solved using a computer program. This is the generally accepted meaning – but it leaves the impression that if you are not using math or a computer, then you are not using a model. This apparent semantic distinction is often made when saying that we should make policy decisions without using climate models because they are not to be trusted. Some even say that we should make decisions based on data, not models. On November 27, 2018, White House press secretary Sarah Huckabee Sanders dismissed the findings of a government report on climate change saying that "it's not data-driven. We'd like to see something that is more data-driven. It's based on modeling, which is extremely hard to do when you're talking about the climate." The last part of the statement is certainly true: Climate modeling is extremely hard to do! But the rest of it is based on common misconceptions about what constitutes a model.

Suppose you say, "I don't believe that the climate is changing, and I expect it to remain the same in the future." You may think your statement does not use a model – but in fact it is very much based on a model. Mathematically, you are saying that the rate of change of climate is zero. Meteorologists even have a special name for such a model, which is routinely used to assess the skill of weather forecasts; it is called the *persistence model*.[11] Statisticians would call it the *stationarity model*.[12] You may instead say, "I don't believe in models. So, I'll assume climate will change over the next 30 years at the same rate that it has changed over the last 30 years." Mathematically, you are saying that the rate of change of climate is constant. Scientists have a name for this too: It is the *linear regression model*, where you draw a straight line through temperatures over the last 30 years and extrapolate for the next 30 years. You may then throw up your hands and say, "I have absolutely no idea what the future is going to be like – the climate is going to change randomly." Yes, scientists have a name for that, too! It is called the *stochastic model*. The only way not to use a model is to completely ignore the future.

So when we talk about the future, we must choose between different types of models.[13] We want to pick the best one. But what criteria do we use to decide which model is the best model for prediction? This is one of the most important questions in modeling. For weather models, the answer is easy: We test each weather model on thousands of weather events in the past, and we pick the one that produces the most accurate forecasts. For climate models, it is much harder to answer that question. We could pick the model that simulates our current climate best, but that does not guarantee it will make the best prediction for our future climate (Section 10.2). We need to consider additional criteria.

We discussed the distinction between simple and complex models. Perhaps we could borrow the guiding principles of fundamental physics and invoke Occam's Razor, or the principle of parsimony, to justify a preference for simpler models. But simplicity is hard to define, and it is not automatically a virtue in models used to predict the behavior of a complex emergent system (Chapter 8). Simple models are useful for improving our qualitative understanding of climate but not for quantitatively predicting long-term climate change.

What about the distinction between statistical (data-driven) and physical (hypothesis-driven) models? This is an important distinction, one worth discussing in some detail. Usually, this is the distinction people have in mind when they say they "prefer to make decisions based on data rather than based on models." They implicitly use the word "data" to refer to (inductive) statistical models and the word "model" to refer to (deductive) computer models based on hypotheses.

The distinction between inductivism and deductivism is useful as an analytic tool, but when it comes to predicting climate, there are no purely data-driven or purely hypothesis-driven models. If we decide to use a statistical model, we have to decide on the details: Do you fit a straight line, sinusoid, or exponential curve to the data? How far back in time do you fit the data? And so on.

If they could, statisticians would use temperature data from many past events of emissions-driven global warming to train statistical models so that they could select the best-fitting model for predicting the rate of future global warming. But we only have measurements of temperature for one incipient emissions-driven global warming event – the one that is happening right now. (There is indirect information, called paleoproxy evidence, for climate change occurring tens of thousands to millions of years ago, but it is not accurate enough to train a statistical model. Also, past global warming events were not driven by emissions.) Since we do not have enough training data, we would end up having to make simplistic hypotheses to build a statistical model: about what curve is the best to fit, how to deal with geographic variations, and how to adjust it for future changes in emissions. Unlike the physically motivated hypotheses used to build a comprehensive climate model, these simplistic statistical hypotheses are essentially untestable because we are trying to predict an unprecedented event that is the emergent property of a very complex system.

In other fields that rely on empirical modeling, such as economics or epidemiology, there are often fundamentally different approaches to modeling within the field, each founded on a different set of assumptions.[14] In climate modeling, however, there is no question about the basic framework: It starts with the well-tested deductive approach used for a weather model, dividing the globe into a collection of grid boxes and solving the physical equations governing climate using a computer.[15] Any approach – such as a purely statistical approach – that ignored the fundamental equations would lose the deductive and constraining power provided by the laws of physics. The differences between climate models are in the details that build upon the basic framework, like how to represent clouds or aerosols that are much smaller than the grid box, and how to formulate chemical and biological processes that are not well understood. Each climate model has poorly constrained parameters and formulas that try, in different ways, to capture the behavior of such processes. The values of these parameters are estimated inductively, by calibrating the climate model to features of the current observed climate.

Our complex climate models and their predictions are not perfect, but they are better than the alternatives. As Swiss climate scientist Reto Knutti writes, if we want to plan our future path, "it would be smarter to have an imperfect map than no map at all."[16] Since inaction is also a form of action, we have no choice but to act to the best of our knowledge in dealing with climate change. These imperfect models encapsulate the best of our imperfect knowledge. A common criticism of climate models is that they are highly "tuned," with "fudge factors" that are adjusted to fit data, rather than being purely deductive. However, if we wish to use human intuition as the alternative to climate models for predicting the future, as some contrarians suggest, we must recognize that our intuition is also highly tuned – to our own lifetime of experiences and biases.

14.2 What Is Data?

All measurements are right; some measurements are useful.

A play on George Box's aphorism that *all models are wrong*

We have argued that any statement regarding the future is essentially based on a model. What about statements regarding the past? All quantifiable statements of what happened in the past can be described as measurements, or raw data. Raw data alone are not necessarily useful in determining whether an event of interest happened in the past or is happening in the present. To answer questions of this nature, we need to distinguish between that which is of interest – the *signal* – and everything else – the *noise*.

Raw data from certain accurate measurements, like Charles Keeling's carbon dioxide measurements on Mauna Loa, are easy to interpret directly because the signal (in this case, increasing carbon dioxide) is strong. When the signal is weak, as in the early stages of a drought, an El Niño event, or global warming itself, the raw data are not very useful; even experts can disagree on whether the event is actually happening. As the signal becomes stronger, there is less disagreement. Any statements about a weak signal in the past are no longer pure data because such statements require interpreting or adjusting the data to isolate the signal, which can only be done using a model for the noise.

This book is about climate models and their predictions, but climate predictions made by models are not the primary evidence for global warming. Data – specifically, records of surface temperature – show that the Earth has a warming trend (Figure 7.2). It is not a completely unambiguous signal, due to the background noise of natural climate variations like El Niño events that affect global temperatures, but the warming signal has gotten much stronger in the last 50 years.

To measure climate trends accurately, we try to use the same instrument at every location of interest. To calculate the global-average temperature, for instance, instruments such as thermometers need to be calibrated for accuracy, and measuring stations need to be located around the world. If a thermometer in a measuring station fails and is replaced by a differently calibrated thermometer, we may see an artificial jump in the station's measurements. If the number of measuring stations changes over time, that can appear as a trend in the data if the averaging procedure does not correct for it.

Scientists carefully examine every surface measurement of temperature, beginning with the first such measurements in the mid-nineteenth century. They check the details of the measuring stations and correct for any measurement errors. For instance, the effects of economic growth near measurement stations, such as the higher temperatures associated with the growth of cities (known as the *urban heat island effect*), can cause trends unrelated to global climate change to appear in the data. If these effects are not accounted for, the global warming trend computed from raw, uncorrected data may appear larger than its true value. To make these corrections, scientists have to use models of the urban heat island effect. This model could be as simple as fitting a trend line to the measuring stations close to a city and comparing it to the trend for the measuring stations farther away from the city, but, as simple as it is, it is nevertheless a model.[17] Without such a model, there would be no way to correct for this distortion of the climate change signal.

Stations using thermometers have provided simple and accurate measurements for more than a century and half, but these stations are almost all located on the continents, so there is little coverage of oceanic regions. Satellites, which have measured atmospheric temperatures from space since the 1980s, can provide truly global coverage, but they are only good at measuring temperatures several kilometers above the surface. Climate models simulate warming trends in surface temperatures, as well as warming trends above the surface, which are actually stronger than the surface trends associated with increasing greenhouse gas emissions – so we ought to be able to validate them.

In 1990, two scientists from the University of Alabama in Huntsville (UAH), Roy Spencer and John Christy, analyzed 10 years of satellite temperature data and reported that the data showed no warming trend.[18] This was inconsistent with the predictions of climate models. These claims were the start of a controversy regarding satellite temperature trends: Media headlines essentially said that models had been contradicted by data. But had the models been contradicted by raw data, or by a signal estimated from the raw data? In the latter case, we must look at how the signal was estimated.

Images from satellites are very good as weather snapshots; they can provide incredibly detailed information, crucial to providing more accurate initial conditions for and thus improving weather forecasts (Section 2.2). However, in trying to compute the signal of climate trends from satellite data, we run into a number of roadblocks. To separate the signal in raw data from the noise, we need at least a simple model for the noise. Satellite data are particularly difficult to deal with in this regard.

First, we need data over multiple decades to compute a climate trend. We cannot use data from a single satellite to do that, because satellites have a limited lifetime, typically less than a decade. We need to splice together measurements from multiple satellites, in order to create a time series with which we can compute trends. Data from different instruments, calibrated at different times, has to be blended together, which requires the creation of a statistical model to bridge the transition in time.

Second, even measurements from a single satellite suffer from a problem known as orbital decay: Due to friction, the altitude of the satellite orbit slowly decreases over time.[19] A satellite cannot use a simple thermometer to measure temperature because it is hundreds of kilometers away from the air whose temperature needs to be measured. Satellite instruments actually measure the amount of energy for each wavelength of light emitted by the air, from which they can calculate the temperature of the air. This method is more error-prone than using a thermometer because all factors that affect the transmission of radiation through the atmosphere, such as humidity, need to be corrected for.

The complicated process of calculating air temperature from energy measurements is called "satellite retrieval" and is essentially a physics-based radiation model. Orbital decay will affect the retrieval calculation because a decrease in the satellite's altitude not only alters the amount of air the radiation must travel through but also affects the local time of measurement, adding to the error.

Correcting for measurement errors associated with a single satellite and splicing together data from multiple satellites is a fairly involved modeling exercise. We can think of this modeling exercise as a "virtual instrument" used to measure long-term trends. Different research groups carrying out satellite retrievals build this virtual instrument in different ways and get somewhat different answers.

Spencer and Christy made one set of assumptions to splice together data from multiple satellites to estimate temperature trends. Soon after the satellite-trend controversy erupted, other research groups used different sets of assumptions to analyze and splice together the satellite data, coming up with alternative estimates of the trends that agreed much better with climate model predictions. But Spencer and Christy have not accepted these other estimates, and both sides have continued to spar. Climate scientist Andrew Dessler, who has studied this controversy, explained in 2016 that the satellite record was not of sufficient quality to validate or invalidate climate models because new issues were being continually discovered in the satellite record of temperature trends.[20] We might think that another set of data could have helped resolve this controversy, that from the radiosonde balloon measurements used in weather forecasting (Section 2.2). But statistical corrections also need to be applied to radiosonde temperature measurements to estimate long-term climate trends, reducing their utility in resolving this issue.[21]

Contrarians often invoke the satellite-trend controversy to claim that the climate models "systematically overestimate the observed warming of the globe and, specifically, the upper tropical troposphere."[22] It is true that the climate models are imperfect and have structural uncertainties, as discussed in Chapters 11 and 12. But, as we have argued in this section, even what is referred to as "observed warming" in the upper troposphere is not directly observed; it is inferred from raw data using simple models, which have their own uncertainties.

14.3 Models versus Data: The Problem of Induction

if we had observations of the future, we obviously would trust them more than models, but unfortunately observations of the future are not available at this time.[23]

Tom Knutson and Bob Tuleya, GFDL scientists, responding to the question "Should we trust models or observations?"

The primary reason to worry about human-induced global warming is that it is a completely unprecedented event in Earth's history. Warm temperatures have occurred in the past, and carbon dioxide concentrations have been higher in the past – but typically the former drove the latter, as was the case during the ice age cycles (Figure 5.3). Also, past warming events unfolded over thousands of years; the current warming event is unfolding over tens of years. That means slow climate feedbacks likely played a bigger role in past warming events than they may be playing in the current warming event. We can learn from the warm climates of the distant past, but those insights may not provide a complete description of the unprecedentedly rapid global warming event occurring right now.[24]

Temperature measurements are key to understanding global warming. However, there is a signal-to-noise problem with temperature estimates from the past. There are no direct temperature measurements available from when the warming signal was strong, as it was thousands to millions of years ago, and we have to estimate temperatures indirectly from other physical evidence known as climate proxies, such as tree rings, sediments, and ice cores. Where we do have direct temperature measurements, as in the last 150 years, the signal is quite weak until near the end of the data record. We have had comprehensive data since the 1980s, when satellites started measuring the Earth's energy balance, but that is a rather short and noisy data record from which to extract the signal.[25]

We need models because we can't make measurements in the future, but we also need models because we can't make direct temperature measurements in the distant past. Extracting the temperature signal from indirect measurements involves the use of models. As discussed in the previous section, using a model this way is akin to using a virtual instrument to make measurements, and the virtual instrument error needs to be explicitly accounted for in the uncertainty estimates. With physical instruments, we use measurements made independently of the instrument to verify its accuracy. But it is not possible to verify these virtual instruments when dealing with events that occurred many thousands of years ago. Quantitative estimates of past climate properties can and do change over time, as new data are discovered or new analysis techniques are invented.[26] This adds unquantifiable uncertainty to observational estimates of past climate change signals,[27] very similar to the structural uncertainty that bedevils complex climate models. Often, observational estimates using multiple lines of evidence are combined statistically to generate a single observational estimate. The implicit reasoning behind this is the same Anna Karenina principle that we previously applied to climate models (Section 11.1): If the structural errors in the different observational estimates are sufficiently random, then the combined estimate should be more accurate than each individual estimate.

The most comprehensive study to date that consolidates observational estimates of climate sensitivity is the 2020 assessment by the World Climate Research Program (WCRP).[28] The assessment analyzes many different types of observational data, with some indirect input from climate models, and concludes that it is extremely unlikely that climate sensitivity is below 2°C, and also that it is not likely to be above 4.5°C. But several of the latest generation of climate models, which have substantially improved parameterizations, predict substantially higher climate sensitivity values – although there are many other models that predict lower sensitivity values, more consistent with observational estimates (see Figure 12.2b). This discrepancy is much larger than in the 2013 IPCC assessment, where the high model estimates were closer to the observational estimates.

As some of the latest generation of climate models appear to be "running hot" or predicting stronger warming,[29] a schism has opened up between data and models on the issue of climate sensitivity. This is not a schism between "pure data" and "pure models." As historian Paul Edwards argues in his history of climate modeling, *A Vast Machine*, neither pure data nor pure models exist in climate science.[30] This new schism is more of a conflict between a collection of many functionally different simple models used to analyze data inductively[31] and a collection of many functionally similar complex models that share a deductive framework, albeit one with crucial inductive components.

In the past, IPCC assessments have typically followed the lead of climate models in assessing the range of climate sensitivity.[32] But the sixth IPCC assessment may follow the lead of the WCRP assessment and not extend the high-end range beyond that assessed by the fifth assessment, even though models show a higher sensitivity range. Such a decision would be an implicit rebuke of the latest generation of models with higher climate sensitivities. We can debate the wisdom of sidelining comprehensive climate models in favor of an assessment based on a combination of process models and statistical methods with numerous plausible assumptions, but there is no objective way to combine divergent estimates of climate sensitivity between data and models. We should also consider the possibility that this divergence is a symptom of models having become overly complex, and therefore allowing too many degrees of freedom in their parameterizations (Section 13.2).

In any case, we cannot simply walk away from this latest generation of models because climate sensitivity is not the only metric of importance. The newest models perform better than previous models in some other important metrics relating to the fidelity of their climate simulations.[33] Also, observational estimates of climate sensitivity only tell us what may happen to global-average temperatures in the future; they do not tell us anything about regional climate change. We need shorter-term regional predictions to plan for mitigation and adaptation, and only comprehensive models, which include processes such as ocean heat uptake, can provide them.[34] Furthermore, the inductive nature of the observational estimates cannot account for the unprecedented pace of the warming and the possibility that such warming could trigger unexpected feedbacks.[35]

This is a fundamental problem with purely inductive reasoning based on data. It can tell us a lot about the properties of white swans that we are very familiar with, but much less about the new type of swan that we are glimpsing more and more frequently. It is said that those who cannot remember the past are condemned to repeat it. Inductive reasoning can prevent us from repeating the mistakes of the past; deductive reasoning, though fallible, can help us avoid making new mistakes in the future. For all their flaws, comprehensive models are essential for climate attribution, mitigation, and adaptation. We will need to use the new insights from data to improve representations of poorly known processes to create the next generation of models. And so, the Red Queen's race must go on.

14.4 What is "Business as Usual"?

Climate predictions hinge on the meaning of a common phrase, "business as usual." The definition of this phrase has profound implications for assessing the impacts and risks of climate change, with trillions of dollars of monetary value at stake. Climate prediction requires specifying how the boundary conditions of greenhouse gases like carbon dioxide will evolve in the future (Section 7.6). Different scenarios of carbon emissions, each based on a consistent set of socioeconomic assumptions, provide these boundary conditions. To bracket the impacts of climate change, we try to identify best- and worst-case scenarios among those plausible. Typically, "business as usual" corresponds to the "no policy" scenario; it assumes that without new policies to mitigate carbon usage, emissions will simply continue to increase with expected economic growth.

What does "business as usual" mean? Since this is such an important question for predicting climate change, multiple international committees have attempted to answer it (Section 7.6). The IPCC's first assessment, released in 1990, used four scenarios labeled A, B, C, and D. Scenario A, explicitly labeled Business-as-Usual, assumed that few or no steps would be taken to limit emissions; scenario D corresponded to a world with substantially reduced emissions, with the other two scenarios falling in between. The initial scenarios were based primarily on world population projections. As they were updated in 1992 and again in 2000, more elaborate storylines and socioeconomic models were used to define the scenarios.

When the four RCP scenarios were introduced in 2007, they were meant to be an interim measure. A narrow, energy-focused approach was used to create scenarios urgently needed for the next IPCC assessment, while a slower but more comprehensive and updated socioeconomic approach was being developed by another IPCC committee. The RCP8.5 scenario (Figure 7.4a) was proposed as a "high-emission business as usual scenario," implying that there could be other business-as-usual scenarios with lower emissions.

However, these subtle caveats were lost in translation and RCP8.5 was adopted as *the* "business as usual" scenario soon after its introduction. An influential 2020 McKinsey report on climate risk was based solely on the RCP8.5 scenario, saying "[w]e have chosen to focus on RCP8.5, because the higher-emission scenario it portrays enables us to assess physical risk in the absence of further decarbonization."[36] Many other societal impact assessments have also used this scenario to make dire predictions of very high global warming (4–5°C), by the end of the century (Figure 7.4c). (To put these predictions in context, the last time a warming of such magnitude occurred was at the end of the last ice age. It melted two giant continental ice sheets that covered Canada and Scandinavia.)

The wide usage of RCP8.5 as the business-as-usual scenario to make these business-as-usual predictions has been criticized,[37] especially since it wasn't intended to be the "no policy" variant of the other RCP scenarios.[38] Socioeconomic models, unlike climate models, do not have to obey many strict conservation laws (Section 9.1), which means that emission scenarios tend to be poorly constrained. It has been pointed out that RCP8.5 involves some unrealistic assumptions about emission growth, such as a fivefold increase in coal usage, which may not even be possible given the total estimated coal reserves in the world. Coal usage has been falling in many countries, making the RCP8.5 scenario seem less and less likely.[39] As the baseball-playing philosopher Yogi Berra observed, "the future ain't what it used to be."

The ambiguities inherent to defining the "business-as-usual" scenario mean that we should perhaps not treat RCP8.5 as the singular vision of a "no policy" future. However, this does not mean that we should stop worrying about climate change, nor does it mean that we should delay the strong mitigation steps that are urgently needed.[40] Simply knowing that we may not be on the trajectory to face an extremely high level of warming isn't exactly comforting;

even under a less-extreme scenario like RCP6.0, climate change will be disastrous. And extreme scenarios like RCP8.5 are still of interest to climate scientists, even if they become less relevant to real-world planning, because such scenarios can amplify the signal of global warming and make it easier to study different climate processes.

Emission scenarios are not meant to be deterministic predictions (Section 7.6), and treating them as such may be responsible for much of this confusion. Emissions will appear to track a recent scenario (by the scenario's construction), but that is no guarantee that they will stay on the same track many decades into the future. Some scientists argue that higher-emission RCP scenarios may actually be more realistic because the current concentration pathway approach neglects increased carbon emissions caused by amplifying carbon cycle feedbacks in a warmer world.[41] That may be true, but it would be more appropriate to handle these feedbacks explicitly using comprehensive Earth System Models, rather than trying to tweak scenarios to compensate for them.

The debate over the usualness or unusualness of the business assumptions that underlie extreme scenarios like RCP8.5 can never be fully resolved. Unlike in the natural sciences, there is no underlying truth that all parties can hope to eventually converge on. In 2015, a long-awaited set of new scenarios, called Shared Socioeconomic Pathways (SSP), offered five scenarios to complement those of the RCP. These were more comprehensive, based on narrative storylines about socioeconomic development, and they considered different obstacles to mitigation and adaptation.[42] One of the scenarios, SSP3 ("regional rivalry – a rocky road"), was based on a storyline of resurgent nationalism and was considered unlikely when it was in development (Section 7.6).[43] But, in the year following its release, events such as Brexit and the 2016 US presidential election made SSP3 seem much more likely!

The IPCC explicitly referred to the high-emission scenarios as "Business-as-Usual" in its First Assessment released in 1990,[44] but it no longer encourages that terminology, noting that "the idea of business-as-usual in century-long socioeconomic projections is hard to fathom."[45] It prefers "baseline scenarios," which are "not intended to be predictions of the future, but rather counterfactual constructions that can serve to highlight the level of emissions that would occur without further policy effort." These baseline scenarios are contrasted with "mitigation scenarios," each of which corresponds to a different set of policies for reducing emissions. Does this relabeling make it clear that the "baseline" is not about "business as usual"? Maybe. But the term "baseline" can also be ambiguous, especially if it extrapolates current trends into the distant future.

14.5 What Is Prediction?

The fact that we predict eclipses does not, therefore, provide a valid reason for expecting that we can predict revolutions.

Karl Popper

We have argued that statements about the past are data, and that statements about the future always use models, explicitly or implicitly. These statements about the future made using models are called predictions. There are various flavors of predictions, with distinctions that are not always appreciated or understood. When we discussed the difference between weather prediction and climate prediction, we noted the distinction that Karl Popper made between conditional scientific predictions and unconditional prophecies (Chapter 5). A scientific prediction is usually a statement of the form "if X is true, then Y will happen." The validity of the prediction is conditional on the validity of the assumptions (which may include the implicit assumption that the model is accurate for the type of prediction being made, if that has not been verified).

A prophecy simply says, "Y will happen." It is an unconditional statement about what will happen in the future, *no matter what*. Some scientific predictions are as good as prophecies. When a model is used to make numerous predictions that are verified to a high degree of accuracy, its predictions can acquire the status of prophecies. For Popper, an astronomical prediction that a solar eclipse will happen on a certain date qualifies as a prophecy, because the model used to predict the eclipse is celestial mechanics, which is well-tested. Additionally, the initial positions and future trajectories of the sun and the planets required to make the eclipse prediction are accurately known. And, crucially, no actions taken by humans can materially alter the trajectories of the planets between the present time and the time of the predicted eclipse.

On the other hand, consider a societal revolution. Such a revolution is usually the result of a cascading sequence of human actions, many of which cannot be predicted (rather like a societal Butterfly Effect). In 1979, five leaders of the Iranian revolutionary student movement voted on whether they should occupy the American Embassy or the Soviet Embassy in Tehran.[46] Two leaders, including one named Mahmoud Ahmadinejad, voted to occupy the Soviet Embassy because they considered Marxism a bigger threat to the revolution. They were outvoted by the three other leaders, and the students occupied the American Embassy instead. A single vote bifurcated the path of history. (Ahmadinejad later became a stridently anti-American president of Iran.) The trajectory of a society is hard to predict – and that applies equally to emission scenarios for greenhouse gases, which depend on the political, economic, and technological evolution of societies around the world.

Making a prediction is analogous to ordering a steak at a restaurant. The word "steak," by itself, is imprecise – we need to select from one of the six degrees of doneness for steak as well as the precise cut. The word "prediction" is also imprecise,[47] as it can correspond to many different degrees of "prophe-ticness." To better understand this, we consider the following set of predictive statements:

1 Next year, the average January temperature will be 0.3°C (32.6°F), and the average July temperature will be 24.7°C (76.5°F) in New York City.[48]

2 Tomorrow, a cold front will pass over New York City, lowering temperatures by 10°C (18°F).

3 Tomorrow, there is an 80 percent chance of rain over New York City.

4 In the next eight months, there is a 65 percent chance of a warm El Niño event in the tropical Pacific.[49]

5 Over the next five years, global-average temperature is expected to remain high and is likely to be between 0.17°C and 0.43°C above the recent long-term average.[50]

6a In 80 years, the global-average temperature will rise by an additional 3.7°C if we take no further mitigation steps to curb carbon emissions, resulting in carbon dioxide concentrations that continue to rise throughout the twenty-first century (RCP8.5 scenario; Figure 7.4).[51]

6b In 80 years, the global-average temperature will rise by an additional 2.2°C if we take some mitigation steps and carbon emissions peak around 2080 (RCP6.0 scenario; Figure 7.4).

6c In 80 years, the global-average temperature will rise by an additional 1.0°C if we take aggressive mitigation steps and carbon emissions peak in 2020, declining rapidly thereafter (RCP2.6 scenario; Figure 7.4).

Some of the above predictions are essentially prophecies; some are not. As a chef can cook the same cut of steak to different degrees of doneness, all of the above predictions can be made, in principle, by running the same climate model for different lengths of time.[52] (All good climate models must also be capable of making decent weather and El Niño forecasts, because climate is the average of weather. But they are not "tuned" to maximize their short-term forecasting skill.)

1 Next year, the average January temperature will be 0.3°C (32.6°F) and the average July temperature will be 24.7°C (76.5°F) in New York City.

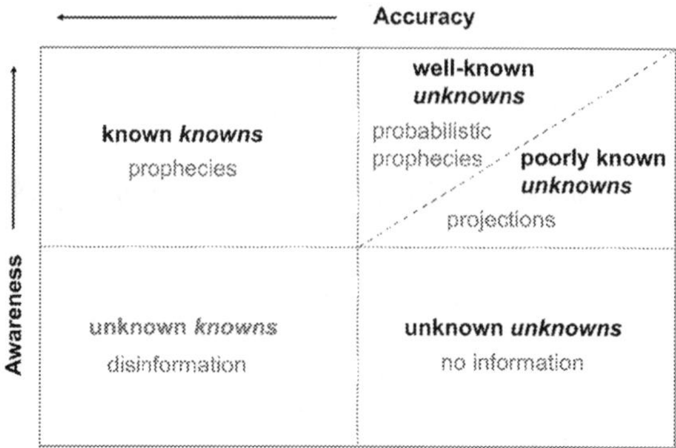

Figure 14.1 The Propheticness-of-Prediction Matrix: a rough mapping of the spectrum of predictions to the modified Rumsfeld uncertainty matrix (Figure 13.1b).

Prediction 1 is a prophecy. Models can predict the large temperature difference between January and July very well. We know the boundary condition – the position of the sun – accurately, and neither initial conditions nor any human actions will modify the prediction. But Prediction 1 is not a deterministic prophecy because it is a statement about the general statistics of weather. A *deterministic prophecy* would be an exact and detailed statement about the future – in this case, an exact temperature prediction for each day of the month.

We can combine the notion of prophecy with the concept of probability. A prophecy is a pronouncement of Laplace's Demon, the epitome of determinism: It can predict exactly and indefinitely into the future. Suppose that we are not able to predict the future exactly, but we are able to predict the exact probability of a set of events happening in the future, as can be done by an ensemble of Weather Demons. We introduce the notion of a *probabilistic prophecy*, where we predict the probability of an event accurately, *no matter what.* (This is not a new concept; it is just a new name for an old concept, usually known as *objective probability*.) A probabilistic prophecy is certain about the event's uncertainty, which can be considered a well-known unknown (Figure 14.1). Prediction 1 is a probabilistic prophecy about the statistics of temperature. It may seem like a trivial prophecy, but it is extremely useful.

2 Tomorrow, a cold front will pass over New York City, lowering temperatures by 10°C (18°F).

Prediction 2 is conditional on our knowledge of the initial conditions of the atmosphere, which are unaffected by any human actions (in the future). The boundary conditions are either well known (solar heating) or almost constant (carbon dioxide). Since 24-hour forecasts of temperature are quite accurate, with less than 1.7°C error, Prediction 2 is pretty close to a deterministic prophecy.[53] Forecasting temperature requires following the motion of very large-scale air masses, which computer models can do well. This means that different models are usually in close agreement for their 1-day forecasts of temperature, although they may disagree on their 10-day forecasts (due to model errors and the Butterfly Effect).

3 Tomorrow, there is an 80 percent chance of rain over New York City.

Prediction 3 is less certain than 2. Rainfall predictions tend to be probabilistic, not deterministic, because fine-scale processes like clouds and rain are not as well simulated by weather models as air temperature. But the probability of rain is simulated quite well, and it can be verified:[54] We can count the number of days that a model predicts an 80 percent chance of rain and verify if it actually rained during 80 percent of those days. Therefore, Prediction 3 can be considered close to a probabilistic prophecy: Human actions will not affect the probability. A similar example of a probabilistic prophecy would be a short-term hurricane forecast (Figure 2.1b).

Probabilistic weather prophecies can be informative and useful for estimating risk. However, a conditional probabilistic weather prediction – such as saying that there is a 70 percent chance of cooler temperatures than average, *if* the polar vortex becomes stronger next week – is less useful, since we may not be able to predict the strength of the polar vortex.

4 In the next eight months, there is a 65 percent chance of a warm El Niño event in the tropical Pacific.

Prediction 4 is an example of a short-term climate forecast. There is significant uncertainty in El Niño forecasts, due to model errors as well as the Butterfly Effect associated with initial conditions in the ocean (Section 7.5). Only limited verification of models against data is possible because there have been only about 20 El Niño events over the last 70 years for which we have good observations, and each event tends to be a bit different from the last. The El Niño prediction will not be affected by human decisions, which makes this a weak probabilistic prophecy.

5 Over the next five years, global-average temperature is expected to remain high and is likely to be between 0.17°C and 0.43°C above the recent long-term average.

Prediction 5 is the sort of prediction typically referred to as a decadal climate prediction, although this particular prediction only covers the next five years. Like Prediction 4, it is conditional on our knowledge of the initial conditions in the ocean and the accuracy of our model. Due to their longer time period, there is even less data to verify decadal predictions over the last 70 years of observations than there is for El Niño events. Decadal predictions are also conditional on there being no major volcanic eruption(s) during the five-year prediction period because volcanic eruptions can cool global temperatures significantly – but such events are unknown unknowns that cannot be predicted.

6a. In 80 years, the global-average temperature will rise by an additional 3.7°C, if we take no further mitigation steps to curb carbon emissions, resulting in carbon dioxide concentrations that continue to rise throughout the twenty-first century.

6b. In 80 years, the global-average temperature will rise by an additional 2.2°C if we take some mitigation steps and carbon emissions peak around 2080.

6c. In 80 years, the global-average temperature will rise by an additional 1.0°C if we take aggressive mitigation steps and carbon emissions peak in 2020, declining rapidly thereafter.

Predictions 6a,b,c are typical of long-term predictions made using a climate model. They are conditional; each depends on a particular socioeconomic scenario to predict carbon dioxide concentrations in the future. The Climate Demon hears three different plausible storylines of carbon emissions from the Socioeconomic Demon and then tells three different temperature stories. The three predictions are essentially unverifiable, because they are centennial timescale predictions of a single unprecedented event.[55] They are also affected by model errors; different models will predict different degrees of global warming for the same scenario (Figure 7.4c).

Long-term climate predictions like 6a,b,c are technically referred to as "projections" (a term borrowed from economics), to emphasize their conditionality (Section 7.6).[56] But the semantic distinction between predictions and projections is often blurred in popular discussions of climate change.[57] We will likewise refrain from making such a distinction (for the most part). Unlike Prediction 3, the set of conditional Predictions 6a,b,c does not constitute a probabilistic prophecy because there is no way to uniquely assign unconditional probabilities to emission scenarios.[58] Sometimes probabilities are assigned to scenarios because risk analysis requires numbers, but that is not motivated by scientific evidence. Any statements along the lines of "there is a 30 percent chance of an additional 3.7°C warming by the end of the century" are always subjective. Therefore, the set of Predictions 6a,b,c should always be treated as a group; they are only valid in a collective sense, not individually, because the future is not constrained to track a single scenario.

The Weather Demon and the Climate Demon are essentially twins, sharing the same DNA of computer code. But the prophetic natures of their predictions are quite different. How can the same computer code make the whole gamut of predictions, from accurate prophecies to uncertain projections? On the scale of 24 hours, the strong signal is the weather phenomenon; the weak noise is climate change, which can essentially be ignored. On the scale of decades or longer, the roles are reversed: The weak signal is climate change, and the strong noise is the weather phenomenon. Climate is the average of weather, and averaging the weather over time weakens the weather noise relative to the climate signal. But it also amplifies the cumulative effect of the systematic errors in the approximate formulas that represent fine-scale weather processes like clouds, which compounds the uncertainty in the climate prediction.

Note also that the most prophetic (1) and the least prophetic (6a,b,c) statements are both climate predictions! The skill of climate prediction depends on our knowledge of the boundary conditions. Like eclipse predictions, the prophetic nature of Prediction 1 arises from fundamental physics that determines the tilt of the Earth's axis of rotation, which provides the boundary condition for the seasonal cycle prediction. The boundary condition for centennial predictions like 6a,b,c is the future trajectory of carbon emissions, which is subject to the "societal Butterfly Effect" of inherently unpredictable human actions.

***

The words and terms we use to talk about different aspects of the climate change problem matter a great deal. Confusion about the meaning of basic terms like "model" and "data," and the distinction between the two, persists. There is no "pure model" versus "pure data" comparison in climate science;[59] it is not possible to talk meaningfully about the future of climate without using a model of some kind, and it is difficult to analyze climate data without at least a simple model. Models are used to make "predictions," which is another misunderstood term, due to conflation of many different kinds of prediction; some predictions are more prophetic, or less conditional, than others. Probabilities associated with predictions can also be confusing: It is possible to associate objective probabilities with short-term weather forecasts, but not with long-term climate projections.

Uncertainty in model predictions is sometimes used to justify inaction: Climate contrarians invoke a version of the precautionary principle to argue that as long as there is uncertainty about climate change in the future, it is safest to do nothing in the present.[60] But should we ignore a credible threat simply because it is hard to quantify very accurately? Even if the quantification provided by numerical models is "bogus," as contrarians suggest, does this "stick-to-the-status-quo approach" avoid the quantification problem? We should also note that "doing nothing" is an ambiguous term. It could be argued that we are already "doing something" – emitting large amounts of carbon dioxide – even if there is some uncertainty as to the exact amount of warming it will cause. Does "doing nothing" mean that we continue to emit carbon dioxide? Or that we should stop emitting?

The argument for "doing nothing" implicitly assumes a model, one that predicts that things won't change enough to cause serious harm in the future. The assumed model is the persistence or stationarity model, which predicts that the trend in a quantity of interest, such as global temperature, is small, or essentially "zero." But this "zero" is merely a number that implies very low climate sensitivity. Why should a numeric "zero" trend prediction made by the stationarity model be considered inherently superior to another numeric trend prediction made by a different model? The proper question to ask is: Which model is most consistent with past data, while respecting all applicable constraints? The model that best satisfies these criteria is the cautious choice. Using the best central estimate for climate sensitivity obtained from a variety of models and data sources, even if uncertain, is far better than using the arbitrary default value of zero.[61]

Predictions of future climate made by models are often detailed and occasionally dramatic. But the scientific uncertainties associated with these predictions are not always preserved when these predictions are translated by the media for public consumption, as we discuss in Chapter 15.

15

Taking Climate Models Seriously, Not Literally

One often comes across articles in the news about how we have only X years left to fix climate change, or how global temperatures rising by Y degrees will doom us all, and so on. In what sense are these numbers X or Y true, and how much should we trust the exact numerical values? The media often takes these numbers from climate model predictions literally, but, as we have seen, there is a great deal of uncertainty surrounding these predictions. Climate predictions are not prophecies; they are highly conditional pronouncements that can only come true once all their conditions have been satisfied. The details and the plausibility of those conditions are often poorly described in the news.

A typical media narrative is that, if we carry on with "business as usual," we will, as predicted by a climate model, experience Y degrees of global warming (with an error bar of Z degrees). The scientific provenance of climate models lends gravitas to this numeric prediction and gives it authority.[1] But a more accurate narrative might be the following: If we continue with "business as usual," we will, according to socioeconomic models, emit a certain amount of carbon dioxide, which climate models translate into a certain degree of global warming with harmful consequences (Figure 7.4). The most uncertain aspect of this statement may not be the numeric error bar provided by the climate models, but the unquantifiable error of the socioeconomic prediction, which depends crucially on the assumptions used to deconstruct "business as usual" (Section 14.4). When we take the original media narrative literally, we downplay this nonscientific uncertainty because there are no numbers attached to it.

There are times when a model prediction needs to be taken literally. A weather forecast saying that a Category 5 hurricane will make landfall tomorrow around 7 a.m. near Houston, Texas, should be taken both seriously and literally; it is close to a prophecy (Section 14.5). Climate predictions, however, are based on a range of plausible scenarios, which should be treated as collective information. Therefore, it is not worth fixating on any single scenario or prediction: The phrase "business as usual," for instance, can describe more than one scenario. Climate models are metaphors, and their predictions should be taken "seriously, but not literally,"[2] as has been suggested by climate scientist Richard Somerville.

The problem of insufficient context in media narratives is particularly acute in areas of science that can directly impact human lives. Climate prediction is one such area. Medical research is another. The following are three headlines that appeared recently in the media:

Scientists Create Antibody That Defeats Coronavirus in Lab
Motor Neurone Disease Researchers Find Link to Microbes in Gut
Study on Cannabis Chemical as a Treatment for Pancreatic Cancer May Have "Major Impact"

At first glance, you might think these represent major advances that will soon improve human lives. But these and many other such headlines are based on animal models – in this case, mouse models. Animal models are used in biological research to study human diseases, and the findings described in these articles are often interesting and important research. It's a long way from the mouse model to the human body, though; many promising findings never complete the journey.

As a reaction to the misleading media articles in medicine, an enterprising Twitter user set up the account @JustSaysInMice. All the account does is retweet science articles adding IN MICE to the headlines, like so:

Study on Cannabis Chemical as a Treatment for Pancreatic Cancer May Have "Major Impact" IN MICE

This draws attention to what journalist Andrew Revkin calls the "tyranny of the news peg" – the tendency of the media to report dramatic-sounding scientific results without the appropriate context and caveats.[3]

Due to its human-impact potential, the field of climate prediction is also a fertile source of ominous media coverage. For example, here are three headlines:

Hothouse Earth is Merely the Beginning of the End
Earth's Climate Monsters Could Be Unleashed as Temperatures Rise
Ending Greenhouse Gas Emissions May Not Stop Global Warming[4]

The first two headlines are from articles[5] inspired by a 2018 paper on a runaway "Hothouse Earth" scenario that the paper's authors suggest could happen if global warming exceeds 2°C.[6] The articles highlight dire climate scenarios, even if they note the speculative nature of the "Hothouse Earth" paper – its climate predictions are based on amplifying feedbacks simulated by highly simplified rather than comprehensive climate models.[7] The third headline is from a 2020 press release[4] issued by an academic journal, *Scientific Reports*, about a study which found that the "world is already past a point-of-no-return for global warming."[8] The press release was later revised in response to criticisms from climate scientists[9] to add the caveat that the study used "a reduced complexity model."

In medicine, results from simple fruit fly models need to be confirmed using more complex mouse models, and then finally using experiments with real humans. Similarly, results from a simple climate model (or a single climate model) need to be confirmed with results from more complex models (or multiple models). It is a long scientific road from simple models to the real climate system.

Just to show that it is not only medical researchers and climate scientists whose work generates alarming headlines, we end this section with a headline from July 2020:[10]

> *Theoretical Physicists Say 90% Chance of Societal Collapse within Several Decades*

15.1 Deep Uncertainty: When Error Bars Have Error Bars

> *An error bar, a confidence interval, and a p-value walk into The Error Bar . . .*

Humans tend to associate words with subjectivity and numbers with objectivity. There is a stereotype that ordinary people use words, which are known to be ambiguous, while scientists use numbers, which appear precise. In *The Tyranny of Metrics*, historian Jerry Muller bemoans the fetishization of numbers in modern society.[11] Numeric metrics were introduced as an aid to thoughtful judgment, but they have slowly supplanted such judgment in many walks of life – rather like the proverbial camel that first stuck its nose in the tent and finally took over the entire tent. Before computers and the internet, it took time to collect and calculate numbers; now numbers can be generated instantly, at the touch of a button. Muller is concerned by the obsession with numeric metrics outside science, but the fixation on numbers can be an issue within science itself. Human geographer Mike Hulme writes about how "[g]overning complex phenomena through managing a single number is alluring" – be this single number the warming target for global-average temperature, or the testing positivity rate during a pandemic.[12]

The manner in which these numbers are presented can further affect our thinking. Consider a digital thermometer that displays a temperature of 37.1 and another digital thermometer that displays a temperature of 37.139. Absent other information, you might think the second thermometer is more accurate because it appears to be more precise. But precision and accuracy are not the same. Displaying more digits in the measurement, and being able to repeat that measurement, makes a thermometer more precise. But to determine accuracy, the thermometer readings need to be verified or calibrated using an independent device that also measures temperature. It is entirely possible that the measurement 37.1 is more accurate than the measurement 37.139, if the first thermometer is a better instrument, as determined by the calibration. In that case, the second thermometer would be providing false precision.

Measurement accuracy of temperature is characterized by an error bar, which typically corresponds to one standard deviation (or sometimes two standard deviations) of the measurement. The first thermometer may have an error bar of +/– 0.1, and we can express the measurement as 37.1 +/– 0.1. This means that there is a 68 percent chance that the true value lies in the range 37.0 to 37.2,[13] which is referred to as the *confidence interval*. The second thermometer may have an error bar of +/– 0.15, even if it displays more digits. We can express its measurement as 37.1 +/– 0.15, after rounding.

We are taught in school that physical measurements should be reported with error bars to indicate the level of accuracy. This thinking permeates all quantitative fields of study. While the value of the error bar associated with a measurement receives attention, the accuracy of the error bar itself is rarely examined carefully. Our assumptions about error bars associated with physical instruments do not always translate well to fields in which the instruments used are virtual – meaning that they are models of varying degrees of complexity, built using different assumptions.

As physical instruments like thermometers become more sophisticated over time, we expect their accuracy to improve. Their measurements' error bars should get smaller. This expectation holds for certain virtual instruments: As weather models have become more sophisticated over time, the accuracy of their forecasts has improved. In fact, the cone of uncertainty for hurricane track prediction has narrowed by a factor of two over the past 15 years (Figure 2.1). If the cone of uncertainty for hurricane forecasts had instead widened as weather models became more sophisticated, the widening would have implied a higher-level uncertainty, associated with the estimated cone of uncertainty itself.

What if the error bar becomes wider, or fluctuates up and down in value, as a measuring instrument becomes more sophisticated? That would go against the normal notion we have about error bars. But this happens with "error bars" or the "cone of uncertainty" associated with emergent climate properties such as climate sensitivity. As we have seen, the estimated range of climate sensitivity provided in the IPCC fifth assessment (1.5–4.5) is wider than the one in the fourth assessment (2.0–4.5). The lowering of the lower bound is attributed to an improved knowledge of historical estimates of climate sensitivity.[14] Climate models also show a fluctuating spread in estimates over time (Figure 12.2b). Even if the next IPCC assessment narrows the range, as seems likely, the previous cycle demonstrates that greater knowledge can sometimes lead to greater uncertainty. We can quantify the uncertainty of things we know about, of well-known unknowns; we cannot quantify the uncertainty of poorly known or unknown unknowns.[15] As we continue to learn more about the climate system, a future assessment could widen the range once again.[16]

Our estimates of climate sensitivity are uncertain because we use different virtual instruments, and somewhat different definitions, to measure it. The error bars fluctuate because we cannot truly calibrate these virtual instruments. Weather forecast models are directly calibrated against hundreds to thousands of observed weather events. Complex climate models used to predict climate sensitivity in future scenarios, as well as simpler models used to estimate climate sensitivity from past data, have no comparably accurate datasets against which to calibrate. We have a good idea of the mean value of climate sensitivity, but there is an implicit uncertainty associated with its error bar. We don't know what that uncertainty is; in other words, there is uncertainty about the uncertainty.[17] This type of higher-level uncertainty, difficult to accurately define and quantify, has been called "deep uncertainty."[18] In economics, this is sometimes referred to as *Knightian uncertainty* (as distinguished from Knightian risk, which is the quantifiable uncertainty associated with a probabilistic prophecy).

Most published studies on climate sensitivity include some numeric measures of uncertainty. But using a one-size-fits-all approach such as error bars or standard probabilistic confidence intervals across all types of uncertainty estimates can misrepresent the deeper uncertainties.[19] Uncertainties are not always quantifiable using numeric probabilities, so structural errors could add unquantifiable error to certain types of uncertainty estimates. It may be our "physics envy" which drives us to believe that we can ascribe accurate error bars to every climate parameter we report in our figurative "lab notebook" of climate prediction.[20]

To describe the process of climate model improvement, we introduced the category of poorly known unknowns to the Rumsfeld uncertainty matrix

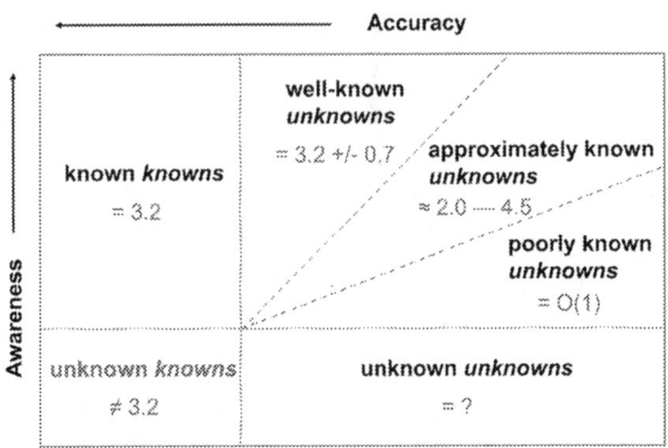

Figure 15.1 The deep uncertainty matrix: a rough mapping of the spectrum of estimates of a parameter to the modified Rumsfeld uncertainty matrix (Figure 13.1b), with a new subcategory of known unknowns. The single numeric estimate 3.2 for the parameter indicates a high degree of certainty of knowledge; the estimate with an error bar, 3.2 +/– 0.7, indicates that we know the exact statistical properties of the error for this "well-known unknown," whereas the approximate range 2.0–4.5 indicates that our error estimates are themselves uncertain for this "approximately known unknown." For a "poorly known unknown" parameter, we may only be aware of the parameter's order of magnitude, as indicated by the mathematical notation O(1) meaning of the order of 1.

(Figure 13.1). We wield our analytic knife once more to carve out another knowledge category, creating a new deep uncertainty matrix (Figure 15.1). We call the new category, in between the well-known and poorly known category, "approximately known unknowns." An approximately known unknown is a property that is less certain than something for which we would provide a standard error bar, but more certain than something about which we would say "we have no idea" or "we just know the rough order of magnitude."[21]

An approximately known parameter may be expressed as a value range, but with no implication of precise numeric probabilities for values outside the range, due to the deeper uncertainty. Climate sensitivity is such an approximately known unknown. Our knowledge of climate sensitivity is by no means exact, even probabilistically, as demonstrated by the fluctuating error bars. But we can constrain it better than we could if we knew only its order of magnitude or the sign.

Introducing more categories of uncertainty can help resolve some of the confusion about climate prediction. However, there is some reluctance among climate scientists to discuss specific uncertainties because these uncertainties are occasionally misrepresented with the intention of casting doubt on the whole enterprise of climate prediction.[22] But if the potential consequences of ignoring predictive information are severe enough, this information should not require absolute or even probabilistic certainty to be actionable.[23] As carbon dioxide concentrations continue to increase, the knowledge that climate sensitivity is approximately in the range 2.0–4.5°C, or even higher, is sufficiently scary. From our approximate knowledge of past climates and our approximate model predictions of future climate, we know that a 4.5°C change in global temperatures represents a massive disruption of the climate system. Our inability to rule out higher sensitivities is an even stronger argument for action, and our inability to completely rule out low sensitivities does not alter that conclusion.[24] The approximate range for climate sensitivity may narrow (or widen) as our knowledge increases, but we need to act on the basis of the best estimates we have now.

15.2 Precision versus Accuracy: Degrees of Warming

Uncertainty in error estimates can affect even the most precise of sciences. In September 2011, the world of physics was stunned by the announcement of an extraordinary discovery.[25] Researchers working with the OPERA underground particle detector in Gran Sasso, Italy, announced that particles known as neutrinos, coming from the CERN particle physics laboratory in Geneva, Switzerland, had arrived at their detector traveling at a speed exceeding the speed of light.

Ever since Einstein's theory of special relativity was confirmed, it has been a fundamental tenet of physics that no particle can travel faster than the speed of light. But these neutrinos had arrived at Gran Sasso 60 nanoseconds earlier than they would have if they had traveled at the speed of light. The measurements of the time of arrival were so precise that the OPERA team claimed they had a 6-sigma statistical significance (where *sigma* denotes standard deviation). A research finding with a 5-sigma statistical significance is considered the gold standard for a "discovery" in physics: It means that there is only a 1 in 1.7 million chance that the result is a fluke or a statistical fluctuation. A 6-sigma result is even better; the odds of being wrong are about the same as the odds of winning a jackpot, 1 in 500 million.

This apparent violation of the cosmic speed limit would not only prove Einstein wrong, but also upend our understanding of the universe and nullify our notions of causality. It could have been a paradigm-shifting result, the start of a scientific revolution. But other research centers were not able to replicate the result, and in February 2012, the OPERA team retracted its finding. What happened? Why retract a finding characterized as having a 6-sigma level of statistical significance? Did the OPERA team "win" the 1-in-500-million jackpot of being wrong? OPERA researchers expressed high confidence in their original result because they had checked and rechecked their measurements multiple times. Their error estimates, however, assumed that their equipment was in working order; they did not account for a hidden structural error in their experimental setup. It turned out that a fiber optic cable between a computer and a GPS unit had come loose and was to blame for the original result. The lesson we can learn is that all statistical error estimates, however precise they sound, are conditional. In the absence of replicability, they are only as good as the structural assumptions they are based on.

The purpose of highlighting this faster-than-light fiasco is not to pick on physics in particular or science in general. Structural or instrumental errors are par for the course in science, especially when the experimental setup is complex, as it was for the OPERA detector. In this case, replicability (or the lack thereof) saved the day. Other research centers did not have the same structural problem with cable connections, and therefore could not replicate OPERA's results.

As we have seen, replication is hard for climate models, which are also highly complex "instruments." We do not want all climate models to be structurally identical because we do not know the "correct" structural approximations for some very small-scale processes; we have to sacrifice replicability to preserve model diversity (Section 11.4). Furthermore, the statistical confidence levels in climate science are far lower than the 5-sigma levels of physics. A 2-sigma or 95 percent level of confidence (in other words, a 1 in 20 chance of being wrong) is usually considered pretty good.

Given these caveats, let us consider the different degrees of uncertainty associated with an important climate parameter. Which of the following *sounds like* the most accurate statement about our current knowledge of climate sensitivity (in °C)?

S1 Climate sensitivity lies in the range 2.0–4.5, with lower values extremely unlikely and higher values not likely.

S2 Climate sensitivity lies in the range 2.6–3.9, with a likelihood of 66 percent and a mean estimate of 3.2.

S3 Climate sensitivity lies in the range 2.2–3.4 (66 percent confidence limits), with a central estimate of 2.8 and with the probability of exceeding 4.5 being less than 1 percent.

S4 Climate sensitivity lies in the range 2.0–4.5, with a likelihood of 66 percent, and a best estimate of about 3.0.

S5 Climate sensitivity lies in the range 1.5–4.5, with a likelihood of 66 percent.

All of the above statements are paraphrased estimates of climate sensitivity from recent studies. S1 is a paraphrasing of the plain language summary of the 2020 WCRP assessment[26] mentioned in Section 14.3. S2 is also taken from the baseline estimate of the same assessment.[26] S3 is paraphrased from a 2018 paper that estimates climate sensitivity by combining observations and model simulations.[27] S4 is paraphrased from the Summary for Policymakers in the IPCC's fourth assessment report, where the number 66 percent has been substituted for the term *likely* used in the actual summary, as recommended by the summary document itself.[28] Similarly, S5 is paraphrased from the corresponding summary for the IPCC fifth assessment report.

Although the ranges in S2–S5 appear to be close to each other in their central estimates, they imply quite different probabilities for very low or high climate sensitivity values. Even though the IPCC does not provide a continuous probability distribution for climate sensitivity, scientists routinely fit standard probability functions to the parameters provided by the IPCC assessment (and other assessments of climate sensitivity, such as S2 and S3).[29] As we see in Figure 15.2, the fitted probability distributions look quite different for the different estimates, especially for the extreme, or "tail," values. For risk assessment, the high tail is much more important than the low tail, because it is associated with potentially catastrophic climate change.[30] We can estimate the tail probability for climate sensitivity values exceeding a high threshold, say 5.5°C: S2 implies this probability is about 1 percent whereas S4 and S5 imply that the probability is more than 10 percent. (These tail probability values are highly dependent on which probability distribution is fitted,[31] but that just emphasizes the uncertainty.)

The IPCC arrives at estimates for climate sensitivity like S4 and S5 by combining estimates that use different types of data and different methodologies. These estimates are based on paleoclimatic data from the distant past, instrumental data from the recent past, and model predictions of the future. The data-derived estimates also rely on models of different kinds; for example, some of them use climate models to estimate the radiative heating.[14] As the IPCC notes:[32]

> [t]here is no well-established formal way of estimating a single [probability distribution function] from the individual results, taking account of the different assumptions in each study. Most studies do not account for structural uncertainty, and thus probably tend to underestimate the uncertainty.

Although the IPCC narratives emphasize the deep uncertainty,[33] associating specific numeric likelihood values with categories such as "likely" and "very likely" can make the uncertainties appear shallow and purely probabilistic to the end user.

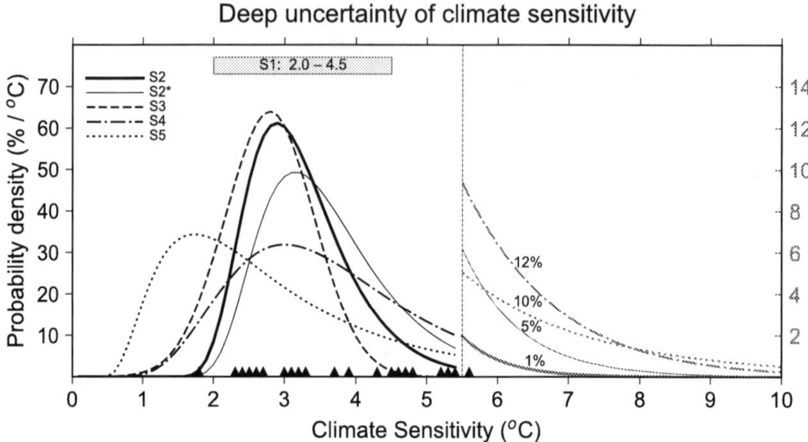

Figure 15.2 Probability density (% per °C) of equilibrium climate sensitivity corresponding to S2 (thick solid), S3 (dashed), S4 (dash-dot), and S5 (dotted) estimates (see text). The thin solid line displays a variant of the S2 estimate (S2*) based on a different statistical assumption (called *uniform S prior*). The gray rectangle on top denotes the S1 range of climate sensitivity. The solid triangles on the x-axis denote climate sensitivities estimated by the latest generation of climate models (for IPCC AR6). *Note that an expanded y-axis scale is used for climate sensitivities greater than 5.5°C.* Percentage numbers show the tail probability of climate sensitivity exceeding 5.5°C, that is, the area under the corresponding curves (except for S3, which implies essentially zero-tail probability). (A lognormal probability distribution was fitted for all cases, except for S3, where a normal fit was used.)

Even data-derived estimates of the range of climate sensitivity can change over time, with new data and different analysis techniques.[34] This means that uncertainty estimates associated with climate sensitivity should perhaps be interpreted as a lower bound:[35] any downstream application, such as a risk assessment model, should take into account a possible increase in the uncertainty estimates.[36] However, the uncertainty in the physical parameters of climate change will be compounded by the uncertainty associated with the choice of emission scenario (Figure 7.4)[37] and potentially dwarfed by even greater uncertainties associated with economic measures of long-term climate impacts.

We can try to reconcile the differing estimates of uncertainty by invoking the distinction between accuracy and precision. To illustrate this distinction, consider a game of darts. The objective of the game is to hit the center of the dartboard, the bullseye. Precision is the ability to hit the same location on the dartboard each time, whether or not that location is the bullseye. Accuracy is the ability to hit the bullseye each time. The darts of a precise player will all land close to one another, even if none of them are near the bullseye. The darts of an accurate player will cluster around the center, even if they are not as close to one another as the darts of a precise player. (In the case of the "faster-than-light" neutrinos we discussed, the OPERA team was landing its darts extremely precisely, but away from the bullseye, as later established by independent measurements.)

Estimating climate sensitivity is like playing a game of darts, except that we don't know exactly where the bullseye is. The Anna Karenina principle (Section 11.1), the assumption that the errors associated with different estimates are random and therefore tend to cancel out, provides a heuristic basis for estimating the average value of climate sensitivity, but that simple cancellation argument does not necessarily extend to estimating the uncertainty in climate sensitivity. Climate data, as interpreted using data-analysis models, provide a rough idea of where the bullseye may have been in the past,[38] and climate predictions, using more complex models, provide a rough idea of where it may lie in the future – but the two may not quite agree (Section 14.3).

The fact that statements S2–S5 differ from one another means that they cannot all be accurate estimates of uncertainty. But they could all be conditionally precise – each reported confidence interval being conditional on the assumptions it is based on. Such conditional estimates of precision can be very useful in comparing a particular study to other studies that make similar assumptions, to see if the precision of that specific approach has improved. It is less meaningful to compare these estimates across different approaches[39] because each approach may have different types of unquantifiable structural uncertainties.[40]

The association between verbal quantifiers such as "likely" and numeric probabilities such as "greater than 66 percent" was introduced by the IPCC to avoid uncertainty in how the verbal quantifiers were interpreted. But if there is real uncertainty about the numeric probabilities, then this association may serve to obscure that uncertainty. It may be preferable to keep statements about climate sensitivity as minimalistic as possible, like S1, associating no specific numbers with qualitative assessments of probability (such as "extremely unlikely"). This is an Occam's Razor argument for displaying data simply, without the complexity of adding uncertain error bars. If numeric uncertainty estimates are provided, their conditionality should be emphasized strongly. If this conditionality is not emphasized, the danger is that the uncertainty estimates may be misinterpreted as absolute measures of accuracy (in other words, as probabilistic prophecies) by those outside the field of climate science. This shouldn't matter for the central estimate of around 3°C, which ought to be taken seriously. But a literal interpretation of the error bar would assign exact numerical values to the tail probability of climate sensitivity exceeding a threshold value (such as 5.5°C), which could lead to underestimation or overestimation of risk in impact assessments.[30] In dealing with the climate crisis, we may need to accept the deep uncertainty and try to use risk-assessment methodologies that do not rely heavily on numerical estimates of the tail probabilities of climate change.

15.3 Pitfalls of the Uncertainty Trough

> *If you like laws and sausages, you should never watch either one being made.*
>
> Otto von Bismarck

Sausage-making is a messy process, as workers at sausage plants know. But what customers see is a neatly packaged product they can take home, cook, and eat. Those who are intimately aware of how something is created often have a different perception than those who only deal with the final product.

This phenomenon has been studied in the context of missile technology.[41] The actual designers of the missile, the "insiders," were more aware of the uncertainties in the missile than were those who simply worked with the finished product, such as the managers at the company making the missiles, who expressed confidence in the product. But those further removed (the "outsiders"), including rival missile-making companies or those simply opposed to the use of missile technology, expressed less confidence in the product than the managers did.

As the social distance from knowledge production increases from "insiders" to "outsiders," uncertainty initially decreases, reaching a minimum, before increasing again. This means that there is a trough in the uncertainty. This is commonly (and somewhat confusingly) referred to as the "certainty trough" because in many fields, such as missile technology, the discussion revolves around certainty rather than uncertainty. However, we will refer to it as the "uncertainty trough" because the minimum does occur in the uncertainty. As noted at the beginning of this chapter, the media often ascribes more certainty to individual model predictions, especially for simplified models, than does the climate science community as a whole. (Developers of individual models often defend their models vigorously, though.[42]) Contrarians, on the other hand, tend to be dismissive of the results from any and all climate models.

In their book *Climate Shock*, economists Gernot Wagner and Martin Weitzman[43] highlight the uncertainty in economic modeling related to global warming and make the case for strong climate action despite, or because of, that uncertainty. But amidst all that economic uncertainty, they say, "one thing we know for sure is that a greater than 10 percent chance of eventual warming of 6°C or more – the end of the human adventure on this planet as we now know it – is too high. And that's the path the planet is on at the moment." They state that they are "taking the IPCC probability distributions literally" because the (fifth) IPCC[44] assessment ascribes "a chance of anywhere between 0 and 10 percent" to the possibility of a 6°C eventual warming.

These economists are doing what is fairly common – assuming that the IPCC assessment provides a probabilistic measure of uncertainty, a probabilistic prophecy, without worrying about the uncertainty of the uncertainty itself. Experts often take error estimates for parameters from areas outside their domain of expertise at face value;[45] they treat them as well-known unknowns, not fully appreciating the uncertainties. The IPCC assessment includes numerous caveats about how its uncertainty estimates are computed, but these are not always meaningful to nonexperts.

The baseline error estimate for climate sensitivity from the 2020 WCRP assessment, which corresponds to statement S2 in Section 15.2, has a much narrower range than the previous IPCC assessments. If we interpret S2 literally, as quantifying probabilities, the probability of climate sensitivity exceeding 5.5°C decreases to less than 1 percent, an order of magnitude lower than in the fifth IPCC assessment (Figure 15.2). The WCRP assessment presents alternative estimates that assign higher probabilities to the climate sensitivity "tail," but these alternative values only underscore the uncertainty of tail probabilities.

To complicate matters further, while the WCRP assessment has narrowed the estimated range for climate sensitivity, climate models for the next IPCC assessment have widened the estimated range (Figure 12.2b). Some of these newer climate models have climate sensitivity values exceeding 5.5°C.[46] According to S2, such a high value of climate sensitivity should have a less than 1 percent chance of occurring, and, according to S3, a near-zero chance (Figure 15.2). But we cannot simply dismiss the model results, because the WCRP assessment may not be the last word on data-based estimates of climate sensitivity, and we still need climate models to assess regional impacts and risk.

The practical implication for climate impact assessment is that the probabilities associated with very high values of climate sensitivity, such as those exceeding 5.5°C, may be just as uncertain as many economic parameters: They are poorly known unknowns. However, the midrange estimates of climate sensitivity in the vicinity of 3°C are approximately known unknowns that have stood the test of time.[47] That knowledge should be worrying enough to motivate strong action to mitigate climate change.

15.4 "OK Doomer": The Power of Numeric Prophecies

please stop saying something globally bad is going to happen in 2030. Bad stuff is already happening and every half a degree of warming matters, but the IPCC does not draw a "planetary boundary" at 1.5°C beyond which lie climate dragons.

Myles Allen,[48] Coordinating Lead Author, ch. 1, *IPCC Special Report on Global Warming of 1.5°C*

The climate crisis engenders a wide spectrum of opinions among the population, ranging from those who – motivated by climate denialism or contrarianism – support inaction to those who demand radical action to mitigate climate change. A new strain that has emerged in recent years is climate doomism,[49] which goes beyond radical action to fatalism and thus demands no action, completing the circle back to contrarianism. Climate doomers often take the worst-case numeric climate predictions from complex models too literally – or else they put too much faith in simple, poorly constrained models exhibiting tipping-point behavior. Such doomist beliefs can have a behavioral impact. A recent study[50] quotes a 27-year-old as saying, "I feel like I can't in good conscience bring a child into this world and force them to try and survive what may be apocalyptic conditions." A weaker version of climate doomism is climate anxiety, a form of psychological trauma that "needs a therapeutic response to help people beyond paralysis and into action."[51]

The climate crisis is certainly an extraordinary threat to the environment and society, unlike anything we have encountered in the history of human civilization. But is it also an existential threat? To be more specific: Is it an imminent existential threat to the entire human species, or even to all life on Earth? Some describe the climate crisis as if it were so,[52] and often invoke numeric predictions from climate models as a justification.[53] Pronouncements like "only 12 years remain" have been made to describe the time we have left to act to mitigate climate change. "12 years" refers to a 2018 IPCC report stating that strong action needs to be taken before 2030.

Another example is the 2017 *New York Magazine* article "The Uninhabitable Earth,"[54] which begins with a section titled "Doomsday" and subtitled "Peering beyond Scientific Reticence." The article refers to the latest

IPCC report as the "gold standard" but also says that its "authors still haven't figured out how to deal with that permafrost melt." Dismissing the serious threat of the most confident climate projection as "just a median projection," it focuses on the worst-case climate projection. The article went viral, and such prognostications of impending doom continue to influence activism.

Many climate scientists who are strong advocates of mitigation efforts disagree with such doomsday assessments. One prominent climate scientist wrote, in response to the article: "The evidence that climate change is a serious problem that we must contend with now, is overwhelming on its own. There is no need to overstate the evidence, particularly when it feeds a paralyzing narrative of doom and hopelessness."[55] Another wrote: "It's just too easy to prove [the article] wrong and hence imply that the entire climate change issue is exaggerated."

Doomsday scenarios that are perceived as being more certain than they actually are can make preventive efforts appear impossible, feeding into a self-fulfilling narrative of fatalism rather than mitigation. In 2018, an article by Jonathan Franzen appeared in the *New Yorker*, with the subheading: "The climate apocalypse is coming. To prepare for it, we need to admit that we can't prevent it." It is a well-intentioned piece, but Franzen includes statements such as: "When a scientist predicts a rise of two degrees Celsius, she's merely naming a number about which she's very confident: the rise will be at least two degrees. The rise might, in fact, be far higher."

The implication is that scientists are too reticent, that they lack the imagination to contemplate extreme scenarios. Scientists often explore extreme scenarios and even make bold predictions: Astronomers, for instance, study how the slowly expanding sun will become a red giant that could engulf the Earth in five billion years. But scientific imagination must be constrained by the laws of physics.[56] We could write a script about a large portion of the globe freezing up in a matter of hours, as in the 2004 movie *The Day after Tomorrow*, but physically speaking, the scenario is impossible. Similarly, we could construct a narrative about the Greenland ice sheet completely melting over the next decade, but that too is physically impossible.

Heisenberg bottled it up and Lorenz tried to banish it entirely, but, even today, there is something intoxicating about determinism. When it isn't served to you, there is the temptation to brew it yourself. In attempting to compensate for presumed scientific reticence by treating a conditional prediction as an unconditional prophecy, and then proceeding to embellish it enough to end on a pessimistic note about mitigating climate change, Franzen's essay achieves little. The end result of such climate doomism is exactly the same as that of climate contrarianism: paralysis of certainty in one case, and paralysis of uncertainty in the other.

David Wallace-Wells, the author of the "The Uninhabitable Earth," published a follow-up article in 2019,[57] titled "We're Getting a Clearer Picture of the Climate Future – and It's Not as Bad as It Once Looked." While emphasizing that the climate outlook is still very bad, the new article added that

> given recent drops in renewable pricing, and the positive signs for coal decline in the developed world, as a prediction about energy use RCP8.5 is probably closer to a "worst case," outlier scenario than anything it would be fair to call "business as usual."

This article did not go viral. Extremely dire climate predictions attract far more attention than climate predictions that are merely dire, even when the latter predictions are more plausible.

Media reporting on climate science performs an important function that cannot and should not be left to scientists alone; journalists and writers translate the science of climate change into the public urgency of the climate crisis, turning numbers into narratives and digits into disasters. But that translation effort comes with its pitfalls, especially if the translator misses the scientific nuances of the original text or embellishes it for rhetorical effect.

The articles discussed in this section used predictions from comprehensive climate models, which tend to be less dire than those from some reduced-complexity models. As we saw in the beginning of this chapter, simplified models can produce dramatic predictions, which, if interpreted literally, can further feed the narrative of doomism. The self-published 2018 essay "Deep Adaptation: A Map for Navigating Climate Tragedy," which is premised on the "runaway climate change" predicted by simple climate models, has sparked a movement with thousands of followers.[58] To motivate its prophecy of social collapse,[59] the essay argues that there are climate tipping points associated with amplifying feedbacks from Arctic ice melt and methane release. Scientists use simplified models to study these tipping points because they deal with phenomena occurring on timescales of hundreds to thousands of years, beyond the reach of today's complex climate models. These simplified models can overemphasize the role of amplifying feedbacks because the simplification process may omit stabilizing feedbacks (Section 8.4). Numeric predictions from highly simplified models should not be taken very seriously until they can be replicated with predictions from more comprehensive, better-constrained models.

The contrarian counterpart to climate doomism – namely, the narrative that fighting climate change will be a colossal waste of resources or will devastate the economy[60] – often relies upon numeric predictions as well. The Danish author Bjorn Lomborg accepts that "climate change is a real problem with substantial and growing overall negative impacts toward the end of the century,"[61] but he dismisses emission mitigation efforts by invoking numeric predictions from economic models, such as the estimate (taken from a single study of New Zealand's economy) that reaching zero carbon emissions would optimistically cost "a whopping 16% of GDP each year by 2050." Likewise, the Global Warming Policy Foundation, a contrarian think tank, warns that the cost of decarbonizing the British economy "will be astronomical" and reach more than 3 trillion pounds.[62] Uncertainties in long-term economic modeling mean that other economic studies do not support these dire predictions of economic collapse caused by climate change mitigation efforts.[63]

Climate activists at times criticize climate scientists for being too cautious in their assessments of climate change scenarios and impacts.[64] These activists sometimes defend their playing up of the potential extreme impacts and playing down of the actual deep uncertainty as being necessary to motivate action on the climate crisis: The end, to them, justifies the means. Justified or not, these actions have called attention to the urgency of the climate crisis, which is a positive outcome.[65] But what's sauce for the goose is sauce for the gander: When we quietly acquiesce to those activists who go beyond scientific evidence, we normalize the same practice for "inactivists"[66] and allow them to focus on unrealistically benign climate change scenarios, cherry-pick scientific arguments, and selectively quote numbers from uncertain economic projections. The end result could be the devaluation of science altogether, leading to a "post-science normal" in which decisions are driven by rhetoric, not logic. And if unscientific decision making becomes the norm, we may end up supporting strategies that appear bold and climate friendly but end up being ineffective,[52] risking delay and even failure in confronting the climate crisis. The road to climate hell may be paved with good intentions.

***

Ultimately, science is not about numbers but about systematic reasoning. Science evolved from natural philosophy, or systematic thinking about nature. Logical reasoning is at the core of both philosophy and science. Numbers are merely tools that scientists use to aid their logical reasoning. Instead of simply saying that our numbers are better than their numbers, it would be far more effective for us to argue that our reasoning is sounder than their reasoning. To do this, though, we must carefully examine the logical arguments and assumptions that lie behind the numbers. If we simply accept numbers at face value, they can become a distraction from logical reasoning, rather than an aid to it. When we consider numbers produced by models, we should remember that models can differ greatly from one another in structure. There are simplified climate models, comprehensive climate models, economic models, statistical models, and so on. Some models incorporate more physical constraints and processes than others; some are more verified than others. Depending upon how they are calculated, different model-derived numbers have different types of uncertainty attached to them, and it is not always possible to express that uncertainty in terms of numeric probabilities. Narrowly focusing on reducing the quantified uncertainties can distract attention from the potentially larger unquantifiable uncertainties inherent in models of complex systems.

We must be careful in interpreting the precision provided by numbers in any field of study in which replicability is hard and predictions cannot be verified. When error bars are presented, it is worth asking: What are the error bars of the error bars? This caveat applies to climate science, not because climate research is more cavalier than other types of research, but because climate scientists necessarily grapple with inescapable structural uncertainties. While central estimates for parameters like climate sensitivity are robust despite these uncertainties, it may be wise to take the numbers reported for low probability tails with a large grain of structural salt.

PART III

The Future

16

Moore's Law

To Exascale and Beyond

To discuss the future of climate prediction, we return briefly to two centers of climate modeling: GFDL in Princeton, New Jersey, and NCAR in Boulder, Colorado. For many years, these two research labs housed state-of-the-art supercomputers within the building for use by weather and climate modelers. In 2011, the GFDL computer moved to Oak Ridge, Tennessee, and, in 2012, the NCAR computer moved to Cheyenne, Wyoming. In the early days of computing, climate modelers like Akira Kasahara had to travel across the continent to physically visit a supercomputer in order to access its computing power (Section 4.2). But the advent of the Internet rendered such trips unnecessary. As supercomputers became larger and more power hungry, it became more cost effective to host them remotely in the most affordable locations.

Electric power is a key player in the future of computing and climate prediction. Electric power also played a key role during the Second World War in creating the town of Oak Ridge, Tennessee, the current home of the GFDL supercomputer. Following the bombing of Pearl Harbor in 1941 and the United States's subsequent entry into the war, the federal government secretly initiated the Manhattan Project in 1942 with the goal of building an atomic bomb. This required the conversion of naturally mined uranium to weapons-grade enriched uranium, a process that consumed enormous amounts of electricity and water. The government searched for a discreet site at which to process uranium; the site needed to have access to abundant electric power and copious amounts of water. The Tennessee Valley Authority, a multi-state power utility, was a major source of both – thanks to some recently completed dams.[1] A secret city was soon created in Tennessee, west of the Great Smoky Mountains, to play host to a uranium-processing facility. This facility, its true purpose known only to the Manhattan Project, was given the innocuous-sounding name "Clinton Engineer Works."

At their peak, the plants at Clinton Engineer Works consumed about 280 megawatts (MW) of electricity, one-seventh of the total power output of the Tennessee Valley Authority, and about 1 percent of the total US electric power consumption at that time.[2] To put that in a modern context, in 2019, the United States consumed an average of 500,000 MW of electricity[3] – 1 percent of that would be 5,000 MW, which is about what was consumed by New York City, home to more than 8 million people.[4] The government directed the equivalent of the power capacity of a major city to just one part of its plan for an atomic bomb.

Clinton Engineer Works later became the Oak Ridge National Laboratory (ORNL), part of the US Department of Energy. The laboratory currently houses some of the fastest supercomputers in the world. These computers, like the uranium-processing plants before them, are also gluttons for electric power. The power needs of a supercomputer are determined by the sum of the power consumed by the thousands of components used to make it. The high-power vacuum tubes of early computers were replaced by low-power computer chips in the 1960s, drastically reducing the power requirement. But, since then, the complexity of these computer chips has grown exponentially, and so have the power needs of supercomputers.

16.1 Making Chips That Don't Fry

In the late 1940s, the transistor was invented to replace the vacuum tube. The transistor modulates the flow of electric current though semiconductors to carry out bit-wise logic and arithmetic operations; it is the basic building block of modern electronics. This development was followed soon after by the invention of miniaturized transistors that could be combined to a make an integrated circuit or a "computer chip." One of the most important concepts in computer chip technology is *Moore's Law*, which states that the number of transistors per square inch of a computer chip approximately doubles every year. This prediction was made in 1965 by Gordon Moore, the cofounder of the chip manufacturer Intel Corporation, based on trends in transistor technology at the time. Since the cost of producing a chip was roughly constant, the implication was that the performance of the chip would double every year for the same price.

Although it is called a "law," Moore's statement was more of an empirical prediction based on trends at the time – and, as an inductive prediction, Moore's Law is not written in stone. In 1975, Moore himself revised his prediction based on more recent trends, increasing the doubling time to two years. Remarkably, that prediction held for more than 40 years, heralding the revolution in personal computing and leading to even more compact devices, including the smart phone and the Internet-of-Things.

It is not just the number of transistors on a computer chip that matters for performance but also the electric power those transistors consume: Too much power consumption can "fry" the chip, that is, heat it beyond its operating temperature and damage it. The area of a computer chip is measured in square millimeters, and the power it consumes is measured in watts. Recall that the solar energy flux that reaches the Earth is of the order of a thousand watts per square meter, which works out to a milliwatt per square millimeter. This energy is a thousand times weaker than chip power magnitudes; in fact, chip power magnitudes are closer to the energy flux at the surface of a star.[5] Efficient heat sinks dissipate this enormous power and prevent the chip from getting fried.

Wouldn't putting more transistors on a chip increase its power consumption? Not necessarily. An IBM engineer named Robert Dennard observed in 1974 that the power consumed by the chip remains proportional to the area. This rule came to be known as *Dennard Scaling*. As transistors become smaller, voltage and current must decrease in order to keep chip power consumption the same. Combining Moore's Law with Dennard Scaling, we can predict that chip performance will double every two years for the same power consumption, that is, per watt of power.

Transistors becoming smaller also meant that electric signals had shorter distances to travel, so their operating frequency could be increased, and they could carry out more calculations per second. Recall that the first computer, ENIAC, carried out 400 calculations per second using vacuum tubes. The first transistor-based supercomputer built in the mid 1960s, the CDC 6600, could carry out 3 million arithmetic operations per second. Modern desktop computer chips, operating at 3–4 gigahertz (GHz) frequency, are a thousand times faster than the CDC 6600, carrying out billions of arithmetic operations per second.

By the mid 2000s, though, Dennard Scaling began to break down. Transistors had gotten so small that leakage of current and its associated power dissipation became a problem.[6] The operating frequency of chips could not be increased beyond about 4 GHz. You may have noticed this yourself; the advertised GHz frequency of computer chips in laptops and desktops hasn't changed much since the late 2000s. What has changed is the number of processing *cores*. Now you can buy computers with four cores or eight cores on the same chip. Each processing core is essentially a separate calculating unit. A multicore computer can carry out multiple calculations at the same time. Manufacturers are putting more cores on each chip to get around the limitations on operating frequency. But often, in order to keep power consumption low, not all parts of a chip are active at the same time, leaving "dark silicon."[7] Having dark silicon means that the chip is not operating at full capacity: An eight-core computer isn't going to deliver eight times the performance of a single-core computer.

Moore's Law slowed down in the 2010s, as putting in ever more transistors per square inch of a chip became harder and harder. The cost of fabricating chips at seven-nanometer scales, the current "bleeding edge" technology, is prohibitively expensive, and there are very few manufacturers with that capability. The performance of individual chips is likely to plateau, and supercomputers will no longer be able to rely on ever-faster chips to speed up computations.

16.2 A Brief History of Supercomputing

The name "supercomputer" is given to members of the fastest class of computers at a given time. In 1945, the ENIAC, which used vacuum tube technology, was the first and only supercomputer. During the 1950s, large computers called mainframes (like the IBM 701 used by Suki Manabe at GFDL), as well as smaller desk-sized computers (like the one used by Ed Lorenz at MIT), were built using this technology. But vacuum tube technology was large, power hungry, and unreliable, making it harder to build more powerful computers using more vacuum tubes. It was only after the invention of the transistor and computer chip that further progress could be made.

The computer market gradually trifurcated into three main categories: at the high end, supercomputers catering to military and scientific users; in the middle, mainframes for business use; and, at the low end, personal computers. Supercomputers used a small number of expensive, custom-designed chips, with elaborate cooling systems, to achieve high performance at a high cost. Personal computers used cheap, bulk-manufactured commodity chips to deliver lower performance at a lower cost. But, as predicted by Moore's Law, the performance of commodity chips increased exponentially over time, doubling the consumers' bang for their buck every two years.

From the 1960s through the 1980s, supercomputers were made by companies, such as Control Data Corporation and Cray, that designed specialized chips and architectures to deliver high performance. To create a supercomputer, a small number of these fast processor chips, typically 4–16, were connected together, with shared memory. This configuration, known as *shared memory architecture*, allowed supercomputers to carry out dozens of operations simultaneously.

In the 1990s, the performance of cheap commodity chips, like the ones made by Intel Corporation, started to approach that of expensive custom-designed chips. Supercomputing switched to a *massively parallel architecture*: Thousands of inexpensive commodity chips were linked together using fast network links. The supercomputer could now carry out thousands of operations simultaneously. But these massively parallel computers created new programming challenges. A calculation which requires a sequence of steps where each step depends on the result of the previous step (called a *sequential* or *serial* calculation) cannot be carried out in parallel. Weather and climate models need to carry out sequential calculations: This is how, for instance, they determine whether a cloud is going to form and whether it is going to rain. Models' reliance on sequential calculations creates bottlenecks that prevent them from being able to fully exploit the power of a massively parallel computer.

Another challenge of massively parallel architecture has to do with the exchange data between the interconnected computer chips, or "nodes," in the parallel computer. Each node in a parallel computer has its own memory. The grid boxes in a climate model can be divided up and assigned to different nodes. According to the laws of physics, air movement in one grid box depends upon the pressure in the neighboring grid boxes. If the neighboring grid boxes are also assigned to the same node, then the node can carry out the calculation on its own. But if one of the neighbors happens to be assigned to a different node, then one or more nodes must communicate in order to determine air motion. Communication between nodes takes time, creating bottlenecks that slow down the calculation. As the number of nodes increases, the bottlenecks become worse because there are fewer grid boxes assigned to each node.

Due to these parallelization and communication bottlenecks, climate models typically end up using only a small fraction of the peak performance of a massively parallel computer. In the 1980s, models could achieve 40–50 percent of the peak performance of supercomputers with shared memory architectures, but today they manage to use barely 5 percent of the peak performance on a massively parallel computer.[8]

From the 1960s to the 1980s, the needs of climate modeling and other high-end scientific applications influenced the design of the custom chips used to make supercomputers. Weather and climate modelers were among the first customers for these boutique products. With the transition to massively parallel architecture using commodity chips, that was no longer the case. The mass-market needs of personal computers were the major driving force for the design of these chips. But starting in the 2010s, supercomputer design began to be influenced by a completely different set of needs.

16.3 Chip Wars: March of the GPUs

Video games and movie animations use some of the most powerful computer chips in the world to generate the most realistic three-dimensional graphics possible. Rendering photorealistic graphics is a very demanding task for a computer. Different views of the action, along with light and shadow patterns, need to be captured at various perspectives. This requires the tracing of millions of light rays and their reflections, which can be expressed as large matrix operations – millions of simple multiplications and additions that follow a predictable pattern.

The normal chips at the heart of a laptop or desktop computer, known as the Central Processing Unit, or CPU, are capable of performing one complex set of operations, which often follow an unpredictable sequence. Central Processing Units display web pages on your browser, protect you from cyber intruders, play music, and notify you of new emails, switching between these different tasks so rapidly that you don't notice the switching. These general-purpose CPUs can also render the graphics for video games using matrix operations, as they did for early video games. But those graphics were primitive because a single CPU could render only so many images while also managing other operations.

Central Processing Units were too sophisticated for the purpose of graphics rendering. Soon enough, much simpler and cheaper chips appeared, specifically intended for the repetitive and predictable operations used to render graphics. These chips were called Graphical Processing Units or GPUs, or simply "graphics cards." Rather than the one complex calculation that a CPU could perform, a GPU could perform many simple calculations simultaneously (in parallel) while consuming roughly the same amount of total power as a CPU. Graphical Processing Units run at substantially lower clock frequencies than the CPU, which means that each GPU addition or multiplication is much slower than the corresponding CPU operation – but since many of the GPU additions and multiplications can be done in parallel, the total number of operations performed by the GPU in a given amount of time is much larger than the total number of operations performed by the CPU in the same amount of time.[9]

With the addition of the GPU, graphics became much more realistic. Modern desktop computers now have a CPU or two, each with a handful of cores (typically less than eight). Each core can perform a single complex calculation at a time. The powerful GPU in the computer can perform numerous, far simpler calculations simultaneously, and it is used as needed to render graphics. When the user plays a video game, the CPU handles the logic of the game and offloads the repetitive task of rendering graphics to the GPU. The CPU tells the GPU what to calculate, and it gathers the results from the GPU to display them on the screen for the user.

It didn't take long for programmers to realize that GPUs could be used for purposes besides rendering video game graphics. They consumed less power per operation than CPUs and were cheaper to produce, which meant any scientific calculation that required predictable and repetitive operations like matrix multiplication could be performed more cheaply and rapidly by GPUs than by CPUs. Configurations of the graphics chips used for non-graphics calculations soon acquired the (somewhat oxymoronic) name of General-Purpose Graphical Processing Units, or GPGPUs, which are usually coupled with a software framework that enables them to be used for general-purpose programming. As chip advances slowed in the late 2000s, computer architects realized that GPGPUs, which were now referred to simply as GPUs, provided a way to build powerful supercomputers that didn't consume outrageous amounts of power.

Chips originally designed for video games thus ended up influencing the architecture of supercomputers. Typically, each node in a supercomputer now has a few CPUs with several GPUs attached to it. The GPUs in a supercomputer perform well when there is more work assigned to them, but they perform poorly for lower workloads with little parallelism. When there is less work to be done, it is faster to assign all of it to the versatile CPUs than to the slower GPUs.

A barrier faced by GPUs in supercomputers is the time and energy it takes to transfer data between the processors and the computer memory. A metric called *arithmetic intensity*, with units of floating-point (or arithmetic) operations per byte (flops per byte), is used to characterize this. It is defined as the ratio of the number of arithmetic operations performed for a computation to the number of bytes of data transferred to and from memory during the computation. GPUs are most efficient when the arithmetic intensity is high because memory access is slow. But climate models tend to have low arithmetic intensity, and memory bandwidth can limit their performance.[10] This makes it harder to use modern supercomputers for climate predictions than for certain other scientific tasks.

16.4 Topping the Computing Charts

Twice a year, in June and November, supercomputer aficionados around the world eagerly await the release of a list known as the TOP500, which is their equivalent of music popularity charts or the Academy Awards. The TOP500, true to its name, lists the 500 most powerful supercomputers in the world. From 2013 to 2018, Chinese computers occupied first place on this list. In June 2018, an American supercomputer regained the top spot. This computer, called Summit, was located at ORNL (Figure 16.1). Summit is a collection of black refrigerator-like cabinets, spread over 9,000 square feet of space and weighing more than 340 tons in total.[11] It occupies about six times the area occupied by the first computer, ENIAC, and is 10 times the weight. But Summit is about 500 trillion times faster than the ENIAC and consumes just 100 times more power. In terms of energy efficiency, computers have come a long way since their vacuum tube days.

Summit consists of 4,608 compute nodes.[12] Each node is like a desktop computer with two chips or IBM Power9 CPUs, each having 22 cores. That makes for a total of more than 200,000 CPU cores. In addition, every node has six NVIDIA GPUs attached. Each GPU has the equivalent of 80 GPU cores, which makes for a total of more than 2 million GPU cores. The theoretical peak performance of Summit is 200 petaflops. (A computer that performs at one petaflop carries out a quadrillion floating-point arithmetic operations per second.)

Summit was tested by running a standard code called Linpack, which is short for Linear Algebra Package. Linpack essentially tests the ability of a

Figure 16.1 Summit supercomputer at Oak Ridge National Laboratory (ORNL). (Photo: Carlos Jones/ORNL/cc-by-2.0)

computer to perform matrix multiplications, which are predictably repetitive operations that can be handled well by GPUs. For this code, Summit was able to deliver close to 150 petaflops, which is about 75 percent of its theoretical peak performance. It used all its GPU cores and consumed 10 MW of power to deliver this level of performance.

Supercomputing suffers from a serious case of the Red Queen Effect. Computers don't stay first in the TOP500 for very long. The next milestone for supercomputers is Exascale: The race is on to achieve 1 exaflop (1,000 petaflops) of performance, which would be a quintillion operations per second. If one scaled the power linearly from Summit, an Exascale machine would consume about 60 MW of power, which would be a little more than 0.01 percent of the electric power consumed by the entire United States. The current generation of climate models is not well-suited to use an Exascale machine efficiently, but the United States's newest climate model, the Energy Exascale Earth System Model (E3SM), is being developed with the goal of making efficient use of emerging Exascale computer architectures.[13]

Further down the 2019 TOP500 list, at number five, was a supercomputer named Frontera, which is located at the Texas Advanced Computing Center in Austin, Texas.[14] Frontera consists of 8008 compute nodes, each with two Intel Xeon chips, each of which in turn has 28 cores. It boasts a Linpack performance of 24 petaflops, using nearly 450,000 CPU cores in total.

Summit had a price tag of $200 million, which is more than triple Frontera's cost of $60 million. Summit's Linpack performance is six times that of Frontera, which sounds like a bargain. But Summit has only half the number of regular CPU cores as compared to Frontera. It is not uncommon that computer code is unable to use the GPU cores, and, for this code, Summit could be six times more expensive to use. The hardware cost of Summit dominates over its energy cost. If all Summit did was run Linpack for five years, its expected lifetime, it would cost just about $40 million (using the ballpark estimate of $1 million per year for 1 MW of electricity consumption[15]).

Linpack performance often does not translate to real-world performance for other scientific code that doesn't rely on predictable matrix-like operations. A different piece of code, High Performance Conjugate Gradient, or HPCG, is used to test the performance of supercomputers for more real-world-like applications.[16] Unlike Linpack, which emphasizes arithmetic operations, HPCG shifts the focus to memory access. Summit achieved only about 3 petaflops in the HPCG test, about 2 percent of its Linpack performance.

State-of-the-art climate models have been run on both Summit and Frontera. Although it is difficult to make apples-to-apples comparisons, due to the optimizations involved, the current generation of climate models appears to run substantially faster on Frontera because it has twice the number of regular CPU cores as compared to Summit. Climate models involve many operations that do not parallelize very well, resulting in poor performance on GPUs.[17] This may change with a future generation of climate models that are better equipped to use GPU cores, but that would require either extensive recoding of the current algorithms used in the models or developing new GPU-friendly algorithms.

In June 2020, a Japanese supercomputer named Fugaku displaced Summit from the pinnacle of the TOP500 list. Its Linpack performance was nearly three times than that of Summit, at 415 petaflops, but it is still not an Exascale machine.[18]

16.5 Care and Feeding of the Growing Climate Demon: Economics and Energetics

At the start of the Cold War, the US Air Force had nearly 1,400 long-range bombers. At the end of the Cold War, around the year 1990, the number had come down to about 400. There are soon expected to be around 140. Not coincidentally, bombers have gotten far more expensive. B-2 stealth bombers cost $2 billion a piece, and even other bombers cost more than half a billion dollars each. The rising price reflects the ever-growing complexity of the modern bomber. Extrapolating trends in aircraft costs compared to overall economic trends, Norman Augustine, the former CEO of the US defense contractor Lockheed-Martin,[19] predicted the following:

> In the year 2054, the entire defense budget will purchase just one tactical aircraft. This aircraft will have to be shared by the Air Force and Navy 3 1/2 days each per week except for leap year, when it will be made available to the Marines for the extra day.

This tongue-in-cheek prediction, also known as Norman Augustine's Law XVI, highlights the growing cost of increased complexity, which eventually leads to questions of affordability.

Top-of-the-line supercomputers have also been increasing in price, even adjusted for inflation. The ENIAC cost $500,000 in 1944 ($7 million in 2019 dollars). Among the most successful supercomputers, the Cray-1, built in 1984, cost $8 million (about $30 million in 2019 dollars). Summit, commissioned in 2018, cost $200 million. The Exascale replacement for Summit expected in 2021, called Frontier, is budgeted at $500 million, about the cost of a new bomber.

As computer costs have soared, so has the complexity of climate models. A version of the model used by NASA scientist Jim Hansen to make his 1988 global warming prediction is still available and is maintained for educational purposes. It consists of about 25,000 lines of code. A state-of-the-art climate model today consists of more than 1.5 million lines of code – a 60-fold increase in 30 years.

To assess climate change using a climate model, we need to carry out simulations extending over many centuries. The IPCC uses a standard protocol describing the minimum set of simulations that need to be carried out.[20] One of these is a 500-year simulation called the *pre-industrial control*, which tests whether the climate model can maintain thermal equilibrium without drifting. Three additional simulations include (1) a historical simulation from 1850 to the present, (2) a simulation assuming a 1 percent increase in carbon dioxide per year, and (3) a simulation assuming a quadrupling of carbon dioxide. Each of these three simulations is carried out for a period of about 170 years each. In total, across all four simulations, that means a minimum of about 1,000 simulation years. For the latter three simulations, an ensemble of three is recommended to account for uncertainty in initial conditions (Section 2.5), which could add another 1,000 simulation years to that minimum.

A common metric used to judge the performance of climate models is Simulation Years Per Day (SYPD)[21] – simply a count of how many years can be simulated by the climate model if the model is given the whole supercomputer for a calendar day. A model that runs at 5 SYPD will take 100 calendar days to carry out the IPCC preindustrial control simulation, and a minimum of another 34 calendar days to carry out the remaining three 170-year (ensemble) simulations, if they can be done in parallel using multiple supercomputer allocations. The model may exhibit errors that need to be fixed, and some simulations may need to be repeated, which could add more development time. So, assuming a generous allocation of time on the supercomputer, it would take a minimum of 180 calendar days to carry out the IPCC assessment simulations at 5 SYPD. If the model's performance were 1 SYPD, carrying out the assessment simulations would take 2.5 calendar years at minimum. Performance below 1 SYPD would make the climate model unusable for the purposes of the assessment.

In 2020, the standard grid size for climate models was a 100 $\times$ 100 km horizontal grid, and the "bleeding edge," or high-resolution grid size for climate models, was a 25 $\times$ 25 km horizontal grid. A recent high-resolution version of the NCAR CESM model (Section 9.3) performs at about 5 SYPD on Frontera while using 10 percent of the machine. For a 1,000-year IPCC assessment calculation at 5 SYPD, we would need to use 10 percent of

Frontera for 200 calendar days, or 1/18 of Frontera over a calendar year. Frontera had a price tag of $60 million with a lifetime of 5 years. The amortized annual hardware cost is $12 million, which means that the IPCC minimum simulations necessary for the assessment would effectively cost $666,000. To that, we need to add energy costs, which we estimate to be about $333,000 (1/18 of $6 million, the annual cost for 6 MW of power usage).

The above calculations estimate the computational cost of a basic IPCC assessment to be about $1 million for the current high-resolution climate models with 25 $\times$ 25 km horizontal grid boxes. A full climate assessment, which would include many additional calculations, could cost 10 times more[7] – and that does not include the computational costs of model development. There is additionally the infrastructure cost of building and maintaining climate models. The yearly budget for a typical climate modeling effort is in the tens of millions of dollars. Such efforts usually involve 100 people or more, including scientists and support staff.[22] By these rough estimates, infrastructural costs are still comparable to the computational costs of climate modeling. Infrastructural costs will increase linearly with increasing model complexity, but the total number of computations for all the grid boxes will increase nonlinearly as the grid box size decreases linearly (Section 12.2). Even with improvements in computational efficiency, we can expect computational costs to grow much faster than infrastructural costs as model complexity and resolution increase.

One of the biggest problems with climate models today is that they don't represent fine-scale processes like clouds very well. What if we were to make the grid box much smaller, increasing the resolution of the model? To answer this question, European researchers in 2019 carried out atmospheric model simulations on Europe's fastest (at the time) computer, Piz Daint.[23] Piz Daint had a price tag comparable to that of Frontera and ranked just below it, coming in at number six on the 2019 TOP500 list. For a 1 $\times$ 1 km grid, with more than half a billion horizontal grid boxes, the model had a performance of 1/23 SYPD, with an average power consumption of 1 MW.

Since the Piz Daint study used only an atmospheric model, it perhaps underestimates the cost by about 50 percent compared to a more comprehensive climate model (which would include oceanic and sea ice models alongside the atmospheric model). Taking this into account, we can optimistically extrapolate that to reach 1 SYPD performance for this fine a grid may require a bigger computer that costs a few billion dollars and, using available technology, has a power consumption of close to 50 MW, which is not that different from the anticipated power consumption of proposed Exascale computers.[24] The extrapolation is optimistic because a bigger computer with more GPU cores will have less work per core, lowering the arithmetic intensity. At some point, allocating more cores to carry out the model simulation will become counterproductive, due to communication and memory bandwidth limitations. Beyond that point, the SYPD value cannot be increased without advances in computing technology that improve the performance of a single core.

In 2021, the European Union started an ambitious decade-long project called Destination Earth, at a cost of about \$9.5 billion.[25] The project aims to build a model with a grid box size of 1 × 1 km; the expectation is that at this finer resolution, better representation of processes such as clouds and rain could significantly increase the prediction skill of atmospheric models.[26] Weather prediction will likely improve,[27] but climate prediction may not see the same level of benefit, due to compensating errors in other climate processes that are still poorly represented (Section 12.2). Even with a 1 × 1 km grid, a model may only begin to capture cloud processes, as climate scientist Kerry Emanuel has argued.[28] It may therefore be hard to reduce prediction uncertainty without using a large ensemble of simulations.

What if we consider an even more difficult problem, making the grid finer by a factor of eight – reducing the grid box size to 125 × 125 m, so that we can truly capture the details of clouds? For each reduction in the size of the horizontal dimension of the grid box by a factor of two, the cost of computation goes up by a factor of eight, or 2^3 (Section 12.2). The calculations using Piz Daint found that when the size of the grid box was reduced from 47 × 47 km to 2 × 2 km (by a factor of 23.5), computational energy usage went up by a factor of 22^3; when the size of the grid box was reduced even further, from 2 × 2 km to 1 × 1 km, computational energy usage went up by a factor of 1.8^3 (about 6).[23] This suggests that GPU cores were being used more efficiently as more work was assigned to them. (Changing the vertical dimension will increase the cost further, but we ignore it, because it is less important than the horizontal grid being made finer.)

Extrapolating the Piz Daint numbers, we estimate that we need up to 6 × 6 × 6 (about 200) times better performance to go from a 1,000 × 1,000 m grid box to 125 × 125 m grid box. Computers running climate models have shown 100-fold improvements in power efficiency over each of the past few decades.[8] But with the breakdown of Dennard Scaling and the slowdown of Moore's Law, that may not be the case in future decades. If a future supercomputer were to simply add more chips to an Exascale computer *without significantly improving its power efficiency*, then a 100-fold increase in power requirements to handle a 125 × 125 m grid would mean that the supercomputer would consume 5,000 MW. This is about 1 percent of current national electricity generation, equivalent to the current energy use of New York City, and proportionally comparable to the energy use of Oak Ridge during the Manhattan Project. The electric bill for such a supercomputer would be about \$5 billion per year, although improved power efficiency could lower this estimate considerably. Hardware costs of top-of-the line supercomputers are also rising much faster than inflation, as discussed earlier, and they appear to be rising faster than energy costs. (It is worth noting that climate modeling typically uses only a small fraction of a top-of-the-line supercomputer, as such a computer is invariably shared among all fields of science.)

The extrapolated estimates of computational costs assume a finer model grid with no changes in model complexity. This is a reasonable assumption for weather models, but not for climate models. The addition of representations of physical processes like the carbon cycle to a climate model can dramatically increase model complexity and computational cost,[8] even if the size of the grid box stays the same. Despite significant increases in computer power during the six years between the fourth and fifth IPCC assessments, the size of the grid box did not change very much.[29] The extra computational power was used up by the added complexity in the models.

In our focus on building models with ever-finer resolutions, there is an elephant in the room that we have thus far ignored. That elephant is big data. Even today's bleeding-edge climate models, running at 25 × 25 km resolution, generate humongous amounts of data. Since the volume of data generated scales directly with the number of grid boxes, a 1 × 1 km model will generate at least 1,000 times more data than a 25 × 25 km model. Current data storage systems are measured in petabytes (a petabyte corresponds to 1 million gigabytes). The 1 × 1 km model would therefore require storage systems measured in exabytes, units of 1,000 petabytes. Storing and analyzing these extremely large volumes of data could be as daunting a challenge as running the fine-resolution models themselves, unless the data can be analyzed while the model is running.[30]

In keeping with the spirit of this book, our crude numerical estimates of future computer costs should not be taken literally. Nevertheless, current trends suggest that the world may be able to afford supercomputers that can run one or two climate models running with a 125 × 125 m horizontal grid box sometime in the future – but not dozens of them. The European research center CERN recently approved the construction of a \$24 billion supercollider, and the International Space Station cost more than \$100 billion. Both are long-term international efforts that support a single facility, and it has been suggested that climate modelers should also pool their resources and launch an international effort to meet their computational challenges.[31] Such an effort, if successful, could indeed allow exploration of climate simulations at the 1 × 1 km resolution and perhaps beyond, but this success might come at the cost of diversity in model structure and simulations. We are still grappling with structural uncertainty in even finer-scale microphysical processes, and there are still chemical and biological processes that we do not understand. As long as this continues to be the case, we will need more than a couple of models to sample structural diversity, and more than a few simulations to understand the processes.[32] To invoke the Anna Karenina principle (Section 11.1), the question is: Can money buy happiness (or model perfection)? If so, we need to support only one happy family (or the one true model). If not, we will still need to support multiple unhappy families (or several imperfect models).

***

The supercomputers used to run the more comprehensive climate models are getting more complex and more expensive. They also consume prodigious amounts of electricity; hardware costs currently outweigh energy costs, but this may not be the case for long. More elaborate climate models require ever more computing power, but they can currently use only a fraction of the peak capacity of a modern supercomputer because they do not parallelize well – leaving a great deal of dark silicon, or underutilized processing power. Dark silicon may not consume energy, but it still costs money to manufacture. The supercomputing industry's narrow focus on a single theoretical Linpack metric, rather than a broad emphasis on practical applications, may have led to this suboptimal situation. Computer scientists recommend the creation of domain-specific architectures geared toward specific scientific applications[33] to get around the problem of the inefficient use of supercomputers. Rather than relying as they do now on the whims of general-purpose CPU and GPU manufacturers, climate modelers may need to take control of their own destiny in computer hardware and build "custom silicon" in order to efficiently run their climate models.[34]

If climate models continue to become more elaborate and more expensive, there will be fewer of these high-resolution models around the world, due to the sheer scale of the resources required to support them. As indicated by the spread in climate sensitivities of the latest generation of climate models (Section 12.2), having more elaborate models may not necessarily reduce uncertainty in long-term climate predictions. In fact, a reduction in model diversity in the face of persistent structural uncertainty would hamper our ability to average out the compensating errors among models. However, increasing the spatial resolution of models can benefit short-term climate prediction as well as weather forecasting. Structural errors in the representation of climate processes have a weaker impact on shorter timescales, and model predictions can be verified for accuracy.

17

Machine Learning

The Climate Imitation Game

> *the question, "Can machines think?" should be replaced by "Are there imaginable digital computers which would do well in the imitation game?"*
>
> Alan Turing

We began the story of climate prediction with the dawn of computing in the late 1930s, when Alan Turing conceived the universal computer, now known as the Turing Machine. Turing left Princeton to return to England in 1938. During the Second World War, he worked in secret alongside British cryptographers to help break the German Enigma code,[1] providing the Allies with a major advantage. Afterward, he returned to research on computing, first at the National Physical Laboratory and later at the University of Manchester; he worked on designing computers and became deeply interested in the philosophy of computing and human thought.

Turing wrote a famous paper in 1950 titled "Computing Machinery and Intelligence," in which he proposed a method to determine if a machine is intelligent. In the paper, Turing addresses the question "Can machines think?" – or, rather, he sidesteps the question as being meaningless, because it hinges on the word "think," which can mean different things to different people.[2] Instead, Turing considers a game called the Imitation Game, where a man (A) and a woman (B) communicate via texting with a third person (C), an interrogator. C's objective is to determine which of the two is the man and which is the woman, based solely on the text; C can see neither A nor B. A's objective is to fool C into misidentifying him as a woman. B's objective is to help C identify her correctly as a woman. If A can successfully pass as a woman via text better than B can, A wins. (The paper also considers a variant of the game where B is replaced by another man trying to fool the interrogator.)

Turing proposed that a machine replace the man (A) in the Imitation Game and see if its success rate in fooling C beats the success rate when the game is played with three humans. If the machine bests humans in this rather specific task of a man pretending to be a woman, then it passes the test. This has come to be known as the Turing Test. Its popular interpretation is more general than that originally proposed by Turing: If a machine can communicate answers to questions via text in a way that is indistinguishable from a human's answers to the same questions, then it has passed the popular version of the test. In either form, the Turing Test is a practical alternative that avoids the philosophical quagmire of deciding what words such as "thinking" or "intelligence" actually mean.

Turing's personal story ended tragically. Homosexual acts were a crime under Victorian-era laws enforced in Britain at that time. Turing was gay, and in 1952, he was convicted of "gross indecency" for having an affair with a man. He chose to be chemically castrated, as an alternative to going to prison – and he died two years later, humiliated, two weeks before his 42nd birthday. His death was ruled a suicide. It took 55 years for a British Prime Minister, Gordon Brown, to issue a public apology for this injustice, and to acknowledge Turing's service to science and the nation.

***

Is imitation possible without thought? The assumption behind the Turing Test is that for certain actions performed by humans, imitation is not possible without thought. Weather and climate models are also imitations; they play the Imitation Game against the real world. We therefore ask a related question: Can we predict weather and climate without understanding?[3]

We have argued that understanding climate models, using the reductionist approach of a model hierarchy, is crucial to building our confidence in them (Chapter 8). An important reason for doing so is because we want to use models to predict an unprecedented event – global warming. We cannot test a climate model to see whether it can "imitate" many global warming events because we barely have one event on which to test it.

Is understanding important for building a good weather model? We have thousands of weather events against which to check if a weather model is good at imitating reality. Perhaps understanding is not important, or perhaps it is at least less important, in the case of a weather model. In any case, the word "understanding," like the word "thinking," is fraught with ambiguity and means different things to different people. Humans who play the Imitation Game many times may become much better at it each time, but they may not always be able to understand or articulate why they become better because the improvement is "instinctive." A more instinctive player may best a more thoughtful player at the game. But if the rules of the game were to suddenly change, the more thoughtful player might be able to adapt more quickly than the more instinctive player.

In this chapter, we further explore the concept of predicting instinctively – that is, predicting without perhaps ever understanding in a reductionist sense.

17.1 Can Machines Learn?

Humans learn to think. For machines to learn to think like humans, they need to learn human activities. Attempts to teach such activities to machines date from the early days of computing, when scientists taught computers to play simple games of strategy, like checkers. As computers became more powerful, the challenge became teaching computers to play complex games of strategy, like chess. This area of research – how we can make computers think like humans – is known as *artificial intelligence*.

Human thinking is an emergent property of the human brain, the result of billions of neurons interacting with each other through trillions of connections known as *synapses*. In 1943, scientists started using a reductionist approach to physically imitate the human brain by constructing a mathematical model of a connected network of artificial neurons. After the invention of the computer, this neuron network could be simulated using software. Thus, artificial neural networks were born. Early artificial neural networks featured a single layer of interconnected artificial neurons, which could be trained (taught to make correct decisions) through a process known as *machine learning* (ML). They received a set of numbers as an input and produced a number as an output. Depending upon whether the output was correct or incorrect, the parameters of the artificial neural network were adjusted to reinforce correct outputs and discourage wrong outputs, in a crude imitation of how the human brain learns.

Scientists soon developed algorithms for neural networks with multiple layers, where the output of one layer fed into the next layer. This allowed for multiple layers of abstraction. For example, a neural network to recognize objects may have a first layer that recognizes edges, a second layer that recognizes shapes, and a third layer that recognizes the object – in effect, imitating the hierarchical process of human visual cognition. These multilayered ("deep") neural networks were initially much more expensive to train than their predecessors, but the development of faster processors and GPU technology in the 2000s provided the necessary computing power (Section 16.3).

The training of deep neural networks, known as "deep learning," is essentially induction on a grand scale. One of the most successful applications of deep learning is in image processing. Preventing objectionable photos from being uploaded to social media sites is a daunting task; millions of users upload photos every day, many of them innocuous but some highly objectionable. Human employees attach labels to millions of photos indicating if they are acceptable, and deep learning uses this labeled dataset of images to train a deep neural network to classify images as either acceptable or unacceptable. The machine learns to imitate an emergent property of the human brain, the ability to recognize objectionable images. Language translation, which can also be considered a high-level emergent property of the brain, is another successful application of deep learning. In the early days of machine translation, the rules-based approach dominated: Rules from linguistics were used by the machine to aid in the translation. But with the advent of deep-learning techniques, the brute-force approach – using a large body of translated text to train the translation software – proved superior to the rules-based approach.[4] Linguistic concepts embody our understanding of natural language, but such understanding was not essential for the machine to learn.

Artificial neural networks are trained using large volumes of available data. A network shown thousands of pictures of white swans in the data will learn to recognize a white swan quite well. But if there are no black swans present in the training data, the network may fail to recognize a black swan. This "problem of induction" is a well-known deficiency of artificial neural networks, and it is referred to as *generalization error.* When these networks are used for the tasks they are trained for, such as recognizing familiar types of images in a dataset, this is not a major concern – but these networks can break if they encounter unfamiliar features in the dataset.

To test the robustness of deep neural networks, researchers use a technique called the *adversarial attack.* In a cheeky adversarial attack, Canadian researchers literally inserted the image of an elephant into a photo of a room that was being analyzed by computer vision software trained to recognize objects. The introduction of the "elephant in the room" threw the software off – not only did it have trouble recognizing the elephant, but it also misidentified some objects that it had previously identified correctly.[5] The Covid-19 pandemic was in some ways a real-world adversarial attack. Businesses rely on supply chains: They obtain components from different suppliers and assemble the components to make the product they sell to consumers. In recent years, many businesses have begun using sophisticated ML-based algorithms, trained using historical data of customer behavior, to optimize their supply chains and minimize the inventory of components they need to stock in their own warehouses. When the buying habits of consumers changed in drastic and unanticipated ways at the start of the pandemic, some of these ML-based predictive algorithms broke. All this is to say we need to be aware of the generalizability problem of ML before we use it to predict the trajectory of an unprecedented event like climate change.[6]

The Covid-19 pandemic is an example of a black swan event. By definition, all models will fail to anticipate a black swan event, as was the case with the ozone hole, for example (Chapter 6). But a well-understood deductive model can often be quickly modified to explain the black swan event after it has happened, and it can even be adapted to predict the event's future course. It will take much longer to modify a poorly understood inductive model, such as an ML model, because there will not initially be enough data to train it anew.

17.2 Model Wars: Inductivism Strikes Back

Climate modeling faces many challenges: the rapidly increasing complexity and computational cost, the difficulty of efficiently using modern computers, and the uncertainty plateau in climate prediction. The recent revolution in the field of ML opens up several interesting possibilities to address some of these challenges.[7]

The technology of GPUs is being driven by the needs of ML because there is a vast market of users.[8] Modern GPUs have specialized processors for efficiently handling ML tasks. Analogous to the Linpack and HPCG benchmarks (Section 16.4) is a new performance benchmark called HPL-AI, which measures the "artificial intelligence" or ML capabilities of supercomputers. In 2019, the Summit computer achieved a performance of 445 petaflops for HPL-AI, about three times its Linpack performance. This means that ML code can run very efficiently on Summit because the GPUs are specifically optimized for it.

If we had a large dataset of climate events, we could train an ML climate model to predict the climate events in that dataset. As we know, we don't have multiple global warming events on which to train an ML climate model. But we could potentially use such an approach for weather prediction. The past 70 years of weather data could be used to train an ML weather model, which would produce a purely data-driven inductive model. The input to the ML weather model would be the weather conditions at the initial time, and the output would be the weather conditions 24 hours later. The model could be trained using over 25,000 daily weather maps generated over 70 years of global observations.

Before the invention of the computer, meteorologists actually used a similarly inductive approach. During the Second World War, a team of forecasters relied upon their prior experience of weather patterns to make the famous forecast for the Allied Landing at Normandy on June 6, 1944 – D-Day.[9] This technique is called *analog forecasting*: Given the weather map for today, a meteorologist essentially flips through all previous weather maps and locates the map from the past – day X, say – that displays the best analog to today's conditions. The weather map for the day after X would serve as the analog forecast for tomorrow. Analog forecasting didn't work very well because, in a global, three-dimensional sense, it is rare that a particular day's weather exactly resembles weather for another particular day in the past. Finding a match for the local weather pattern in the past is insufficient to constrain the future because weather patterns move around the globe quite rapidly.

Today's computer weather models, which are primarily deductive, handily beat human analog forecasting, which was inductive and also (as performed by humans) subjective. But it is possible that an ML weather model using modern deep-learning techniques could learn to imitate nature better than humans could.[10] This is an active area of research, as an inductive weather model could be much more computationally efficient on modern supercomputers than a deductive weather model. Such a model could predict the emergent weather behavior of the atmosphere without ever having to understand how the different components of the atmosphere interact with one another.

But we are more interested in the climate problem, where the data-driven inductive approach to prediction will not work due to the lack of sufficient data to constrain long-term climate change. One approach is to use advanced ML techniques to derive optimal estimates of all the poorly known parameters in climate models, using better simulation of current climate as the optimization criterion. Another approach is to add ad hoc correction terms, estimated using ML, to the model equations to minimize errors in the simulation of current climate. But neither approach addresses the structural uncertainty problem because the optimal parameter estimates, or correction terms, could merely reflect optimal compensation between structural errors and might not be valid in a future climate.

There are potentially two more ways in which we can use ML to improve climate prediction:

I Replace the entire model with a faster imitation created using ML.

II Replace a component of the model, such as the parameterization of clouds, with a faster imitation derived via ML.

Approach I, where an ML model simply emulates the emergent behavior of a deductive model, can be used for both weather and climate prediction to improve computational efficiency. We could run a deductive model for a simulated period of hundreds of years, generating a large simulated dataset that could be used to inductively train the ML model – and to train it much better than is possible with our limited observational dataset. The downside of this approach is that the ML imitation would suffer from the same systematic errors and limitations as the deductive model; it would merely be cheaper to run. Another downside of standard ML algorithms is that they optimize prediction without ensuring conservation of properties such as mass and energy. As we have emphasized before, the constraining power of a climate model may be as important as its predictive power. Machine learning algorithms will therefore need to be modified to respect physical constraints.[11]

Approach II is reductionist. We can focus on a specific process in the climate model that is poorly represented – clouds, for instance.[12] With sufficient observational data, we can train an ML-derived cloud parameterization to replace the physics-based cloud parameterization in a climate model. Without sufficient observational data, we can still use a physics-based cloud model, called a *cloud-resolving model*, to generate synthetic data on which to train the ML parameterization. Cloud-resolving models use very fine grids to capture the detailed evolution of clouds subject to different background conditions like winds, temperature, and humidity. Simulations using this fine-scale model can train an ML cloud parameterization that can then be used in the larger climate model. It is important, though, that the ML parameterization not only capture large day-to-day weather variations in cloud coverage but also accurately simulate weaker long-term climatic trends in the data, if it is to be used for climate prediction.

The hope is that we will be able to apply these ML approaches to predict climate prediction over a range of timescales, from a few months to hundreds of years. In forecasts of short-term climate phenomena like El Niño, up to a year in advance, these approaches may avoid the problem of generalization error because the background climate does not change very much over the course of a year. But both of these ML approaches may suffer from generalization error when they are applied to centennial climate prediction. In Approach I, the ML model may fail when it is used for global warming scenarios associated with conditions that were never encountered by the original deductive model during the generation of training data.

In Approach II, the generalization error problem may be less severe. Clouds, for example, see the equivalent of climate change in their environment across time and across locations. To the extent that naturally occurring changes in our current climate capture the range of possibilities in a future climate, the ML cloud parameterization may avoid generalization error. Recall William Gibson's observation about the future already being here, just not evenly distributed around the globe (Section 7.7). If the training data includes enough pieces of the "future" that already exist sometime or somewhere in the present or the past, then ML can simulate those aspects of the future as they become more widespread.

But if a future climate change scenario presents conditions for cloud processes that are nowhere (or rarely) to be found in the range of current climate conditions, then the ML cloud parameterization may fail due to generalization error. One model-twin research study which used ML to replace the representation of rainfall in a climate model found that the ML rainfall parameterization, when trained on the current simulated climate, did not capture the simulated climate change that well.[13] But when the ML rainfall parameterization was trained on the warmer simulated climate, it was able to capture the current simulated climate because the warmer climate spanned a broader range of training scenarios.

A workaround for the problem of generalization error is to treat the faster ML climate models as "climate emulators," efficiently exploring the parameter space of model uncertainty by trying out different values for adjustable coefficients.[14] If the ML model exhibits an interesting behavior, the behavior can be confirmed with the slower, non-ML version of the model. This avoids the false-positive problem of a spurious interesting behavior in the ML model. It does not avoid the false-negative problem, though, because even a thorough exploration of the parameter space of the ML model may not capture all the interesting behaviors the full model (without ML) is capable of.

Data-driven ML approaches to modeling work best when model predictions can be verified. Therefore, the most promising applications of ML-based climate modeling may not be in predicting long-term climate change but in solving long-standing problems – such as the poor simulation of the tropical Pacific El Niño – whose predictions can be verified. Should ML models turn out to provide a significantly higher prediction skill for El Niño and other short-term climate phenomena, as compared to current deductive models, we can be more confident in the ML approach.

17.3 Mapping the Landscape of Models

a rose is a rose is a rose.

Gertrude Stein, *Sacred Emily*, from *Geography and Play* (1922)

A model is a model is a model – that is how those outside the field of climate science typically view climate models. Nonexperts may not always appreciate that there are very many different types of climate models, and they are not all created equal. Any natural system, simple or complex, can be modeled either by fitting a statistical model to data (an inductive approach) or by using a numerical model derived from testable hypotheses (a deductive approach; Section 1.2). The potential return of large-scale inductivism, via ML, to climate modeling motivates us to revisit this philosophical distinction. Fundamental sciences like physics emphasize the deductive approach; empirical fields of study like economics emphasize the inductive approach. Fields like chemistry and biology fall in between – as does climate modeling, which includes elements of physics, chemistry, and biology.

Somewhat orthogonal to this distinction, we can make a different distinction for complex natural systems. Such systems can be analyzed either as a whole (an emergent approach) or by breaking them down into parts (a reductionist approach; Section 4.4). The predicted global warming is an emergent property of climate; the rain associated with a single cloud can be studied using a reductionist approach.

As noted previously, these binary distinctions are simplistic. They capture neither the overt nor the subtle interconnections between the various approaches to modeling complex systems. Nevertheless, these distinctions can help us map the evolving landscape of climate modeling. To this end, we create the two-dimensional Inductive–Deductive–Emergent–Reductionist (IDER) diagram (Figure 17.1), in an attempt to pigeonhole the many modeling concepts and ideas discussed in this book.

On the Y-axis, we have the Inductive➔Deductive range (Figure 17.1). On the X-axis, we have the Emergent➔Reductionist range (like the xkcd "purity" scale; Figure 11.1).

We can also think of the Y-axis as showing decreasing data-dependence as we move upward and of the X-axis as showing increasing complexity as we move leftward. The hypothetical C-B Limit (Section 12.3), where we solve the complete set of deductive equations governing an emergent system exactly, would be located at the top left corner.

Weather models used to make short-term forecasts would be near the top and to the left in the IDER diagram, on account of their deductive and emergent nature; they model large-scale fluid motions. *Climate models* are below and to the left of weather models; they are more emergent and less deductive, as they are more dependent on parameterizations of fine-scale processes, with coefficients that are adjusted ("tuned") using data (Section 10.1). An *Earth System Model* – a climate model that includes additional processes like the carbon cycle (Section 9.4) – is more emergent but even less deductive than the "vanilla" climate model, as the Earth System Model includes additional parameterizations that must be tuned.

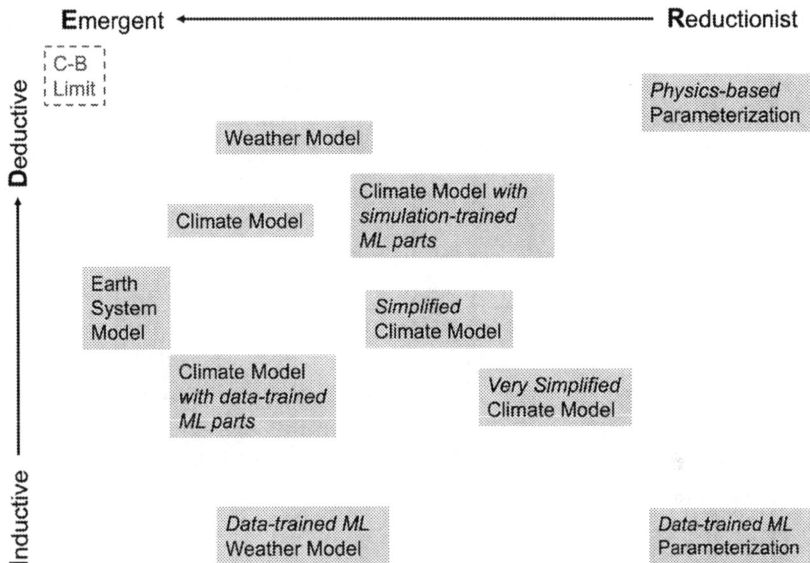

Figure 17.1 Inductive–Deductive–Emergent–Reductionist (IDER) diagram for classifying models.

When we reduce the size of a climate model's grid boxes, we can make it more deductive (and move it upward on our diagram) if a previously parameterized process, such as a cloud, is explicitly resolved by the finer grid. A *simplified climate model* would be shifted to the right of a full climate model, being more reductionist (Section 8.3); the simplified model would also typically fall below the full climate model, as the simplification process often involves replacing complex parameterizations with simpler empirical approximations. A *very simplified climate model* would be yet more reductionist and often more inductive as well.

We can consider parameterizations of fine-scale processes as small models on their own and place them on the diagram accordingly. A pure *physics-based parameterization* is deductive and reductionist, and appears at the top right corner. But if we had to adjust many coefficients of this parameterization using data, it would become more inductive and move downward. A CFD model (Section 12.2), with calibrated coefficients, would fall into this category.

An analog forecasting model or a pure *data-trained ML weather model* (Section 17.2) would be at the bottom and to the left, inductive and emergent. By itself, a pure *data-trained ML-parameterization*, such as a cloud parameterization, would be in the bottom-right corner, inductive and reductionist. But a *climate model with data-trained ML* parameterizations would fall below the original deductive climate model, being more inductive. A *climate model with simulation-trained ML* parameterizations would be shifted rightward (more reductionist) compared to the corresponding deductive model; due to generalization error, some emergent behavior may be missed by this speeded-up version of the climate model, making it behave somewhat like the *simplified climate model.*

Economic and risk assessment models tend to be more data-driven and emergent, which would put them in the bottom left quadrant, if they were to be included. One property not explicitly considered in the IDER diagram is the constraining power or conservation properties of models. Typically, deductive models would automatically include conservations of mass, energy, and so on. More inductive models, especially statistical models, may not necessarily be very constrained. Note that as we move from a weather model to a climate model to an Earth System model in the diagram, the model becomes more elaborate but may not approach the C-B Limit; as the model becomes emergent, it also becomes less deductive and more dependent on empirical parameterizations of biogeochemical and other complex processes.[15]

The two-dimensional IDER diagram also says nothing about the prediction accuracy of the different models. (For that, we would need to add a third dimension to this diagram.) Simplified versions of complex models are typically less accurate than their corresponding complex models, and inductive models generally do not perform as well as deductive models in predicting unprecedented events. Models with simulation-trained ML parameterizations may exhibit spurious behaviors that lead to less accurate predictions than those of the corresponding full models, due to the generalization problem.

***

Graphical Processing Units are helping computers become more intelligent. They are often optimized for ML, which allows computers to create an inductive imitation of a complex system, such as the human brain or the atmosphere – without requiring any "understanding" of the system, at least in a reductionist sense of the word. But the imitation has its limitations because it may not be able to handle unprecedented scenarios like a global pandemic or global warming, due to the generalization problem. Machine learning that uses only simulated data for training, rather than real data, is an imitation of an imitation; since the simulated data will still need to be obtained from a structurally diverse set of models, the computational speed-up provided by the inductive imitation may not be sufficient to break out of the uncertainty plateau inhabited by climate models (Section 12.2).

Computing machines existed even before the 1940s, but they were built for a domain-specific (usually military) purpose. The paradigm shift that occurred in the 1940s was the invention of the general-purpose computer, which could perform many different tasks: It could design bombs one day and predict weather the next. Machine learning is currently in its early, domain-specific stage. Perhaps a paradigm shift will occur when general-purpose ML (or artificial intelligence) is created. But in the meantime, domain-specific ML could be useful for inductively modeling aspects of weather and climate.

18

Geoengineering

Reducing the Fever

"Take two aspirin and call me in the morning." That was doctors' routine advice to the sick back when aspirin was the standard fever reducer and its side effects were not fully known. Should we let a fever run its course, or should we use fever-reducing drugs to lower the body's temperature? This is a dilemma faced at some point by every parent. While the advice of pediatricians is mixed, fever-reducing drugs continue to be widely prescribed and used. Everyone agrees, though, that these drugs do not cure the underlying disease. Use of fever-reducing drugs can provide physical and psychological comfort from symptoms, provided we are taking steps to combat the disease. But these drugs can also have nasty side effects. Aspirin can cause gastrointestinal bleeding, and it is no longer prescribed routinely as a fever reducer. Even the safest fever reducers, acetaminophen and paracetamol, can cause liver damage in very high doses.

Climate geoengineering is deliberate large-scale intervention in the Earth's climate to combat climate change.[1] It covers a range of techniques, some of which resemble MacGyver-like last-minute quick fixes to prevent the imminent disaster of global warming. The most viable forms of geoengineering, which involve reflecting sunlight, tend merely to treat the symptoms of global warming without addressing the causes.[2] Proponents of those forms of geoengineering argue that if the symptoms of a disease become debilitating, treating them with fever-reducing drugs can buy precious time to cure the underlying disease. But many scientists oppose these forms of geoengineering, arguing that, first, they may not benefit everyone; second, they may have unexpected side effects; and third, even if they miraculously work for everyone with just mild side effects, we may become dependent on the drugs and never cure the disease. Aspirin and other fever reducers have been tested on many thousands of patients, and their side effects are well known. Since we have only one Earth to work with, it would be neither ethical nor practical to experiment with a treatment for the symptoms of global warming at varying doses to determine its efficacy and side effects. The only way we can test geoengineering on a large scale before deployment is by using computer models of climate, which, as we know by now, are imperfect.

18.1 How to Train the Climate Demon

If an unfriendly nation gets into a position to control the large-scale weather patterns before we can, the result could even be more disastrous than nuclear warfare.

Howard T. Orville, Meteorologist and Advisor to US President Eisenhower

Many physical attempts have been made to modify weather. For example, clouds have been seeded with silver iodide in an attempt to make them produce rain. This has never been demonstrated to work, but that didn't stop the US military from trying to use weather as a weapon in the Vietnam war. Between 1967 and 1972, the US Air Force deployed more than 47,000 canisters of rain-producing iodides over Southeast Asia in an attempt to extend the rainy season and hinder North Vietnamese troop movements. The *New York Times* quoted one State Department official, who justified the attempt, saying, "What's worse, dropping bombs or rain?"[3]

The basic premise of geoengineering has its origins in John von Neumann's ambitions of weather control. This was shown, by the work of Ed Lorenz and others on deterministic chaos in the atmosphere, to be out of the question (Chapter 2). But climate control was still a possibility: von Neumann had the idea that the Earth's albedo or reflectivity could be altered, which would affect the boundary condition of the atmosphere (Section 5.1) and could therefore be used to control climate. A colleague of von Neumann from the Manhattan Project, Hungarian-American physicist Edward Teller – known as the father of the hydrogen bomb – was also interested in climate control.[4] This interest motivated Teller to support creation of the world's first computer climate model by Chuck Leith at LLNL in the early 1960s. After Leith left Teller's lab in 1965, that climate model was abandoned.

Many years later, in 1997, Teller published a report on the "Prospects for Physics-Based Modulation of Global Change." The report discussed different ways to counter global warming by reflecting sunlight to cool the planet. One the most promising ideas was inspired by the 1991 eruption of Mount Pinatubo in the Philippines. Sulfate aerosols injected into the stratosphere by the volcano stayed there for a few years, and they reduced the global-average temperature by about half the global warming signal at that time. Nature had performed an experiment to demonstrate that if we could somehow inject enough reflective aerosols into the stratosphere, the planet would cool significantly. This idea had originally been put forward in the 1970s by Soviet climate scientist Mikhail Budyko.

Several other geoengineering strategies have been proposed. They fall into two broad categories:

1 *Carbon dioxide removal*: These strategies involve chemical or biological techniques to extract carbon dioxide from the atmosphere and store it underground or in the ocean. They are less controversial, as they attack the root cause of the disease: the increasing amount of carbon dioxide in the atmosphere. But they are also unproven and not economically viable, at least currently.[5] Since they don't directly relate to climate modeling, we won't discuss them any further.
2 *Solar radiation management*: These strategies change the amount of sunlight reaching the planet, by injecting sulfate aerosols into the stratosphere or by putting giant mirrors in space. Compared to carbon dioxide recovery, some of these techniques appear viable and affordable. But they treat the symptoms, rather than the disease, and are therefore much more controversial. We will discuss the use of stratospheric aerosols to combat global warming in some detail because it appears to be the most viable geoengineering technique at this time.

18.2 Sulfates in the Sky

Let us inject millions of tons of sulfates into the stratosphere, creating a cloud of sulfuric acid droplets that envelops the Earth, reflects sunlight, and cools the globe! This idea is not quite as crazy as it sounds, but it may also not be a panacea, as it is sometimes portrayed. It is not so crazy, because major volcanic eruptions actually inject sulfates into the stratosphere, and this does cool the globe (Section 12.1). But the idea is hardly a panacea, either: While the overall effect may be one of global cooling, it may not exactly cancel the harm caused by global warming in particular regions, and it may even exacerbate the harm in others.[6]

The idea of using reflective aerosols in the stratosphere to combat global warming, or *solar geoengineering*, has some pedigreed proponents. One is Paul Crutzen,[7] who shared the 1995 Nobel Prize in chemistry for his research on stratospheric ozone depletion. Solar geoengineering has emerged as one of the less-expensive strategies for geoengineering, and it could be attempted with something close to currently available technology. High-flying aircraft would need to start injecting about 10 million tons of sulfur in the lower stratosphere every year to counter the effect of increasing carbon dioxide concentrations; the annual amount injected would be comparable to that from the Mount Pinatubo eruption of 1991.[8] The estimated cost of this operation would be in the tens of billions of dollars per year – much smaller than the cost of decarbonizing the world's economy, which is usually measured in trillions of dollars.

Supporters present solar geoengineering as a stopgap solution that buys the world time to address climate change in a more measured way. Most climate scientists are wary of solar geoengineering for several scientific reasons. For one, there is limited observational data on the actual effect of a large injection of sulfates into the atmosphere; very few major volcanic eruptions have occurred since the start of the satellite era. We must therefore rely on climate models to assess the efficacy of solar geoengineering. For instance, NASA scientist Jim Hansen successfully used a climate model to predict the global cooling resulting from the Mount Pinatubo eruption in 1991, which is a close analog to what solar geoengineering proposes.

One of the scientists' most important concerns, however, is not a scientific one: It is the "moral hazard" of geoengineering activities potentially resulting in a reduction in efforts to curb greenhouse gas emissions.[9] In other words, if the focus shifts to treating the symptoms of the disease, the root cause of the disease may not be addressed. Society may become dependent on ever stronger doses of the fever reducer as greenhouse gas emissions continue unabated.

In response to increased interest in the topic, the US National Academy of Sciences set up a committee in 2015 to evaluate the feasibility, risks, and rewards of major geoengineering strategies. The committee examined observational and modeling evidence to produce two reports, one on carbon dioxide removal and the other on solar radiation management. The latter report was titled "Climate Intervention: Reflecting Sunlight to Cool Earth."[10] The committee consciously decided not to use the term "geoengineering" in the title to avoid making the strategy sound as simple and straightforward as building a bridge or a building. The term "intervention" better conveys the seriousness of tampering with a complex and powerful system that we do not fully understand.

The report concludes that it is technologically feasible to cool the planet significantly using stratospheric aerosols to counter global warming, at a cost orders of magnitude lower than reducing carbon emissions. Therefore, if global-average temperature is the only metric of success, it should work. But global-average temperature is useful only in academic discussions. It is the regional climate – the temperature and rainfall in your city, or on your farm – that matters. As human geographer Mike Hulme argues,[11] letting a single metric like the global temperature become the focus of climate policy can lead to goal displacement,[12] in which the goal becomes controlling the numeric metric, rather than countering the direct impacts of climate change on people's lives and natural ecosystems.

Importantly, the report states that models "strongly suggest that the benefits and risks will not be uniformly distributed around the globe." Because greenhouse gas-induced warming and stratospheric aerosol-induced cooling have different three-dimensional spatial structures, they will not cancel each other out in every region, even if they tend to cancel each other out in the global average. Many regions will benefit from the cancellation, but some may continue to suffer, and some may suffer even more. If a climate model could predict exactly which regions would benefit and suffer, then we could carry out extreme weather attribution studies (Section 7.3) and a cost-benefit analysis to selectively compensate the regions that would be harmed by solar geoengineering.[13] The climate model would need to predict regional climate change skillfully for the next several decades and longer. But this is especially hard for climate models to do for two reasons:

1 Models exhibit an uncertainty plateau even for global parameters like climate sensitivity (Figure 12.2b), despite cancellation of errors at the regional level when computing the global average. When predicting regional climates, model uncertainty will be further magnified.
2 In predicting short-term climate for the next 10–30 years, the uncertainty of initial conditions in the ocean (the oceanic "butterfly effect") and in the atmosphere will add stochastic noise to the signal of greenhouse gas-induced warming and aerosol-induced cooling (Section 7.5).

These limitations of models make it hard to attribute short-term regional climate changes to specific causes. That is to say, it would be tricky to determine, in a completely model-independent way, if a particular drought or flood was exacerbated as a side effect of solar geoengineering.

Apart from the practical limitation of regional damage attribution, there are additional issues with the solar geoengineering solution as an alternative to drastically reducing emissions.[14] There are well-known side effects: The concentration of carbon dioxide would continue to increase in the atmosphere, which means that ocean acidification and its impact on marine ecosystems would only get worse. Then, there are the poorly known side effects: Dramatically altering the stratosphere by continuously injecting aerosols into it could cause serious damage to the protective ozone layer, for example. And, finally, there are the completely unanticipated side effects, the unknown unknowns. With his background in ozone research, Crutzen was well aware of these possibilities. He said in his 2006 essay: "The chances of unexpected climate effects should not be underrated, as clearly shown by the sudden and unpredicted development of the [A]ntarctic ozone hole."[7]

Moreover, the strategy of mitigating global warming using stratospheric aerosols also risks setting up a single point of failure. Many nations will need to coordinate and pool their resources to maintain the "aerosol shield." If carbon emissions continue unabated, the shield may need to be continually strengthened, at greater and greater cost. Any disruption in maintenance – due to disagreements between nations, for instance – could cause the shield to collapse very quickly, in just a few years, and all the suppressed warming would break loose. As climate scientist Raymond Pierrehumbert puts it,[15] the aerosol shield could become the Sword of Damocles hanging precariously over our planet.

***

Inadvertent climate modification – this was the original name for what we now call global warming or climate change. It should remind us that we have been slowly geoengineering climate for well over a century through greenhouse gas emissions, unwittingly for much of that time. We now use models to understand the global implications of continuing this inadvertent geoengineering, but details of the regional implications are harder to pin down. And we worry about triggering abrupt climate change, if we inadvertently cross a tipping point. Proposed intentional geoengineering options attempt to counter the impacts of greenhouse gas emissions, but on a much shorter timescale.

The paradox of intentional geoengineering is that the most benign options – such as capturing carbon dioxide directly – tend to be prohibitively expensive, while the most affordable options – such as injecting sulfates into the stratosphere – are the riskiest. We have to consider the known and unknown side effects of the latter. As we learned from the succession of engineering fixes that tried to control the lean of the Tower of Pisa (Chapter 10), hacking a complex system maintained by intricate balances can have unintended consequences. But perhaps the biggest danger is that the tenor of techno-optimism surrounding geoengineering will provide cover for those who wish to avoid the real challenge of reducing carbon emissions.

Solar geoengineering may be a poor and even dangerous strategy for actually mitigating or countering global warming, and some would argue this justifies not allocating any public resources to research on that topic. At the end of the day, though, small-scale research into geoengineering is a legitimate form of scientific inquiry, provided real-world experiments are carried out in a limited and safe manner.[16] It may be better for such research to be carried out as part of mainstream climate science than being pushed outside the gates of international regulation.

Climate models are good at producing qualitative insights into climate change, but they are limited in their ability to provide quantitative estimates of regional impacts, as would be necessary to implement geoengineering in an equitable fashion. In Chapter 19, we consider the philosophical issue of dealing with qualitative threats, and what we can do to quantify climate impacts and risk.

19

Pascal's Wager

Hedging Our Climate Bets

We began at the dawn of computational weather and climate prediction, which was driven by the unbounded scientific ambitions of John von Neumann at the IAS in Princeton. In this penultimate chapter, we return to the story of von Neumann, but detour as von Neumann did, away from science. Numerical atmospheric prediction was just one among many fundamental contributions von Neumann made to the fields of mathematics, computing, science, and economics. His contemporaries spoke of his abilities in superlative terms: Physicist Hans Bethe said he had "sometimes wondered whether a brain like von Neumann's does not indicate a species superior to that of man."[1] Superhuman though his mental abilities were, his body was very much human. He was diagnosed with cancer in 1955, around the time the Princeton Meteorology Group disbanded. By 1956, he had been moved to the Walter Reed Hospital in Washington, DC, where dignitaries and friends frequently visited him. As a Cold War hawk, von Neumann was privy to much classified information – so soldiers were always on guard at his bedside, in case he betrayed military secrets while talking in his sleep![2]

Von Neumann had grown up in a nonobservant Jewish family in Hungary. Although he nominally converted to Catholicism before his first marriage, he had essentially remained agnostic up to this point. It was in the hospital that he began to worry about what might happen after his death. Perhaps, as one of the inventors of game theory, he decided to plan rationally for the afterlife and optimize his chances. A few months before his death, he returned to the Catholic Church and spent time with a Benedictine monk. He believed, according to his daughter Marina, that "Catholicism was a difficult religion to live in but the best one to die in."[3] "Terrified of his own mortality," Marina writes, "he found comfort in the promise of personal immortality in an afterlife." Von Neumann passed away in February 1957, at the age of 53.

Biographer Norman Macrae writes that von Neumann's deathbed conversion was motivated by the argument that "so long as there is the possibility of eternal damnation for nonbelievers, it is more logical to be a believer at the end."[4] This echoes Pascal's Wager on the Existence of God, a famous argument advanced by seventeenth-century French philosopher and mathematician Blaise Pascal.

Like von Neumann, Pascal was a child prodigy and a pioneer of computing in his day, having invented the mechanical calculator. As a mathematician, he tried to find a rational answer to the question of God's existence or nonexistence. Since it is not possible to mathematically prove the existence or nonexistence of God, Pascal focused on how a rational person should live. When you choose to live one way or another, you are betting on the existence (or nonexistence) of God. There are four combinations of possibilities, two for existence and two for lifestyle choice, which form a 2×2 decision matrix: (1) God may *or* may not exist. (2) You have a choice of living piously, being virtuous and religious, *or* living sinfully, leading a life of decadence and godlessness.

If you live sinfully, you suffer eternal damnation in Hell if God exists (infinite loss) and enjoy some pleasures otherwise (finite gain). If you live piously, you forego some pleasures if God does not exist (finite loss) but find eternal happiness in Heaven otherwise (infinite gain). According to Pascal, the rational thing to do is to bet the finite against the infinite: to live piously, foregoing finite pleasures for the certainty of avoiding infinite loss and the possibility of winning infinite gain.

Pascal's Wager has been invoked to justify the taking of preventive action to mitigate the threat of catastrophic climate change.[5] It is worth pointing out that this argument only works against credible threats; if we applied it indiscriminately, we would all become Chicken Littles, constantly worrying about and trying to guard against every unlikely catastrophic event, such as asteroid impacts, zombie outbreaks, and even alien invasions. Climate data and models provide scientific evidence that catastrophic climate change is a credible threat, even if there is uncertainty about the numeric measures of its probability and impact. There is no such evidence for zombie outbreaks or alien invasions. (NASA uses scientific models to calculate the threat of asteroid impacts, and will recommend strong action if and when the threat becomes credible.)

In general, Pascal's Wager supports the pursuit of a more sustainable lifestyle to insure against environmental catastrophe or "climate hell." The costs of adapting to climate change after it has happened will be high; the costs of acting now to mitigate climate change are a lot lower. This is sometimes referred to as the "precautionary principle" – acting out of an abundance of caution. But how abundant should we be with our caution? The answer depends, in part, upon how bad the consequences are for a lack of caution. It also depends upon how abundant our economy is now.

19.1 An Actuarial Approach to Long-Term Climate Change Risk

There is no such thing as a bad risk, only a bad premium.

Insurance industry saying

Pascal's Wager is not a radically new concept. Most people who drive a car routinely make a similar wager. You purchase car insurance to protect you and your family from the cost of accidents. You bet a small amount of money, the insurance premium of several hundred dollars, against a potentially huge loss from an accident, in which you could be liable for a million dollars. Since we are concerned with the climate crisis, what is the premium that we should pay to protect against the harmful consequences of climate change? Let us see how far we can push the car insurance analogy.

When we walk into an insurance agent's office, we have a qualitative goal, to purchase liability insurance for accidents that may happen when we drive. We walk out with something quantitative, an insurance policy with a coverage amount and an annual premium. The information we use to decide what kind of policy we want to buy is usually based on our present situation, that is, our annual income and our wealth. We don't personally carry out a comprehensive analysis of the probability of accidents in our city and the cost of damages associated with these accidents; since data is available for millions of accidents that happen each year, an accurate analysis can be done statistically, using the inductive approach of actuarial science. Insurance companies employ actuaries who compute the annual premiums for various categories of customers, ensuring that the amount collected in premiums is sufficient to cover the expected losses from accidents.

We can also insure against bad weather. Farmers and organizers of outdoor sporting events do this routinely. Weather insurance is an actuarially sound problem, meaning that there are many decades of data available with which to make probabilistic prophecies for weather events and compute the absolute risk of loss (the probability of each event multiplied by the monetary damage due to that event). Using this data, an actuary can calculate the premium each customer must pay next year to cover the expected losses.

But insurance against long-term climate change is quite different. It is as if the inventor of the gasoline-powered car, Karl Benz, were to walk into Horse Cart Hazard Insurance in 1885 to buy family liability insurance against car accidents – not for the next year, but for accidents that might happen further into the future, 80 years later, to cars driven by his grandchildren. There would be no data available to inductively apply actuarial science to a scenario so far into the future.

The actuary could construct physics-based models of risk that depend on the speed and mass of the newly invented car, as compared to the speed and mass of the horse cart. It might even be possible to verify this physics-based model with limited data. The actuary would also need a socioeconomic model, with many uncertain parameters, to calculate how many cars would be on the road in the next 80 years, how much these cars would cost, and how frequently and how fast people would drive. An important assumption would be the cost of damages due to an accident, which could be a nonlinear function of the assumed driver speed. If we make enough assumptions, we can always come up with a number and error bar for the risk.

If Karl Benz were to talk to a different insurance company, Stagecoach Security Assurance, their actuary might come up with a very different estimate for the applicable premium. Calculating the long-term accident liability risk for the world's first gasoline-powered automobile is not an actuarially sound problem, which means that any premiums that an actuary calculates would depend more on the assumptions made than on the statistics of the available data.

A hypothetical Insurers Association could have appointed an expert committee to assess the risk using plausible assumptions, which would have produced a single number for the premium. But expertise comes from experience. Experts on past stagecoach performance could have estimated risks associated with newly introduced stagecoach designs. But would any experts in the 1880s have been able to predict actuarially sound losses due to automobile accidents 80 years down the road?

Likewise, calculating the long-term damage risk for the world's first greenhouse-gas-driven warming event 80 years into the future may not be an actuarially sound problem. Recall that Svante Arrhenius, the first scientist to calculate climate sensitivity and arguably the foremost expert on climate change in his time, made plausible assumptions in 1896 to come up with perhaps the first "business-as-usual" scenario for carbon emissions (Section 3.4). He estimated that it would take 3,000 years for human activities to increase atmospheric carbon dioxide by 50 percent – but, by 2020, only 124 years from Arrhenius' estimate, we have already increased carbon dioxide by 40 percent. Assessing the likelihood of carbon emission scenarios for estimating climate risk is difficult. According to economist William Nordhaus, who won a Nobel Prize, in part for his work in calculating climate risk, subjective or judgmental probabilities must be used in models that calculate climate risk because objective probabilities, in the actuarial sense, are not available. The problem is that there is no unique way to determine these judgmental probabilities.[6]

It is said that Lloyd's of London, the famous insurance marketplace, will insure anything! This includes unprecedented one-off events. Lloyd's has insured body parts of famous performers and has reportedly promised cash payouts in case of alien abductions and the zombie apocalypse. Since such risks are essentially unquantifiable, the underwriters at Lloyd's use a narrative-style approach, based on past experience, to estimate the premium.[7] Using this approach, if Karl Benz is the only person buying car insurance 80 years ahead, his insurance company can still sell him a policy, even if the risks are uncertain. Since the coverage is limited to one person, underestimating the risk won't ruin the company, because it will average out against overestimates of risk in other one-off insurance policies. But if thousands or millions of people want to buy policies against the same event, the mistakes made in calculating insurance premiums for the event will not average out. Perhaps this happened to Lloyd's in 2018, when it lost more than a billion dollars[8] due to a spate of claims related to the California wildfires – likely exacerbated by global warming – and other natural disasters. The losses were such that Lloyd's ended up exiting some California insurance markets.

For the global warming problem, the analog of insurance premiums is the *social cost of carbon* – the cost in today's dollars of the long-term impact of one ton of carbon dioxide emitted today. In other words, it is the insurance premium per ton of carbon dioxide to cover the cost of future damages, as the emitted carbon stays in the atmosphere and warms the planet. If we are not willing to pay that premium now for the emissions that we are responsible for, then we are not taking responsibility for the damages our actions will impose on the rest of the world.

The social cost of carbon is calculated using an integrated assessment model, the same kind of model that is used to compute climate change scenarios (Section 7.6). These models make numerous assumptions about economic growth, technological progress, and something called the *discount rate*, expressed in percent per year. The discount rate is the rate by which we discount a future benefit compared to today's price for the same benefit. If we don't spend some money now, we can invest it, and it will be worth a lot more in the future. For example, a lump sum lottery payment is considerably lower than the total annuity payout over a 30-year period because of discounting. In a quantitative sense, the discount rate is rather like an annual return on financial investments, such as a bank interest rate or stock market growth rate. But in a more qualitative sense, it is also a measure of how much we value our own happiness, against the happiness of our children and grandchildren.

Nordhaus and another economist, Nicholas Stern, carried out two integrated assessments using two different discount rates, 3 and 1.4 percent, respectively.[9] They came up with two very different numbers for the social cost of carbon. Nordhaus estimated the social cost of each ton of carbon dioxide to be less than \$10, whereas Stern estimated it to be \$85.[10] Nordhaus later updated his estimate to a higher number,[11] but that only serves to underscore the uncertainties in assessing the monetary impact of climate change. If we are interested in purchasing climate insurance but get wildly different quotes for the premium, what are we to do?

We could survey many different integrated assessment models and come up with an average spread for the social cost of carbon. But this spread may be more a measure of the diversity of assumptions in the models than an estimate of the true error, as we discussed in the case of climate models (Section 11.3). If all integrated assessment models were to copy techniques and assumptions from each other, the spread would be smaller, but that does not mean it would be more accurate in any sense. We could survey multiple economic experts, instead of surveying models, to estimate the average social cost of carbon and its standard deviation. But there really are no experts on what the economy will be like under a much warmer climate. As Nordhaus says, "we cannot [...] estimate the economic impact of a 3°C rise in global temperature from historical data because nothing resembling that kind of global change has occurred in the historical record of human societies."[12]

Since discount rates are basically "guesstimates" on centennial timescales, there is no way to prove that any particular discount rate is the "true" rate. Discount-rate uncertainty can have a bigger impact on estimating the financial cost of long-term climate change than uncertainties associated with scientific parameters such as climate sensitivity. Attempting to assign numerical values to the human and ecological cost of climate change will further compound this uncertainty.[13] As economists Gernot Wagner and Martin Weitzman argue, climate change "belongs to a rare category of situations where it's extraordinarily difficult to put meaningful bounds on the extent of planetary damages."[14] Perhaps instead of arguing over what the "true" social cost of a ton of carbon dioxide is, we ought simply to accept that it is a poorly known unknown of the real world with a broad range of possible values – not unlike the many poorly known unknowns in climate models (Section 13.1).

The goal of mandatory car insurance is to distribute the cost of few accidents among a large pool of drivers, so that people can continue to drive without going bankrupt after an accident. But if the goal were to discourage people from driving altogether, then the mandatory premium would need to be set as high as is economically viable. For climate change mitigation, the goal is indeed to discourage carbon emissions, which we need to do as strongly as is feasible. As Wagner and Weitzman say, "the appropriate price on carbon is one that will make us comfortable enough to know that we will never get close enough to 6°C and certain eventual catastrophe."[14] Therefore, we should perhaps treat the social cost of carbon in real life – the implied carbon price – as a parameter that we can tune to obtain the desired reduction in emissions, much as we routinely tune model parameters to get better climate simulations.[15]

19.2 Blowin' in the Wind: Reasoning through the Fog of Numbers

Not everything in life is quantifiable. Musician and Nobel Laureate Bob Dylan asked some of the most famous numeric questions in literature, such as how many roads a man needs to walk down to be called a man. The answer to all his questions, as he noted somewhat mystically (and meteorologically), was "blowin' in the wind." Focusing solely on the numeric answers to questions can obscure the different types of uncertainty that lie behind different answers. We make decisions involving uncertainty every day of our lives, and we generally do not try to numerically estimate the risk involved in, say, crossing a busy road. Even numeric measures of uncertainty can have their own uncertainty. What is the error bar of the error bar? The answer, sometimes, is blowin' in the wind.

In *Radical Uncertainty: Decision-Making beyond the Numbers*, British economists John Kay and Mervyn King argue that numeric probability estimates produced by risk analysis models provide a false sense of precision. In the aftermath of Donald Rumsfeld's famous remark about known and unknown unknowns, there was a massive intelligence failure in finding the weapons of mass destruction whose purported presence led to the Iraq War. It turned out that there were no such weapons. A directive issued subsequently required that numeric probability values be used in all risk analysis. In 2011, when President Barack Obama had to decide whether to launch the raid that killed Osama Bin Laden in a compound in Pakistan, security advisers presented conflicting numbers for the probability that Bin Laden was actually in the compound. Obama said, "What you started getting was probabilities that disguised uncertainty as opposed to providing you with more useful information."[16] He ended up treating the decision essentially as a coin flip.

Kay and King also argue that the 2008 global financial crisis was triggered in part because the assumptions underlying the models used to make quantitative estimates of risk turned out to be wrong, which had enormous consequences for the world economy.[17] When planning for the future, expressing all risks simply as numeric values allows them to be easily aggregated for the purpose of mathematical analysis.[18] But this leveling of the uncertainty playing field for mathematical convenience ignores the different levels of qualitative uncertainty associated with many of the poorly understood risks.[19] The quantification of all uncertainties, which can be considered a form of "probabilistic determinism," is a contentious issue in climate prediction. This is especially the case for emission scenarios, which are sometimes assigned subjective numeric probabilities because the risk analysis methodology requires those numbers.[20]

An obvious difficulty with assigning a probability or likelihood to a scenario is that our decisions will alter the scenario – an effect known as *reflexive uncertainty*. Some have therefore proposed that socioeconomic human behavior should be directly modeled within climate models.[21] We have already argued that there are limits to the reductionist approach in modeling even the physical aspects of climate, for which we have mathematical equations (Section 12.3). Models' verifiable skill in predicting short-term climate phenomena like the El Niño is fairly modest, and the spread in their long-term climate projections is quite large. Under the circumstances, attempts to extend the reductionist approach to the prediction of individual (or collective) human behavior, which is not very amenable to mathematical description, appear hubristic.

We know from history that the actions of one human or the attractions of one idea can alter the course of the future. The Butterfly Effect (Section 2.4) does not apply to an individual climate projection, which uses a plausible emission scenario as a prescribed boundary condition. But if the atmospheric boundary condition is determined interactively by a predictive socioeconomic model that is coupled to the climate model, then the Butterfly Effect and deterministic predictability limits come back into play. The complex nonlinear numerical evolution of the combined socioeconomic–climate system will exhibit chaotic behavior over time, leading to unpredictable transitions between different scenarios (e.g., see Figure 12.1) and generating brand-new scenarios. Long-term deterministic socioeconomic prediction is not feasible. We can only consider a range of plausible scenarios.

An alternative to numerical models and subjective probabilities, as proposed by Kay and King, is to use a narrative process that asks human experts to assess risks associated with future events, akin to the approach used by Lloyd's of London. The goal of such an approach is to cut through the fog of numbers that can obscure logical reasoning. Such a narrative approach could describe short-term climate change over the next decade or two; events on this timescale would likely not be too dissimilar to events that have occurred in the past, for which there is human expertise available. But such an approach would only solve half the problem when predicting unprecedented events associated with global warming over the next century or two for which there is no human expertise available. The narrative approach could alleviate the problem of false precision associated with probabilistic interpretations, but it would not reduce the underlying uncertainty – and if the human "experts" were less constrained in their predictions than the climate model, the uncertainty might even be amplified.

In the context of climate change, any methodology for determining the course of action, narrative or numerical, must respect all known constraints on the system. These constraints include conservation of energy, momentum, mass, carbon, and the land area available for specific uses, among others. If there were only a few simple constraints that could easily be considered by the human mind, a narrative process could work. But for something as complex as the climate system, a mathematical or computer model becomes a necessary part of any decision-making process. Machines are better than humans at keeping track of small differences between large numbers that determine emergent properties of complex systems, which is why computers handily beat humans at weather forecasting. We therefore have to rely upon complex climate models for long-term planning, despite their imperfections. But we can choose not to interpret their numeric outputs too literally.

Recently, an alternative "storyline" approach to framing the impact of climate change has been proposed.[22] (Note that this usage of "storyline" is not the same as the "scenario storyline.") This is more of a qualitative narrative-style approach, using plausible assumptions and expert opinion, that does not rely upon unquantifiable probabilities. It has the virtue of combining the physical science aspects of climate and the societal impacts of climate change. This "storyline" approach is worth exploring, provided it does not rely upon overly simplified climate models – such as purely thermodynamic models – and provided its conclusions are consistent with the predictions and constraints of comprehensive climate models. For all their limitations and uncertainties, comprehensive climate models are better for prediction than unconstrained ideation.

19.3 Planning for the Future: Spanning the Uncertainties

Plans are worthless, but planning is everything.

Dwight D. Eisenhower, US President and former general[23]

There are two strategies we can deploy in the face of climate change: *mitigation* through emission reduction and *adaptation* to climate impacts. We can debate the relative importance of these two strategies,[24] but both require planning for the future. The preparatory analysis that goes into a plan can be more important than the plan's actual details because a plan may need to be repeatedly modified as new information is received during an evolving crisis. An extreme example of this is the unexpected discovery of the ozone hole (Chapter 6). Atmospheric chemistry models did not anticipate this development, but the models were sophisticated enough that they could be quickly adapted to deal with it. In the climate crisis, as socioeconomic developments and warming temperatures create different impacts, emission scenarios will need to be revised, models will need to be modified, and plans will need to be changed.

To optimally plan for climate change mitigation and adaptation, we need to predict the future to the best of our abilities. For the climate crisis, this requires more than predicting just the physical aspects of climate, such as temperature, rainfall, and winds. We need also to assess socioeconomic impacts: A city planner, for instance, may need to know what the flood danger for a new neighborhood is, especially one near the coast. A hydrologist may want to know how rainfall patterns may change in the future. A navy may want to know where to locate a new base along a coastline. To answer questions of this nature, we need to figure out how the climate of a region may change in the future. Comprehensive climate models provide our best estimates of that information.

Our knowledge of climate sensitivity can be characterized by the range 2.0–4.5°C. Focusing on this range may be sufficient for qualitative global policy discussions, but it is not sufficient for a detailed analysis of impacts. A global property like climate sensitivity does not provide much information about regional climate change because it only captures the properties of a single variable (temperature) for a single time-independent scenario (carbon dioxide doubling) for a single region – the "global average." Since warming does not occur uniformly from region to region, many different patterns of regional climate change can average out to produce the same global warming.[25]

To assess local impacts, we must estimate climate properties for each region of the globe, for each scenario, and for each property of interest (such as temperature, rain, and winds). We need a global model to compute this information, even if we are only interested in a specific region, because regional prediction skill is primarily derived from global prediction skill for the interconnected climate system. For example, rainfall in California is affected by a phenomenon known as the Pineapple Express, an atmospheric wind-flow pattern (or "atmospheric river") that brings with it moisture from the tropical Pacific Ocean. To predict trends in California rainfall, we need to predict how the Pacific will respond to climate change. Of course, it is a good practice to validate regional climate predictions using local knowledge and expertise wherever possible,[26] rather than blindly feeding them into a socioeconomic impacts model. (A regional model with a finer spatial grid, known as a regional climate model, is sometimes used to add detail to predictions from the global model, through a process called *downscaling*.[27])

The only way to compute the statistics of regional climate change in a manner that respects all the physical constraints of the system is to perform a large number of computationally expensive simulations, using a set of comprehensive climate models. This may seem daunting, but it is routinely carried out every six to eight years as part of the IPCC's assessment process. All the output data from different climate model simulations are made freely available for downloading[28] so that individual users can use this data without having to repeat these expensive computations. Sorting through and analyzing the voluminous amount of data can still be a challenge, though.

Let us walk through the different steps in using climate model simulations to assess regional impacts. To quantify the impacts, we need to feed predictions from the climate model into a socioeconomic impacts model that assesses damages. (The choice of impacts model varies from application to application, so we will not discuss it here.)

In predicting the properties of physical climate, there are three main types of uncertainties that we need to deal with.[29] Ranked roughly by how hard it is to quantify them, these are:

I *Scenario uncertainty* (boundary condition error)
II *Model uncertainty* (structural and parametric error)
III *Stochastic uncertainty* (internal variability error)

To predict climate, we first need to predict the boundary condition: the emission of carbon dioxide and other greenhouse gases that warm the planet (Section 7.6). How high do we think emissions will get? How low can they become? This is type I uncertainty, *scenario uncertainty*. It is small over the first decade or so of a climate prediction, but it grows over time, becoming the dominant source of uncertainty over centennial timescales (Figure 7.4c).[30]

Anticipating the societal behavior and economic consequences that will determine carbon emissions over a 100-year period is perhaps the hardest uncertainty to deal with. This uncertainty falls outside the realm of science, at times veering closer to science fiction;[31] some emission scenarios may assume the development of technologies that do not exist yet, such as affordable geoengineering to capture carbon dioxide from the atmosphere.

To bracket the range of climate impacts, we first choose a high-emission scenario – a "business-as-usual" or baseline scenario – to serve as the upper bound of climate risk. This can be a controversial choice since there is no longer a single designated baseline scenario in the new SSP scenario framework (Section 14.4). We need to carefully examine the assumptions behind the different scenarios, and perhaps even choose multiple high-emission scenarios. Once we have made our selection, we can choose a low-emission scenario to serve as the lower bound. This is less controversial because it is more about what humans *should* do than about what humans *will* do. Finally, we should choose a few more plausible scenarios between the high- and low-emission cases, so that the continuum of possibilities is well sampled, and so that our attention isn't exclusively focused on two extreme outcomes, neither of which may come to pass. (As climate scientist Stephen Schneider has noted: "End of the world" or "good for you" are the two least likely outcomes in the spectrum of potential outcomes.[32])

To make any statement about the future, we need to use a model (Section 14.1). But we have many models to choose from, and they are all imperfect, often in different ways. What models do we choose to use? This is type II uncertainty, *model uncertainty* (Figure 7.4c). Since different models are imperfect in different ways, we should choose multiple models to span a range of imperfections (the Anna Karenina principle; Section 11.1).[33] But some models just may not be good enough for us to work with. We need to decide: What is an acceptable model?[34] We could consider a model acceptable if it is sufficiently comprehensive in representing all the important processes and physical constraints of the climate system, without overly relying on adjustable parameters. The model should also be able to simulate the statistical characteristics of recent climate change (global warming since the preindustrial period) and current climate reasonably well. It may not be worth fixating on how accurately the model simulates the past, because there is no clear way to define a single accuracy metric for that – and, in any case, a model's performance in a single metric may simply reflect the effort put into tuning the model to that metric. But we should require models to pass a certain overall simulation quality threshold before we declare them fit for the purpose of prediction. If our focus is on a specific region, any model deemed acceptable should use local expertise to ensure that relevant features and processes are simulated well. Among the models that are deemed acceptable, we should adequately sample model diversity to properly compute the multimodel average (Section 11.3).

Even if we have the perfect boundary condition (emission scenario) and the perfect climate model, we cannot make a perfect prediction of the future. We would need perfect knowledge of the initial conditions to do that (due to the Butterfly Effect). We don't know, and can never hope to know, the precise initial conditions for the atmosphere, ocean, land, sea ice, and every other component of the climate system. But we can quantify this uncertainty using an *ensemble* of climate predictions using the same model but starting from a range of initial conditions, as we did for probabilistic weather forecasting (Section 2.5). Each prediction in the ensemble will be one realization of the future, and our actual future will be (hopefully) yet another realization wandering among the many simulated realizations.

The average prediction of the ensemble, for each model, is the climate prediction (the signal) responding to the boundary condition (Chapter 5). The spread among the different realizations is the noise, or the error bar, which represents type III uncertainty, *stochastic uncertainty*. This type of uncertainty is essentially irreducible because it is an expression of the inherently random nature of the climate system (Section 7.5).[35] Globally, this is the dominant uncertainty over the first decade or two of a prediction, but it becomes relatively less important than the other uncertainties afterwards, even though its magnitude remains constant.[30] In regional climate prediction, this uncertainty remains important for much longer, many decades into the future, because the stochastic noise is not attenuated by spatial averaging.

As we have emphasized, different scenarios and models should be treated as providing collective information, a consilient view of the future that should be taken seriously. Predictions made using a single scenario or a single model should not be taken literally. For type I and type II uncertainties, it may not be necessary to sample a very large number of scenarios or models, because the goal is to span the range of impacts. For type III uncertainty, it is essential to have a large enough ensemble of predictions,[36] typically with 50 or more predictions for each model, to sample the full probability distribution of internal variability. Using a single simulation, or even a small ensemble, can obscure the distinction between stochastic variability and model uncertainty.[37] To estimate stochastic uncertainty accurately, therefore, it may be preferable to use a less sophisticated model that is computationally affordable than a more sophisticated "cutting-edge" model that is too computationally expensive to permit a large ensemble of predictions.

Limitations on quantifying scenario uncertainty apply primarily to long-term climate predictions, looking decades or centuries into the future. For short-term climate predictions looking only a decade or two into the future, type I uncertainty is not as important because the different scenarios will not yet have diverged. The biggest uncertainties for these short-term predictions are of type II and type III.[30] We could use additional models to better sample model diversity for short-term predictions, to try and reduce type II model error through averaging. While type III uncertainty is essentially irreducible, improved knowledge of the oceanic initial condition can reduce it somewhat over the early part of the prediction by delaying the growth of stochastic error.[30]

The ensemble of simulations using multiple scenarios, models, and initial conditions may end up predicting a range for local climate variables that is too broad for meaningful impact assessment. For example, the range of predictions may span both increased and decreased rainfall in a region.[38] How can we work to narrow this uncertainty? Will improving models help? Improvements in climate models will have no effect on type I scenario uncertainty, and, because internal variability is fundamentally irreducible, model improvements will not reduce type III stochastic uncertainty, either. Actually, making climate models more comprehensive (by adding slow processes like permafrost feedbacks and continental ice sheets) could actually increase stochastic uncertainty by quantifying previously unquantified uncertainties (Section 9.5). In principle, improving deficient model components should help with decreasing type II model uncertainty, but, in practice, compensating errors can prevent model improvements from decreasing the spread in emergent properties like climate sensitivity (Section 12.2). This means that type II uncertainties that affect long-term climate prediction may not decrease with model improvements. Instead, model improvements may need to focus on improving the general fidelity of climate simulations and on increasing the skill of short-term climate predictions.

Promising areas for model improvement include the simulation of extreme weather, regional climate, and prediction of short-term climate phenomena like the El Niño.[39] These are often overlooked when the focus is on predicting the long-term evolution of global-average temperature. Even as they become more and more elaborate, many models continue to exhibit large errors in simulating regional climate.[40] Fine-scale maps of predicted regional climate change can provide "the illusion of precision" if they are taken literally rather than seriously, as many uncertainties may not be properly represented.[41] And extreme weather phenomena, such as typhoons or hurricanes, are still poorly simulated by models, in part because the grid is not fine enough. The advantage of focusing on predicting short-term climate phenomena like the El Niño is that these predictions can actually be verified, which allows us to track down and fix errors in model components. Steady improvements in weather forecasting skill (Figure 2.2) suggest that progress is possible in these areas, which means that we may also see improvements in short-term climate prediction.

We have outlined a rather complex procedure to assess regional impacts, involving computationally expensive models and voluminous amounts of model output. Is there a cheaper or quicker way to assess these impacts? Not really. There is no "secret sauce" or shortcut that can accurately provide all the required information.[42] Alternative approaches, such as a simplified climate model or a data-driven statistical model, would involve sacrificing fidelity – they may neglect important processes and feedbacks, and possibly abandon some physical constraints.[43] For example, trying to estimate regional impacts statistically from an assumed probability distribution of global properties like climate sensitivity could lead to very uncertain estimates of tail probabilities (Figure 15.2). Rather than relying on poorly known statistical tail distributions of global properties,[44] a more robust way to assess the impacts of extreme climate change would be to average the statistical properties of extreme weather simulated by the acceptable climate models under a plausible high-emission scenario.

19.4 Assessing Climate Risk: The Difficulties with Probabilities

What humans really want, according to philosopher Bertrand Russell, is not knowledge but certainty. But what climate models provide is not certainty but knowledge. To make decisions regarding mitigating or adapting to climate change, we must convert the knowledge provided by climate models into risk assessments.[45] Risk is defined as the product of damage and probability. Earlier in this chapter, we discussed some of the difficulties, such as the choice of a discount rate, in assessing financial damages many decades into the future. To compute the *absolute risk* of climate change, we also need to ask: What are the probabilities of all the weather and climate events that can cause damage in future scenarios? This is a hard question because the answer can be, in a sense, "blowin' in the wind."

The collection of simulations carried out using the range of plausible emission scenarios and diverse models will provide numbers describing the

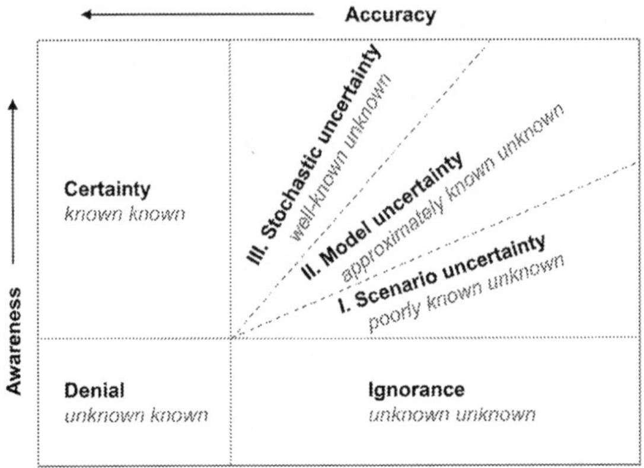

Figure 19.1 Climate impact and risk assessment mapped (approximately) to the deep uncertainty matrix (Figure 15.1). Stochastic uncertainty of impacts can be quantified using probabilistic error bars. For model uncertainty, we can estimate numeric ranges but not probabilistic error bars. For scenario uncertainty, numbers may not provide more than order-of-magnitude estimates for impact and risk.

occurrence of heat waves, heavy rainfall, high wind speeds, and more for each region, model, and scenario. For any one event, such as a heat wave, it is tempting to treat the spread of these numbers as a probability distribution of the event's occurrence – as a probabilistic prophecy. This is not a bad assumption for stochastic uncertainty. Chaotic redistribution of climate states, due to the Butterfly Effect, will make each of the ensemble predictions equally likely. This "certain uncertainty"[46] represents a well-known unknown in the deep uncertainty matrix (Figure 19.1), with a well-defined error bar and probability distribution.

Nonstochastic uncertainties are not easy to characterize probabilistically.[47] This is especially true of type I scenario uncertainty, analogous to the poorly known unknown in the deep uncertainty matrix. Each scenario needs to be treated separately because the IPCC does not assign numeric probability values to individual scenarios.[48] Also, the IPCC's caveat about scenarios being discrete samples from a continuum of possible futures is often lost in translation: Rather than analyze a range of scenarios, analysts "tend to use one of those scenarios in a deterministic fashion without recognizing that they have a low probability of occurrence and are only one of many possible outcomes."[49] (If there is one deterministic prediction that we can make about the future, it is that determinism will never die – it is too deeply ingrained in human nature!)

To get around the problem of the lack of objective probabilities for future emission scenarios, IPCC scientists debated assigning numeric subjective probabilities to these scenarios.[50] It was argued that, otherwise, downstream users of climate predictions might assign their own numeric subjective probabilities,[51] perhaps in an even more arbitrary manner.[52] However, it could also be argued that those closest to the risk being assessed are in a better position to assess and compare impacts from different scenarios, as they have skin in the game. In either case, subjectively quantifying the probability of a scenario creates its own unquantifiable uncertainty.

Even type II uncertainty is not very amenable to a probabilistic description,[53] because the set of acceptable models is just an "ensemble of opportunity"[54] – it is chosen from the models that happen to be available, which means that it is not possible to assign a probability to a particular model. But type II uncertainty is fundamentally different from type I uncertainty: It is not about predicting the societal behavior of the unknown future, but about imperfectly simulating the scientific aspects of the Earth's climate. Therefore, if model diversity is sampled carefully, a simple and transparent approach (such as weighting the acceptable models equally), might be defensible for the purposes of computing the average, because the errors will tend to cancel out (the Anna Karenina principle). Assigning weights is less defensible when estimating the uncertainty or the error bar for climate properties. Therefore, type II uncertainty corresponds to the approximately known unknown of the deep uncertainty matrix (Figure 15.1), where we have confidence in the central range of values but not so much in the tail of the probability distribution.

We should try to use robust risk assessment strategies that do not require the assigning of subjective probabilities to scenarios and do not rely on accurate estimates of tail probabilities.[55] If subjective probabilities are assigned, it should be done in a transparent manner, and the sensitivity of the assessed risk to variations in the assigned probabilities should be clearly quantified. We should also focus risk assessment on shorter-term predictions (for 2050, say, rather than 2100) to reduce the role of scenario uncertainty and minimize the impact of model uncertainties.

So far, we have considered the difficulties in estimating the absolute risk of climate change, due to the lack of scenario and model probabilities. It is more straightforward to use climate models to assess a different type of risk, the *relative risk*, which only requires comparing one scenario to another. For example, we can assess the change in the risk of coastal flooding (due to sea level rise) from a high-emission to a mid-range-emission scenario. The advantage here is that type I scenario uncertainty is removed, and no subjective probabilities are necessary. Type II model uncertainty remains, but assigning equal weights to the acceptable models for averaging might be defensible, as before, so long as we sample model diversity carefully. The impact of model errors may also be mitigated somewhat when we compute the difference between the simulations for two scenarios, because some common errors in the simulations will be subtracted out. Type III stochastic uncertainty can be handled as usual, by using a large ensemble of simulations for each model.

One aspect of climate risk that we have not addressed so far is the risk associated with abrupt climate change events and global tipping points. Evidence from past climates suggests that such events are possible, but simple climate models that simulate such events cannot provide reliable quantitative estimates of the timing and magnitude of such events (Section 8.4). Complex climate models currently do not simulate such events, but that does not preclude their occurrence because some of the relevant climate processes are still not incorporated in the complex models. The risks associated with such events cannot be ignored, but they may need to be considered in a more qualitative fashion, outside the framework that we have outlined.

The data from climate model predictions can be fed into a risk analysis model, keeping in mind the caveats about how the different uncertainties should be interpreted. In this context, it is worth recalling Suki Manabe's version of Occam's Razor: Commenting on adding complexity to a climate model, he noted that buying a $10,000 amplifier to drive a $100 speaker will not result in better sound. But the amplifier will make it more expensive to diagnose and fix problems with the sound system. Given the uncertainties that already abound in climate prediction, and additional nonquantifiable uncertainties that relate to societal and ecological impacts, it may be counterproductive to use very complex mathematical techniques to assess the risk of long-term climate change. To echo how Kay and King deal with radical uncertainty, we should settle for a simple and transparent approach that is "good enough" to minimize risk[42] and does not interpret model-derived numbers too literally. If we are too focused on obtaining the "optimal" solution using a sophisticated but opaque quantitative approach, we may end up in the straitjacket of faux certainty – causing us to ignore other solutions that may turn out to be superior once nonquantifiable impacts and uncertainties are taken into account.

***

This book is about the science of modeling, but it is also about the limitations of modeling, such as the predictability limit for weather (Chapter 2) and potential limits to the reducibility of climate (Chapter 12). Economic modeling also has its limits: It presents us with a range of choices based on various assumptions about our socioeconomic future, but it is up to us to choose. Science and economics can force us to ask moral questions relating to climate change, but we cannot rely on them to provide final answers. The climate scientist Gerard Roe has said that "when the quantitative analyses are uncertain, it raises the relative importance of the moral arguments."[56]

Unconditional predictions of the future apply to the whole human–climate system. Like any complex system with nonlinear interactions, the human–climate system is subject to the deterministic predictability limit identified by Ed Lorenz and others. When we separate the climate system from the socioeconomic system, or distinguish between predictions and projections, we are wielding the analytic knife in order to make the human–climate prediction problem more tractable. But the Butterfly Effect is not eliminated by this reductionist approach – it is merely partitioned. Stochastic uncertainty explicitly represents the quantification of the physical Butterfly Effect associated with the initial conditions of the climate system. Scenario uncertainty implicitly represents the unquantifiable societal Butterfly Effect associated with the boundary conditions of the climate system, that is, the uncertainty in socioeconomic predictions of the actual trajectory of carbon emissions.[57]

For the socioeconomic system, each anticipated scenario presents the appearance of a deterministic trajectory of emissions. But any violation of the assumptions underlying that scenario could result in emissions following a new and unanticipated scenario.[58] The unproven expectation, therefore, is that making climate projections for a selected range of "deterministic" scenarios is sufficient to bracket the range of the actual chaotic evolution of the future trajectory of climate flitting amongst many scenarios (like those in Figure 12.1).[59] Hypothetically speaking, even building the perfect human–climate model would not eliminate the uncertainty due to deterministic chaos. Practically speaking, model imperfections arising from the limits to reducibility of the complex human–climate system add to this uncertainty.

The deep uncertainties that we enumerated in this chapter may seem overwhelming because they appear to be beyond our control: Stochastic uncertainty is fundamentally irreducible, and model uncertainty appears to be hard to reduce, at least in the near term. But we should keep in mind that the predictions of transient climate response (TCR) averaged across models have remained fairly steady over the years (Figure 12.2b): This transient warming due to doubled carbon dioxide concentrations – a situation that could come to pass in the next 70 years, without mitigation – is in the range of 1.3–3.0°C. We could perhaps learn to live with warming on the lower end of that range, if we implement aggressive adaptation measures, but an additional 3°C of warming on top of the current 1°C of warming will have catastrophic consequences for many regions. To put this number in perspective, when the Earth was *cooler* by about 5°C during the peak of the last ice age (Figure 5.3), the climate was radically different (Section 7.7).

Author and former trader Nassim Nicholas Taleb, who popularized the notion of black swan events, argues that asymmetry in the predicted range of climate impacts, and the uncertainty itself, are strong reasons to motivate precautionary measures, writing in a coauthored essay that:

> [it] is the degree of opacity and uncertainty in a system, as well as asymmetry in effect, rather than specific model predictions, that should drive the precautionary measures. Push a complex system too far and it will not come back.[60]

Uncertainties inherent in scientific predictions are not a barrier to urgent action on climate change; it is more the lack of political will to make difficult but necessary socioeconomic choices. Our actions to reduce carbon emissions will limit the range of future warming. This is not so much a scientific prediction as an expression of the scientific constraints on the human–climate system. Although we will never have the ability to predict next century's climate as accurately as next day's weather, we still have the power, through our actions, to narrow the brackets of future climate change. As climate scientist Zeke Hausfather told Bloomberg, the "uncertainties that we can't control really reinforce the need to control the one uncertainty we can, which is our own future emissions."[61]

20
Moonwalking into the Future

In the language of the Aymara people in the Andes, the future is behind us, not in front of us. The word for "in front of" (*nayra*) is the same as the word for "past"; the word for "behind" (*qhipa*) is the same as the word for "future." Older Aymara speakers point forwards when referring to the past, unlike younger Aymara speakers bilingual in Spanish, who tend to point backwards.[1] This metaphor for the passage of time equates it with the source of knowledge. We can see the past in front of us, albeit less clearly as we get further away from it, but we cannot see the future.

NASA climate scientist Kate Marvel has said that "climate change isn't a cliff we fall off, but a slope we slide down."[2] Think of climate as a car we have been driving backward, at greater and greater speed, down the slope of climate change. The slope could end in a precipitous drop-off – we don't know, and we can't look behind to check, but we can change the angle and speed at which we drive backward. The only direct knowledge that we have is what we see through the windshield and the side windows: the past and the present. Models are like the hazy rearview mirror, providing an imperfect view of the future behind us. We need to combine all this information and adjust our driving accordingly.

Global warming is real, and climate predictions must be taken seriously. It is hard to precisely quantify the uncertainty associated with predictions of next century's climate, but we do have enough qualitative and quantitative information about global warming to warrant action. The deep uncertainty that precludes us from ruling out very low or very high warming is not an excuse for inaction: In fact, it should motivate us to take stronger action than we would take if we were certain of a precise degree of warming.[3] But how should we act if the future behind us, viewed only through the rearview mirror, is too blurry for us to know how worried we should be?[4]

The answer is staring us in the face, as the Aymara might say. The best defense against dire warming scenarios and scary climate tipping points, the steeper slopes and cliffs that may lie behind us, is to stay as close as we can to the preindustrial conditions that we see before us. We know that our climate was relatively unchanged for a period of 10,000 years.[5] This knowledge comes not from computer models of climate but from the fields of history, archaeology, and paleoclimatology. They tell us that we are currently near a stable climate equilibrium: The more mitigation steps we take to stay close to that equilibrium, the less likely we are to reach catastrophic levels of warming or tipping points. If we act weakly or give up completely, allowing emissions to increase and warming to accelerate, we are more likely to encounter any tipping points that may exist, and we may end up in a very different equilibrium.

We can't turn back time, but we can roll back our emissions of carbon dioxide and other greenhouse gases. Like moonwalking – a dance move in which the body slides backward while the legs appear to be walking forward – we can continue to move back to the future while proceeding forward with lowering carbon dioxide emissions closer to the past values we see in front of us. Models cannot tell us precisely how much to roll back our emissions, but they tell us it is crucial that we roll them back as much and as quickly as possible.

Climate contrarians distrust the rearview mirror of model predictions, preferring to drive blindly with the assumption that all will be fine behind them just as it is before them. Climate doomers claim to clearly see a cliff's edge in the rearview mirror, but they are convinced that hitting the brakes is futile because the car is going too fast to stop in time. In between are those who accept that model predictions are uncertain but disagree over how much the car needs to slow down. Agreeing on a specific limiting value for global warming, such as 2°C, can be useful as a motivational target for climate change mitigation, just like the numeric thresholds we set for cholesterol levels to minimize health risk.[6]

Dealing with the climate crisis is not really about science anymore. It is about the steps we are willing to take today to secure the future of our children and grandchildren in our own nations, and to ensure the welfare of vulnerable populations in all nations. As the climate scientist Katharine Hayhoe puts it, "[t]he poor, the disenfranchised, those already living on the edge, and those who contributed least to this problem are also those at greatest risk to be harmed by it. That's not a scientific issue; that's a moral issue."[7]

The pressing question is: How can we rapidly reduce greenhouse gas emissions to zero? Although zero emissions is the ideal goal, a more practical short-term goal is "net zero," where truly unavoidable emissions are compensated by *verifiable* offsets. Reaching net zero will essentially stop global warming within a decade or so,[8] although it will take many centuries and negative emissions for temperatures to get back close to their cooler preindustrial values. We should note, though, that reaching net zero will not quickly lower the carbon dioxide concentration, because the natural removal of excess carbon dioxide already in the atmosphere will take many centuries.

The net zero emissions target addresses the quantification problem associated with Pascal's Wager: Just how pious should we be to avoid "environmental damnation"? In this case, we should try to be as pious as we used to be. The piety of preindustrial conditions (near-zero carbon emissions) is our target. Preindustrial piety was possible with primitive energy technology because overall energy needs were low. But relying upon personal sacrifices to reduce emissions is not likely to be sufficient; those who desire to be pious may be unwilling to make the necessary personal sacrifices,[9] especially if they feel that the burden of the sacrifices is not being equitably shared. Therefore, we need preindustrial piety without the preindustrial asceticism. We must achieve this piety at the infrastructural level – through regulatory actions, carbon pricing, and targeted investments to achieve mitigation – while relying on modern "zero emission" technology to service the much larger societal energy needs.[10]

Some view the climate crisis as a fundamental conflict between humanity and nature, but we can frame it more narrowly as a dissonance between climate change and economic change. Both kinds of change feature models and uncertainty, known and unknown unknowns. But there are important differences. Climate models are subject to physical constraints, and, from those, we know that excess carbon dioxide will stay in the atmosphere for centuries. This means that climate adjustment timescales are very long, whereas economic upturns and downturns occur on timescales of a decade or so. The climate car has an extremely long braking distance; the economy car can brake much more quickly.

"Objects in the mirror are closer than they appear" is a common warning on passenger-side mirrors of cars. As we drive backwards, the dangerous warming scenarios in the climate car mirror may also be closer in the future than they appear. Or they may be further than they appear. We can't really be sure. There are two reasons to hit the brakes on our climate car – the uncertainty of how far away the danger is, and the certainty that the braking distance is very long. If we knew exactly how far away the danger was, we could perhaps delay when we started to brake. Since we do not know, we need to start braking as soon as possible. The uncertainty adds to the urgency.[11] Applying the brakes on the climate car – implementing strong and immediate measures to reduce carbon emissions – will absolutely transform the economy. The transformation should be combined with measures to assist those affected by higher energy costs and job losses.

We can make an analogy between poorly known unknowns in climate modeling and poorly quantified costs of future climate change in mitigation policies. Climate scientists tune or calibrate poorly known unknowns in their models to achieve a stable climate (Section 10.1). The numbers required for such mitigation policies as carbon pricing, cap-and-trade quotas, renewable energy mandates, and dollar amounts of technology subsidies or infrastructure investments can be treated as poorly known unknowns – tunable parameters. Due to all the uncertainties we have highlighted, there may be no a priori way to accurately determine optimal values of these parameters. We may therefore need to "tune" the real world to determine their optimal values and compensate for implementation errors that we will likely encounter along the way.

The history of the Tower of Pisa teaches us humility in the face of complexity, and the story of the ozone hole shows us that models cannot always anticipate the complexity of nature. We need to adaptively determine the best policy to mitigate climate change, by implementing a strong policy and adjusting its components based on their efficacy.[12] The lack of long-term certainty in the policy may be frustrating, but it is better than being stuck with an inflexible policy that turns out to be ineffective, or with no policy at all – both of which would lead to more disruptive climate change.

20.1 The Paris Potluck

Models and their predictions can help us plan how to address the climate crisis. But scientists can only provide conditional answers to the question, "How many degrees is it going to warm?" The more urgent question is, "What can we do now to prevent too many degrees of warming?" The answer to the first question depends crucially on the answer to the second question.

In December 2015, almost all the nations of the world came to an agreement in Paris, France, on how to tackle the climate crisis. The goal of this agreement, known as the Paris Agreement, is to keep global-average temperature increases well below 2°C, and ideally below 1.5°C. The Paris Agreement lays out carbon emissions reduction targets for each nation. Every five years, these targets must be revised higher to have more emissions reductions. Each nation gets to pick its own targets, and compliance with these targets is not mandatory; there are no real penalties for not complying. Countries are free to leave the accord, giving only a few years' notice. (The United States gave notice in June 2017 that it would exit the Paris Agreement, but it rejoined in 2021.)

Global agreements to fight climate change are negotiated by an entity called the United Nations Framework Convention on Climate Change (UNFCCC), the political counterpart to the scientific IPCC. In 1997, the UNFCCC negotiated the Kyoto Protocol, which, unlike the Paris Agreement, mandated emission reductions. But it didn't work very well: Canada formally withdrew from the Kyoto Protocol in 2011, and Japan and Russia eventually stopped implementing it, causing the protocol to collapse. The Paris Agreement was designed to be more robust, to be able to survive even without the participation of large nations such as the United States.

We should remember that the Paris Agreement was the bare minimum that could be agreed upon. It is not clear how effective its voluntary targets and voluntary compliance will be, but, at the moment, there appears to be no better way forward for internationally coordinated action to address the climate crisis. Hayhoe describes this situation using a potluck analogy.[13] Reducing global carbon emissions requires each nation to do its part, much like attendees at a potluck dinner. A potluck of friends and acquaintances relies on a combination of cajoling and shaming to assure that there is enough food for everyone, but this may not exactly work for a potluck of nations.

This potluck aspect of climate change mitigation is not always well understood. It has been pointed out, for instance, that California's pioneering climate change mitigation efforts did little to prevent global warming from aggravating the spate of wildfires that tore through the state in 2020.[14] California may bring more food to the potluck, but that does not mean it will get more to eat. If the other states (and countries) don't volunteer to bring enough dishes to the table, or if they bring less than they promised, then everyone goes home hungry. Although the climate crisis has regional impacts, it can only be solved through international coordination. It is very different from regional environmental problems like pollution, which can be successfully addressed through regional action.

A problem with many of the mitigation plans for staying below warming targets like 2°C is that they rely, to a significant extent, upon "negative emission technologies" that will actively remove carbon dioxide from the air.[15] These have been referred to as "carbon unicorns"[16] because these carbon-removal technologies are as yet unproven at global scale and may not be affordable.[17] Extending the potluck analogy, this is like the potluck organizer promising that a new, inexperienced catering service will ensure that there is enough to feed everyone at the potluck.

It is useful to compare this "potluck" situation with another climate-related problem that was solved through coordinated international action: the problem of stratospheric ozone depletion (Section 6.4). Admittedly this was a much more focused issue, and it was much easier for nations to agree upon a solution: the Montreal Protocol. But there are still some important lessons to be learned from the Montreal Protocol, which was more like a defined cooking plan than a potluck.

Ozone loss was perceived by the public as a clear and present danger. The increased risk of skin cancer and cataracts felt personal, and it helped that the science of ozone loss was much more straightforward to explain than the science of climate change. The very name "ozone hole" was evocative and scary – like a hole in the roof of the planet. It was also key that the culpable industries eventually bought into the existence of the problem. Du Pont, a major manufacturer of refrigerants that had previously resisted regulation, accepted in 1987 that CFCs would need to be phased out. In part, this was due to the discovery of the ozone hole and the increased public wariness of CFCs that followed. There also may have been a less altruistic motive: CFCs were not protected by patents, and developing patentable replacements was a wise business move. Indeed, new refrigerants without the harmful ozone-depleting properties of CFCs were soon developed. The use of these alternatives would be transparent to the end user, with no change in functionality from CFCs and little difference in cost. This provided an additional incentive for countries to join the Montreal Protocol: Countries that had signed the protocol could only trade refrigerants among themselves, which meant that non-signatories could not trade with signatories. There were also extra incentives to encourage less developed countries to participate.

The Montreal Protocol is adaptive. Its drafters were very much aware that scientific understanding of the ozone depletion problem was incomplete. The protocol included a mechanism to update the substances that needed to be regulated, and it stipulated that periodic scientific assessments would be carried out to amend it as needed, based on new scientific knowledge. The success of international environmental regulations like the Montreal Protocol, as well as national and regional environmental regulations addressing acid rain due to sulfur emissions,[18] inspired the structure of the Kyoto Protocol. But the scale of the greenhouse gas problem is vastly bigger than that of the CFC problem, and it touches upon every aspect of the global economy, including energy, transportation, and housing. Not to mention that the sacrifices required to drastically reduce carbon emissions are comparatively immense; it was the legally binding nature of the sacrifices required by the Kyoto Protocol that appears to have doomed it.

The Paris Agreement is a starting point, and its voluntary nature makes it less likely that it will collapse in the way that the Kyoto Protocol did. Recently, many countries, including major carbon emitters like China, have made impressive pledges to reduce their emissions and reach net zero.[19] But the voluntary nature of the agreement also means that there is no strong mechanism to ensure that these pledges are fulfilled, or to prevent pledges from being diluted or revoked by a nation due to unanticipated political or economic developments. There is a need for a stronger, but still adaptive, agreement that effectively reduces emissions, one that incentivizes joining and perhaps disincentivizes those not in compliance or those staying out.[20] Even businesses may be willing to support creating such an agreement if it is done in a transparent and globally coordinated manner.[21]

It is worth comparing the scale of the resources needed to achieve the goals of the Paris Agreement to those that were mobilized to deal with another global crisis, the Covid-19 pandemic. According to a 2020 essay in *Science* magazine, "low-carbon investments to put the world on an ambitious track toward net zero carbon dioxide emissions by mid-century are dwarfed by currently announced Covid-19 stimulus funds."[22] Despite the chaotic and frenetic nature of the early response to Covid-19 in many countries, vast amounts of resources were mobilized when the seriousness of the threat became apparent. However, due to the slowly evolving nature of the climate crisis and its uneven global impacts, communicating the urgency of allocating even comparatively modest resources remains a challenge. Unless we remedy this situation, we may end up waiting until the last minute to act, as the quote from the 2008 remake of *The Day the Earth Stood Still* reminds us: "[It] is only on the brink that people find the will to change. Only at the precipice do we evolve."

Epilogue

the only thing we have to fear is fear itself
Franklin D. Roosevelt, First Inaugural Address, March 4, 1933

Each one of us is an author of the climate change story. How should we ensure that it does not end as a tragedy? How do we deal with the problem formerly known as climate change, and now referred to as the climate crisis, the climate emergency, the climate catastrophe, and the climatocalypse? How optimistic or pessimistic should we be in tackling global warming? The answer depends on personal beliefs and values – it is beyond the realm of pure science.

Above all, our actions should be motivated by calm thinking, not desperation. "Fear is the mind-killer," wrote Frank Herbert in his 1965 science fiction novel *Dune*, one of the earliest fictional works to explore the theme of planetary ecology and climate change – fear that if we do not act, our species is doomed; or the fear that if we do act, our economy will be ruined.[1] Fear of an inevitable climate catastrophe promoted by some is motivating a few young people to refrain from having children.[2] Fear of a potential economic catastrophe touted by others is hampering even pragmatic mitigation measures.

Fear is a natural and essential instinct that helps us survive. But fear as a motivator works best when we know how to respond to a threat.[3] If we're not sure how to respond, fear can result in a rash and ineffective response – or it can paralyze us, resulting in no response at all. We need to channel the fear into rational thinking. Acting rationally, sometimes contrary to our instincts, was what allowed us to leave our nomadic lifestyles to build human civilization. Acting rationally is the best way to save human civilization from the climate crisis. Climate models are imperfect instruments of rationality, not infallible oracles of destiny; they cannot predict the future with certainty. That said, we have sufficient information from models, corroborated by other scientific evidence, to motivate urgent action. Perhaps the scariest thing about climate change is not the known unknowns (which we can plan for) or the unknown unknowns (which we can prepare for), but the unknown knowns – the misconceptions and disinformation that prevent us from planning and preparing rationally.

A central message of this book is that climate models are metaphors of reality; their predictions should be taken seriously, but not literally.[4] The quantitative details of a climate prediction are less important than the qualitative message that things can get really bad. This is a subtle distinction, but it has some important implications: Yes, we should act strongly now to reduce carbon emissions. No, humans aren't likely to become extinct. Yes, it is OK to have children. No, we don't have only X years to fix the climate crisis. And so on.

Predictions from climate and economic models should be treated more like a compass reading than a detailed map of the future. They tell us which way we should go, but not what our destination will be. If we encounter any unexpected obstacles on the way, we may need to take a detour. What should we choose for our destination if the planet and the economy are in disharmony? There is no Planet B, but there can be an Economy B. We need to trust reason, not fear, to get us there.

Glossary

Boundary condition: A factor or condition external to the weather or climate model that affects the atmosphere, such as solar heating, greenhouse gas concentrations, and volcanic eruptions.

Butterfly Effect: A metaphor for the unpredictability of complex systems. Small perturbations in the initial conditions for a weather model grow exponentially over time, leading to large errors in the weather forecast. Theoretically, a butterfly flapping its wings in Brazil could alter the initial atmospheric conditions to cause a tornado in Texas several days later.

Climate: The statistics of weather, averaged over many years. The climate of a region is characterized by average temperature, average rainfall, and other statistical properties.

Climate model: A computer program that is similar to a weather model but includes additional components, such as the ocean, that are required for predicting climate.

Climate prediction: A prediction of climate for the next season all the way through the next century, computed by running a climate model with appropriate boundary conditions, such as emission scenarios.

Climate sensitivity: The predicted amount of global warming that will occur if the amount of carbon dioxide in the atmosphere is doubled, as compared to preindustrial conditions. Usually expressed in °C, it is more precisely referred to as *equilibrium climate sensitivity*, since it is the long-term warming when the climate system reaches equilibrium after many centuries.

Deductivism: Reasoning from a set of basic assumptions or hypotheses to make testable predictions. More precisely known as *hypothetico-deductivism*.

El Niño: The occurrence of unusually warm temperatures in the tropical Pacific Ocean, which follows an irregular cycle with a period of three to five years. El Niño events affect rainfall and weather patterns in many regions of the globe, from Australia to California. The warmest temperatures occur around Christmastime off the coast of Peru – hence the name, given by Peruvian fishermen, which means "the little boy" (or "the Christ child") in Spanish.

Emergentism: Treating a complex system as more than just the sum of its parts to study properties that emerge only through interactions between parts.

Emission scenario: A projection of future greenhouse gas emissions based on assumptions about socioeconomic factors such as future population growth, economic development, government policies, technological innovation, and so on. Different scenarios assume different policies to avert climate change, ranging from no policy to an aggressive mitigation of emissions.

Fine-scale process: A process, such as clouds or aerosol particles, that occurs on spatial scales finer than a model grid box. Also known as *unresolved process.*

Ice age: A much colder climate. For the last one million years or more, the Earth's climate has fluctuated between "normal" climate – like what we had before global warming began – and "ice age." These fluctuations repeat in a quasi-regular fashion every 100,000 years or so. The last ice age ended about 12,000 years ago, and the Earth's climate entered the current warm period. During the peak of the last ice age, which occurred about 20,000 years ago, global temperatures were about 4–5°C cooler on average; giant continental ice sheets covered much of northern Europe and northern North America, and the sea level was about 120 meters (400 feet) lower than it is today. Ice age fluctuations are believed to have been caused by naturally occurring periodic changes in the Earth's orbit.

Inductivism: Reasoning from past observations of a phenomenon to draw general conclusions about it.

Initial condition: A set of initial values required to solve the equations of time evolution for a system. For the atmosphere, the initial condition is a snapshot of pressure, temperature, humidity, and winds around the globe.

Model: When used in a general scientific context, a representation of a real system, similar to a theory. In the context of weather and climate, a model specifically means a set of mathematical equations, solved using a computer, that represent the atmosphere and other parts of the Earth system.

Parameterization: An approximate formula used to represent the effects of a fine-scale process, such as clouds. The formula typically has adjustable coefficients or parameters, hence the name.

Reductionism: Studying a complex system by reducing it to its simplest parts.

Resolution: The number of boxes or "pixels" in the spatial grid or mesh of the computer model. The typical resolution of current global climate models is 360 × 180 (longitude × latitude) along the horizontal dimensions.

Resolved process: A process such as a weather system that is represented well because it has spatial scales larger than a single grid box of the model.

Theory: A scientific explanation of real-world phenomena. The explanation could be qualitative – a set of logical arguments – as in the theory of evolution. Or it could be quantitative – a set of mathematical equations – as in the theory of relativity in physics.

Transient climate response (TCR): The predicted amount of global warming that will occur if the amount of carbon dioxide in the atmosphere is doubled slowly, over a period of about 70 years, by increasing it at a rate of about 1 percent per year. It will be lower than climate sensitivity because the warming has not yet reached equilibrium.

Weather model: A computer program that subdivides the atmosphere into a three-dimensional grid and solves the equations governing the motion of air and transport of water in each grid box.

Weather prediction (or forecast): A prediction of what the weather will be like for the next 1–10 days, given the current initial condition of the atmosphere. It is computed by running a weather model.

Notes

Preface

1 Carroll (2013)

Introduction

1 Trumbla (2007), Roker (2015), Rohr (2016)
2 Blake and Zelinsky (2018)
3 Rainey (2017)
4 Woetzel et al. (2020), p. 13
5 Colman (2021)
6 Legates (2020)
7 Silver (2012), pp. 185–197

Chapter 1

1 Almanac.com (2016), p. 186
2 ibid., p. 217
3 NCEI (2017)
4 Graham (2018)
5 Batterson (2007)
6 Dyson (2012a), pp. 23–52
7 IAS (2013), p. 4
8 Veisdal (2020)
9 Dyson (2012b)
10 Edwards (2012a)
11 IAS (2013), pp. 21–22
12 Thompson (1983)
13 Finding aid for director's office: electronic computer project files. Institute for Advanced Study,

www.ias.edu/sites/default/files/library/page/DO.ECP_.html

14 Tropp (2003). See also John W. Mauchly papers, 1908–1980, Ms. Coll. 925. Kislak Center for Special Collections, Rare Books and Manuscripts, University of Pennsylvania, p. 4
15 Mauchly (1973), pp. 1, 13–14, 21
16 Tropp (2003)
17 Weik (1961)
18 Electronic computer project. Institute for Advanced Study, www.ias.edu/electronic-computer-project
19 Das (2008), Lynch (2008)
20 Holloway et al. (1955)
21 Mill (1843)
22 Popper (1935); Godfrey-Smith (2003), ch. 4
23 Godfrey-Smith (2003), ch. 3; Henderson (2018)
24 Overbye (2017)
25 Pirsig (1974), ch. 6
26 Carroll (2013)
27 Ayala (2009), Miner (2018)
28 Frisinger (1973)
29 Benjamin et al. (2019)
30 Bauer et al. (2015)
31 Silver (2012), p. 115; Flynn (2018)
32 Dyson (2012a), pp. 154–174
33 Phillips (1982)
34 Benjamin et al. (2019), p. 13.16
35 Thompson (1983), app. B (Letter from Jule Charney to Philip D. Thompson, Feb. 12, 1947)
36 Shaw (2019), Thompson (2019)
37 Smagorinsky (1983), p. 8
38 Witman (2017)
39 Haigh et al. (2014)
40 Lynch (2008)
41 Platzman (1979)
42 Phillips (2000)
43 Smagorinsky (1983), p. 10
44 Lynch (2007)
45 Smagorinsky (1983), p. 13
46 Persson (2005)
47 Harper et al. (2007)
48 As quoted in Fleming (2010), p. 194
49 ibid., ch. 7
50 Shalett (1946)
51 Emanuel (1999)
52 Zworykin (1945)
53 Dyson (1988), p. 182

Chapter 2

1 WMO Bulletin (1996), pp. 8–9
2 Sokal (2019)

3 Silver (2012), p. 120
4 Emanuel (2011), pp. 9–11
5 Platzman (1987)
6 Palmer (2008)
7 Lorenz (1963a)
8 NASA (2004)
9 Laplace (1814), p. 4
10 Almanac.com (2016), p. 217
11 Schrödinger (1935), Trimmer (1980)
12 The notion of lesser demons has been explored by Frigg et al. (2014) in the context of different sources of prediction error.
13 Canales (2020), ch. 2
14 Novak (2011)
15 Kuhn (1962); Godfrey-Smith (2003), ch. 5–6
16 Oreskes (2019), p. 39
17 Poincaré (1899), Oestreicher (2007)
18 Thompson (1957)
19 Smagorinsky (1983), p. 13
20 Benjamin et al. (2019), p. 13.18
21 Rasmussen (1992)
22 Ingleby et al. (2016), Ferreira et al. (2019)
23 Dabberdt et al. (2003)
24 Benjamin et al. (2019), p. 13.20
25 Lynch (2011), pp. 3–17
26 Bauer et al. (2015)
27 Hilborn (2004), Lorenz (2006)
28 Lorenz (1963b), p. 431
29 Lorenz (1993), p. 181
30 Palmer (2009)
31 Palmer et al. (2014)
32 Lorenz (1969), Lorenz (1993), pp. 182–183
33 On a related note, climate scientists are experimenting with the use of sound to represent climate data (Gardiner, 2019).
34 Edwards (2010), p. 141
35 Dessler (2015), ch. 1; Gettelman and Rood (2016), pp. 3–7
36 Silver (2012), p. 125; Bauer et al. (2015)

Chapter 3

1 Goldilocks principle, Resource Library, *National Geographic*, www.nationalgeographic.org/encyclopedia/goldilocks-principle/
2 Dessler (2015), ch. 3; Gettelman and Rood (2016), pp. 14–16
3 Pierrehumbert et al. (2007), p. 150
4 Dessler (2015), sect. 5.1
5 Trenberth et al. (2009), fig. 1
6 Fourier (1822)
7 Archer and Pierrehumbert (2010), pp. 3–4
8 Foote (1856), Huddleston (2019)
9 Jackson (2020)

10 Ortiz and Jackson (2020)
11 Scientific American (1856), p. 5
12 Sorenson (2011)
13 Kolbert (2006), p. 40
14 Rodhe et al. (1997); Weart (2008), ch. 1
15 Rodhe et al. (1997)
16 Arrhenius (1908), p. 63
17 Ramanathan and Vogelmann (1997); Archer and Pierrehumbert (2010), p. 50; Manabe and Broccoli (2020), p. 19
18 Allen and Ingram (2002)
19 Walker et al. (1981); Dessler (2015), ch. 5
20 For a broader discussion of forcings, see Kolbert (2006), p. 104
21 Hausfather (2020b)
22 Crane (2018)

Chapter 4

1 Lewis (2008)
2 Prono (2008)
3 Smagorinsky (1983), pp. 4, 11, 20, 26, 29; Lewis (2008)
4 Smagorinsky (1983), p. 33
5 Ceruzzi (1998), pp. 13, 15, 25, 34
6 Smagorinsky (1983), p. 32
7 AIP–Kasahara (1998), pp. 2–5
8 ibid., pp. 6–7
9 ibid., pp. 10–12
10 ibid., p. 10
11 AIP–Manabe (1998a), pp. 3, 4, 7, 8
12 Manabe and Broccoli (2020), p. 41
13 Smagorinsky (1983),p. 36; AIP–Manabe (1998a), p. 21
14 Kolbert (2006), p. 99; Silver (2012), p. 114; Gettelman and Rood (2016), pp. 38–42
15 Interestingly, one of the paper's coauthors was J. Leith Holloway, a programmer who was hired by Smagorinsky at GFDL. Holloway had previously worked with John Mauchly on analyzing sunspot correlations (Holloway et al., 1955).
16 Weart (2008), ch. 5
17 René Descartes, *Stanford Encyclopedia of Philosophy*, https://plato.stanford.edu/entries/descartes/
18 Pirsig (1974), ch. 6
19 Schmidt (2007b); Schmidt (2014); Gettelman and Rood (2016), p. 47; Vallis (2016)
20 Manabe and Broccoli (2020), pp. 22, 42
21 Weart (2008), ch. 1
22 AIP–Manabe (1998a), pp. 3, 4, 7, 21
23 Strictly speaking, Manabe's original parameterization did not use the actual dew-point temperature but assumed a fixed moist adiabatic lapse rate of 6.5°C/km, based on observations.
24 Manabe (2019)
25 Density of air at room temperature and pressure = 1 kg/m^3. Density of water = 1,000 kg/m^3. Specific heat of air = 1,004 J/(kgK). Specific heat of water = 4,182 J/(kgK).

26 Heat capacity of atmosphere = 1,004 × 100,000/9.81 = 10^7 J/(Km^2). Heat capacity of ocean = 4,182 × 4,000 × 1,000 = 1.6 × 10^{10} J/(Km^2). Taking into account that ocean occupies 70 percent of the surface area, Ocean/Atmosphere ratio = 0.7 × 1,600 = about 1,000.
27 Bryan (2006), pp. 31, 32
28 AIP–Bryan (1989), Griffies et al. (2015)
29 Bryan (2006), p. 32
30 Manabe and Broccoli (2020), ch. 8
31 AIP–Manabe (1989), p. 24

Chapter 5

1 NCEI (2017)
2 McAlpine (2015), Stormfax (2020)
3 Fried (2014)
4 Karl Popper, *Stanford Encyclopedia of Philosophy*, https://plato.stanford.edu/entries/popper/
5 Lorenz (1975)
6 Live Science (2020)
7 Harris (2010)
8 There is an additional source of noise in Mauna Loa, though, because it is an active volcano. Occasionally, it emits carbon dioxide from fissures at the summit. These events are detected and filtered out from the data (Harris, 2010, p. 7867).
9 Weart (2008), ch. 2
10 Since there is more land area in the Northern Hemisphere, it dominates the global seasonal cycle of carbon dioxide.
11 AIP–Manabe (1998a), p. 30
12 Peterson et al. (2008)
13 Manabe and Broccoli (2020), ch. 5
14 Somerville et al. (1974); AIP–Manabe (1998a), p. 17
15 Kolbert (2005)
16 Weart (2008), p. 100; Rich (2018)
17 Charney et al. (1979)

Chapter 6

1 Mill (1843), p. 323
2 www.bas.ac.uk/about/about-bas/history/british-research-stations-and-refuges/halley-z/
3 NRC (1996)
4 Clark (2013)
5 Rowland (2009)
6 Harvey (2013), Vitello (2013)
7 Shanklin (2010)
8 Farman et al. (1985)
9 https://earthobservatory.nasa.gov/features/UVB
10 www.acs.org/content/acs/en/education/whatischemistry/landmarks/cfcs-ozone.html

11 Sect. 3.5, Stratospheric ozone – An electronic textbook. NASA www.ccpo.odu.edu/SEES/ozone/oz_class.htm
12 Rowland (2009), p. 25
13 ibid., pp. 36–37
14 Molina and Rowland (1974)
15 NRC (1996), p. 6
16 Bojkov and Balis (2009), p. 87
17 Christie (2000), p. 44
18 U.S. scientists tackle mystery of seasonal antarctic ozone depletion, *Antarctic Journal of the United States*, June 1988, www.coldregions.org/vufind/Content/ajus
19 Q10. How severe is the depletion of the Antarctic ozone layer?, www.esrl.noaa.gov/csl/assessments/ozone/2018/downloads/twentyquestions/Q10.pdf
20 Schmidt (2017)
21 Solomon (1997), p. 10
22 ibid., pp. 12–13
23 ibid.
24 Schoeberl and Rodriguez (2009)
25 Solomon (1997), p. 23
26 Weiss (2009)
27 Rowland (2009), p. 56
28 https://treaties.un.org/pages/ViewDetails.aspx?src=TREATY&mtdsg_no=XXVII-2-a&chapter=27&lang=en
29 www.epa.gov/ods-phaseout/phaseout-class-i-ozone-depleting-substances
30 https://esrl.noaa.gov/csl/assessments/ozone/2006/chapters/Q16.pdf
31 Solomon et al. (2016)
32 Amos (2019)
33 Christie (2000), p. 90
34 Oppenheimer et al. (2008)
35 Christie (2000), p. 45
36 ibid., ch. 10

Chapter 7

1 https://pulitzercenter.org/sites/default/files/june_23_1988_senate_hearing_1.pdf
2 Rich (2018)
3 Oppenheimer (2007)
4 Agrawala (1998), Ravilious (2018)
5 Le Treut et al. (2007), p. 113, fig. 1.4
6 Flato et al. (2013), p. 753
7 Mitchell et al. (1990), p. 135; Albritton et al. (2001), p. 67; IPCC (2007), p. 12
8 IPCC (2013), p. 16
9 Mitchell et al. (1990), p. 135,
10 IPCC (1990), p. xxiii; Collins et al. (2013), p. 1037
11 IPCC (1990), p. xxx
12 ibid., p. xii
13 IPCC (1996), pp. 4–5
14 IPCC (2001), p. 10
15 IPCC (2007), p. 10
16 IPCC (2013), p. 17

17 Pearce (2010)
18 Keeling (1979); Weart (2008), ch. 2; Dessler (2015), pp. 81–84
19 www.nasa.gov/twins-study, Burrakoff (2019)
20 Bindoff et al. (2013), p. 874
21 Miller et al. (2014)
22 Dessler and Cohan (2018)
23 Hausfather (2017)
24 Soon (2005), Hulme (2007)
25 Donnelly (2009)
26 Philip et al. (2020)
27 Allen (2003a), NASEM (2016)
28 Hansen et al. (1988)
29 Fischer and Knutti (2015), van Oldenborgh et al. (2017)
30 Fountain (2017)
31 IPCC (2014a), p. 53
32 Worland (2021)
33 Vlamis (2021)
34 van Oldenborgh et al. (2019), Blackport and Screen (2020), Cohen et al. (2020)
35 Abatzoglou and Williams (2016)
36 Voiland (2010); Dessler (2015), pp. 95–101
37 https://earthobservatory.nasa.gov/images/44250/clouds-and-global-warming
38 IPCC (2013), pp. 13–14
39 Yan et al. (2016)
40 Feldstein (2000), Clement et al. (2015)
41 Hawkins and Sutton (2009)
42 Deser (2020). This is also referred to as aleatory uncertainty (Hora, 1996).
43 Smith et al. (2020)
44 Maher et al. (2020)
45 Collins et al. (2013), p. 1036
46 Carbon Brief (2018)
47 The scenario called "SSP3: Regional rivalry – A rocky road" (with "high challenges to mitigation and adaptation") is described in Riahi et al. (2017).
48 Gettelman and Rood (2016), pp. 131–132, 174
49 ibid., pp. 187–188
50 Collins et al. (2013), p. 1106
51 Taylor et al. (2012); Collins et al. (2013), p. 1035
52 Collins et al. (2013), p. 1037
53 ibid., p. 1107
54 IPCC (2013), p. 25
55 IPCC (2014a), p. 7
56 Plumer and Schwartz (2020)
57 Giorgi and Bi (2009), Hawkins and Sutton (2012)
58 Shepherd and Sobel (2020)
59 IPCC (2014b), p. 17
60 Allen (2019)
61 Betts et al. (2018)
62 Hawkins and Sutton (2009); Collins et al. (2013), p. 1037; Doyle (2013)
63 Palmer and Stevens (2019), Smith et al. (2020)
64 IPCC (2013), p. 25; Rahmstorf (2013)
65 Mengel et al. (2018)
66 Melting sea ice does raise the sea level by a small amount, due to the dilution of salinity after the melt (Noerdlinger and Brower, 2007).

67 Grossman (2016)
68 Collins et al. (2013), pp. 1084–1085
69 Sherwood and Fu (2014)
70 Alcamo et al. (2007), p. 554; Palmer (2012)
71 Knutti and Sedlaček (2013), Palmer and Stevens (2019)
72 Deser et al. (2012), fig. 1a, bottom
73 Deser et al. (2020), fig. 3a, CESM no. 35 vs. no. 8

Chapter 8

1 Salvia (2017), quoted with permission from Springer Nature/*Physics in perspective*.
2 https://solarsystem.nasa.gov/news/307/galileos-observations-of-the-moon-jupiter-venus-and-the-sun/
3 But Kepler had more observations to explain, compared to Ptolemy (Ball, 2016).
4 Borowski (2012), Spade and Panaccio (2019)
5 Jeevanjee et al. (2017)
6 Held and Soden (2006)
7 Markow (2015)
8 Held (2005), Held (2014)
9 Polvani et al. (2017)
10 Held and Suarez (1994)
11 Rogers (1984), Knutti (2018)
12 Smith (2002)
13 Held (2005), Knutti (2008a), Baumberger et al. (2017), Maher et al. (2019)
14 McSweeney (2020)
15 There is the danger that slow climate instabilities may be suppressed or weakened by this model-selection process (Valdes, 2011).
16 Wagner and Eisenman (2015), Boos and Storelvmo (2016)
17 Dudney and Suding (2020), Hillebrand et al. (2020)
18 Valdes (2011)
19 Dunne (2017)
20 Cubasch et al. (2013), p. 129

Chapter 9

1 Edwards (2010), p. 162
2 AIP-Washington (1998), p. 17
3 ibid., pp. 13, 70
4 AIP-Kasahara (1998), pp. 13, 14, 25
5 Washington (1970)
6 Gettelman and Rood (2016), pp. 45
7 Mauritsen et al. (2012)
8 Kerr (1994)
9 Edwards (2010), p. 165; Washington and Kasahara (2011)
10 Manabe et al. (1975)
11 Kerr (1997)
12 Kay et al. (2016)

13 Ramaswamy et al. (2001), p. 356
14 Dessler (2015), ch. 5; Gettelman and Rood (2016), pp. 121–125
15 Glato (2011)
16 Doyle (2013)
17 Dessler (2015), sect. 5.5
18 Dessler (2015), sect. 5.6–7
19 MacDougall (2020)
20 Hausfather and Betts (2020)
21 Maslin and Austin (2012)
22 Trenberth (2010); Cubasch et al. (2013), p. 141
23 Lane (2004)

Chapter 10

1 Why doesn't the Leaning Tower of Pisa fall over? *TED-Ed Lesson* 2019, ed.ted.com, www.youtube.com/watch?v=HFqf6aKdOC0; see also www.history.com/this-day-in-history/leaning-tower-needs-help
2 Borland et al. (2003)
3 Leonhardt (1997), Pisa Committee (2002)
4 Leaning Tower of Pisa. *Madrid Engineering Group,* http://madridengineering.com/case-study-the-leaning-tower-of-pisa
5 NOVA (1999)
6 Edwards (1999), Lenhard and Winsberg (2010), Schmidt and Sherwood (2015)
7 Hourdin et al. (2017)
8 Schmidt et al. (2017)
9 Gettelman and Rood (2016), pp. 79–80
10 Schmidt and Sherwood (2015), Hourdin et al. (2017)
11 Gettelman and Rood (2016), p. 194; Baumberger et al. (2017)
12 Oreskes and Belitz (2001), Bony et al. (2013), Schmidt et al. (2017)
13 Gettelman and Rood (2016), p. 193
14 Voosen (2016)
15 Gettelman and Rood (2016), p. 168; Hourdin et al. (2017), p. 595; Wang et al. (2021)
16 Golaz et al. (2013), Winsberg (2018a), p. 166
17 Emanuel (2020)
18 Oreskes et al. (1994), Lloyd (2010)
19 Winsberg (2018a), ch. 10
20 Hargreaves and Annan (2014)
21 Schmidt and Sherwood (2015)
22 O'Raifeartaigh (2017)
23 Lenhard and Winsberg (2010), Baumberger et al. (2017)

Chapter 11

1 Mirowski (1992), Sokal (2015)
2 Imagining the universe as a computer simulation is done in science fiction movies, such as *The Matrix*, and also in philosophy: Bostrom (2003), Carroll (2016).
3 Anderson (1972)

4 Parker (2006), Knutti et al. (2019), Schmidt (2020)
5 Diamond (1997)
6 Shirani-Mehr et al. (2018)
7 Stainforth et al. (2007)
8 Lenhard and Winsberg (2010)
9 Gettelman and Rood (2016), p. 194
10 Parker (2006), Hourdin et al. (2017)
11 Knutti et al. (2010), Schmidt and Sherwood (2015)
12 Stainforth et al. (2007), Knutti (2010)
13 Gleckler et al. (2008), Sanderson and Knutti (2012)
14 Hagedorn et al. (2005), Schmidt and Sherwood (2015), Barnston et al. (2019)
15 Sanderson and Knutti (2012)
16 Silver (2014)
17 Edwards (1999), Allen (2003b), Winsberg (2018b)
18 Schmidt et al. (2017)
19 Yong (2015)
20 Schmidt and Sherwood (2015)
21 Pipitone and Easterbrook (2012)
22 SIMIP Community (2020)
23 Parker (2006); Knutti (2010); Collins et al. (2013), p. 1036
24 van der Sluijs et al. (1998), Huybers (2010)
25 Allen and Ingram (2002), Stainforth et al. (2007), Brown and Wilby (2012), Schmidt and Sherwood (2015)
26 Allen and Stainforth (2002), Knutti et al. (2010)
27 Collins et al. (2013), pp. 1036, 1040
28 Knutti and Sedlaček (2013), Knutti et al. (2019)
29 Allen and Stainforth (2002)
30 Collins et al. (2013), p. 1036

Chapter 12

1 Schmidt (2018)
2 Michaels and Maue (2018)
3 Hausfather et al. (2020)
4 Knutti (2008b)
5 Dayton (2016), Freedman (2016), Grant (2016)
6 Scientists (2016)
7 Palmer and Stevens (2019)
8 Emanuel (2020)
9 Schmidt and Sherwood (2015)
10 Nissan et al. (2019), Fiedler et al. (2020), McKenna et al. (2020), Balaji (2021)
11 Hawkins and Sutton (2009), Doyle (2013), Voosen (2020b)
12 Hargreaves and Annan (2014)
13 Hansen et al. (1992)
14 Rasool and Schneider (1971), Peterson et al. (2008)
15 Oreskes (2007)
16 Parker and Risbey (2015)
17 Schmidt (2020)
18 Smagorinsky (1963)

19 Guerrero et al. (2020)
20 Silver (2012), p. 117
21 Panosetti et al. (2019)
22 Smith (2002), Knutti (2008a), Bony et al. (2013)
23 Knutti and Sedlaček (2013)
24 Bock et al. (2020)
25 Jackson et al. (2008)
26 Hausfather (2019), Meehl et al. (2020)
27 Fiedler et al. (2020), McKenna et al. (2020), Wild (2020)
28 Rich (2018)
29 Lenhard and Winsberg (2010)
30 Borges (1975)
31 Limits to reductionism are also topics in other sciences; see, for example, Hossenfelder (2019).
32 McWilliams (2007)
33 A similar term, "irreducible uncertainty," is used in climate modeling, but it usually refers to stochastic uncertainty or internal variability; for example, see Deser et al. (2012) and Hawkins et al. (2015). The structural instability of models has also been referred to as the Hawkmoth Effect, to contrast with the Butterfly Effect, in the context of predictions in low-order chaotic systems that depend upon initial conditions (Thompson, 2013; Frigg et al., 2014; Winsberg and Goodwin, 2016).
34 There is an interesting philosophical discussion about whether the existence of a reducibility limit implies that our universe is a computer simulation (Carroll, 2016).
35 Palmer and Stevens (2019), Panosetti et al. (2019)
36 Kozlov (2021)
37 Russian proverb frequently used by US President Ronald Reagan
38 Pierrehumbert (2014) and Revkin (2014), responding to Koonin (2014)

Chapter 13

1 A common distinction is between Knightian risk and Knightian uncertainty, or between aleatory and epistemic uncertainty; for example, see Knight (1922) and Fox and Ülkümen (2011).
2 Rumsfeld did not come up with this phraseology or classification of knowledge, and he credits its origins to NASA. But his use of it in a high-profile setting was key to this phraseology entering the vernacular and being forever associated with his name.
3 A somewhat analogous knowledge matrix, called the Johari window, has been used in cognitive psychology since 1955, with columns corresponding to self-knowledge and rows corresponding to knowledge of others (Luft and Ingham, 1955).
4 The well-known and poorly known unknown categories here can be thought of as being somewhat analogous to aleatory and epistemic uncertainty, in the jargon of risk analysis (Hora, 1996; Dessai and Hulme, 2003). The analogy is imperfect, though, because aleatory uncertainty is considered irreducible, but model improvements can convert well-known unknowns to known knowns. A similar distinction is also made in risk analysis between (probabilistically) quantifiable risk and unquantifiable uncertainty (Knight, 1922; Stirling, 1998).
5 From Box (1976), reprinted by permission of Taylor & Francis Ltd.
6 Palmer and Stevens (2019)

7 AIP-Manabe (1998b)
8 Subramanian et al. (2019)
9 See also the Hawkmoth Effect (Thompson, 2013; Frigg et al., 2014)
10 Voosen (2019)
11 Rich (2018)
12 Jehl (2004)
13 Žižek (2004)
14 Rayner (2012)
15 Oreskes and Conway (2011)
16 Supran and Oreskes (2017)
17 Moore (2018), Qiu (2018)
18 Schmidt and Sherwood (2015), Trenberth and Knutti (2017)
19 Weart (2008), p. 7
20 Lahsen (2013)
21 Waldman (2019)
22 APS (2015)
23 Dyson (2007)
24 Schmitt and Happer (2013)
25 Ghosh (2017)
26 Phillips (2018)
27 Goenner (1993)
28 Friedlaender, S., 2008. Kant gegen Einstein
29 Oreskes (2019), pp. 142–144
30 Hargreaves and Annan (2014)
31 Held (2011)
32 Allen (2003b), Stainforth et al. (2005)
33 Karmalkar et al. (2019)
34 Held (2005)

Chapter 14

1 Yoder (2019)
2 Conway (2008)
3 Pinker (1994)
4 Korten (2015)
5 Goodell (2021)
6 O'Neill and Nicholson-Cole (2009), Hayhoe (2018), Ambrose (2020)
7 Risbey (2008)
8 AIP–Farman (2009), p. 2
9 Robert Frost
10 Borenstein and Johnson (2020)
11 Silver (2012), p. 131
12 Milly et al. (2008)
13 Gettelman and Rood (2016), p. 8
14 Hutson (2020)
15 Collins et al. (2013), p. 1036; Trenberth and Knutti (2017)
16 Scientists' reactions to the US House Science Committee hearing on climate science, *ClimateFeedback.org*, April 10, 2017, https://climatefeedback.org/scientists-reactions-us-house-science-committee-hearing-climate-science/

17 Hansen et al. (2010). See also https://climate.nasa.gov/faq/38/how-do-scientists-deal-with-these-changes/
18 Spencer and Christy (1990); Edwards (2010), pp. 413–428
19 Wentz and Schabel (1996)
20 Rebuttal testimony of Dr. Andrew Dessler for the Minnesota Public Utilities Commission,
www.edockets.state.mn.us/EFiling/edockets/searchDocuments.do?method=
showPoup&documentId=%7BC36B70CA-5848-4A60-A1D0-1664F0E5250C%
7D
21 Thorne et al. (2011)
22 Legates (2020)
23 Knutson and Tuleya (2005), Schmidt (2014)
24 Gettelman and Rood (2016), p. 195; Bjordal et al. (2020)
25 Dessler (2018)
26 Crowley (2000), Tierney et al. (2020), Bova et al. (2021)
27 Collins et al. (2013), p. 1111
28 Sherwood et al. (2020)
29 Berardelli (2020), Forster et al. (2020)
30 Edwards (2010), p. 282
31 Bintanja et al. (2005)
32 Sutton and Hawkins (2020)
33 Bock et al. (2020), Fasullo (2020), SIMIP Community (2020), Wild (2020),
34 Calel et al. (2015), Knutti et al. (2017),
35 Marvel et al. (2018)
36 Woetzel et al. (2020), p. 25
37 Hausfather and Peters (2020)
38 Moss et al. (2010)
39 Burgess et al. (2020), Hausfather (2020a), Wheeling (2020),
40 Mann (2020)
41 Schwalm et al. (2020)
42 Riahi et al. (2017)
43 Tollefson (2020)
44 IPCC (1990), p. xviii
45 IPCC (2015), p. 543
46 Mackey (2009)
47 Sarewitz and Pielke Jr (1999)
48 www.weather.gov/okx/CentralParkHistorical
49 El Niño/Southern Oscillation (ENSO) diagnostic discussion, June 14, 2018, www.cpc.ncep.noaa.gov/products/analysis_monitoring/enso_disc_jun2018/ensodisc.pdf
50 Decadal forecast 2013, www.metoffice.gov.uk/research/climate/seasonal-to-decadal/long-range/decadal-fc/2013
51 Collins et al. (2013), pp. 1031, 1055 (tab. 12.2)
52 Trenberth and Knutti (2017), Schulthess et al. (2019)
53 Shaw (2017)
54 Allen and Stainforth (2002); Silver (2012), p. 134; Shaw (2018);
55 Betz (2010), Parker (2010),
56 IPCC (2015), Trenberth and Knutti (2017)
57 Trenberth (2007)
58 Dessai and Hulme (2003); Collins et al. (2013), p. 1036; Shepherd et al. (2018)
59 Edwards (1999)
60 Koonin (2014), Stephens (2017), Crowe (2019)
61 Wagner and Weitzman (2015), pp. 67–68

Chapter 15

1 Hulme (2013)
2 Somerville (1996), p. 97 as quoted in Lahsen (2005)
3 Revkin (2005)
4 Scientific Reports (2020)
5 Goodell (2018), Readfearn (2018)
6 Steffen et al. (2018)
7 Betts (2018)
8 Randers and Goluke (2020)
9 Webster (2020)
10 vice.com/en_us/article/akzn5a/theoretical-physicists-say-90-chance-of-societal-collapse-within-several-decades
11 Muller (2018), pp. 39, 40, 47
12 Hulme (2020)
13 Ninety-five percent chance, if the error bar corresponds to two standard deviations
14 Collins et al. (2013), p. 1111; Knutti et al. (2017)
15 Trenberth (2010)
16 Pielke Jr (2001), Maslin and Austin (2012)
17 Silver (2012), p. 390
18 Kandlikar et al. (2005), Curry (2011), Curry and Webster (2011), Heal and Millner (2014), Wagner and Zeckhauser (2016), Nissan et al. (2019)
19 Funtowicz and Ravetz (1990), Smith and Stern (2011), Parker and Risbey (2015), Curry (2018)
20 Thinking of all risk solely in terms of probabilities is also referred to as "probabilism" (Betz, 2010).
21 Risbey and Kandlikar (2007) propose a similar progressive scheme to express IPCC uncertainty. They suggest a "Bounds" measure similar to the approximately known unknowns category, except that their Bounds are still associated with percentiles. They also propose further categories of uncertainty that represent knowledge of the orders of magnitude or just the signs of parameters. See also Spiegelhalter and Riesch (2011) for more on categorizing uncertainties.
22 Edwards (2010), p. 409; Palmer and Stevens (2019)
23 This is known as the precautionary principle, (United Nations, 1992). See also, Lewandowsky et al. (2015), Revkin (2017)
24 Rogelj et al. (2014), Lewis and Curry (2015)
25 Cho (2012)
26 Sherwood et al. (2020)
27 Cox et al. (2018a)
28 Solomon et al. (2007), pp. 22–23
29 Using a lognormal fit for the probability distribution is a common way to characterize climate sensitivity (Rogelj et al., 2014).
30 Calel et al. (2015), Curry (2018)
31 Wagner and Weitzman (2018)
32 Meehl et al. (2007), p. 799; Solomon et al. (2007), p. 65
33 Kirtman et al. (2013), pp. 1004, 1009
34 Tierney et al. (2020)
35 Stainforth et al. (2007)
36 Oppenheimer et al. (2008), Weaver et al. (2017)
37 Baker and Roe (2009)
38 Stocker et al. (2013), p. 83

39 Meehl et al. (2007), p. 799; Collins et al. (2013), p. 1111
40 For example, see the comments generated by the Cox et al. (2018a) paper which S3 was taken from, and the response: Brown et al. (2018), Po-Chedley et al. (2018), Rypdal et al. (2018), and Cox et al. (2018b).
41 MacKenzie (1998)
42 Lahsen (2005)
43 Wagner and Weitzman (2015), pp. 78–79, 51, 52
44 Collins et al. (2013), p. 1111; Stocker et al. (2013), p. 84
45 Shackley and Wynne (1995)
46 Meehl et al. (2020)
47 Otto et al. (2015)
48 Allen (2019)
49 Mann et al. (2017)
50 Schneider-Mayerson and Leong (2020)
51 Ambrose (2020)
52 Hulme (2019)
53 Hallam (2019)
54 Wallace-Wells (2017)
55 Scientists explain what the *New York Magazine* article on "The uninhabitable earth" gets wrong. *ClimateFeedback.org*, July 12, 2017, https://climatefeedback.org/evaluation/scientists-explain-what-new-york-magazine-article-on-the-uninhabitable-earth-gets-wrong-david-wallace-wells/
56 Schmidt (2019)
57 Wallace-Wells (2019)
58 Hunter (2020)
59 In July 2020, the article was revised to add the caveat that the "paper does not prove the inevitability of such collapse," https://jembendell.com/2019/05/15/deep-adaptation-versions/
60 Crowe (2019), Lomborg (2020b)
61 Lomborg (2020a)
62 GWPF (2020)
63 Andrijevic et al. (2020)
64 Hulme (2006)
65 Wallace-Wells (2020)
66 Revkin (2016)

Chapter 16

1 TVA goes to war, *Tennessee Valley Authority*, www.tva.com/about-tva/our-history/built-for-the-people/tva-goes-to-war
2 Reed (2015)
3 2019, tab. 1.1. Net generation by energy source, www.eia.gov/electricity/monthly/epm_table_grapher.php?t=epmt_1_01
4 Electricity outlook 2017: powering New York City's future, *New York Building Congress*, www.buildingcongress.com/advocacy-and-reports/reports-and-analysis/Electricity-Outlook-2017-Powering-New-York-Citys-Future/The-Electricity-Outlook-to-2027.html
5 For example, the sun's surface, with a temperature of 5,500°C, emits 63 million watts per square meter or 63 watts per square millimeter. (The dark silicon problem

and what it means for CPU designers, www.informit.com/articles/article.aspx?p=2142913)
6 Feldman (2019a)
7 Balaji (2021)
8 Schulthess et al. (2019)
9 Thompson and Spanuth (2018)
10 Fuhrer et al. (2018), Schulthess et al. (2019)
11 Lohr (2018)
12 www.ornl.gov/news/ornl-launches-summit-supercomputer and www.top500.org/system/179397/
13 e3sm.org
14 www.tacc.utexas.edu/systems/frontera
15 Halper (2015)
16 Dongarra et al. (2016)
17 Loft (2020)
18 Clark (2020)
19 Augustine (1987)
20 Eyring et al. (2016), tab. 2
21 Schär et al. (2020)
22 Easterbrook (2010)
23 Fuhrer et al. (2018)
24 Feldman (2019b)
25 Voosen (2020a)
26 Palmer and Stevens (2019), Panosetti et al. (2019)
27 Bauer et al. (2015)
28 Emanuel (2020)
29 Kusunoki and Arakawa (2015)
30 Schulthess et al. (2019), Voosen (2020a)
31 Goddard et al. (2009), Easterbrook (2011), Palmer and Stevens (2019), Palmer et al. (2019)
32 Knutti (2010), Lenhard and Winsberg (2010)
33 Hennessy and Patterson (2019)
34 Duben et al. (2015), Schulthess et al. (2019)

Chapter 17

1 Copeland and Proudfoot (2011), Dyson (2012b)
2 Turing (1950)
3 Palmer (2016) discusses a different application of the Turing Test to weather and climate modeling, arguing that high-resolution, limited-area weather forecasts pass the Turing Test for predicting weather, whereas coarse-resolution global climate simulations fail due to large errors in their regional simulation.
4 The Economist (2017)
5 Hartnett (2018)
6 Balaji (2021)
7 Schneider et al. (2017b), Reichstein et al. (2019)
8 Feldman (2020)
9 Lipman (2014)
10 Sima (2020)

11 For example, see Beucler et al. (2021)
12 Palmer and Stevens (2019)
13 O'Gorman and Dwyer (2018)
14 Karmalkar et al. (2019)
15 For an alternative mapping of models, using system complexity versus model abstractions, see Bony et al. (2013), fig. 4

Chapter 18

1 Keith (2000)
2 Kiehl (2006)
3 Hersh (1972)
4 Edwards (2012b)
5 McLaren and Markusson (2020)
6 Robock (2008); NRC (2015), p. 40
7 Crutzen (2006)
8 NRC (2015), pp. 7, 79
9 ibid., p. 152
10 ibid.
11 Hulme (2020)
12 Muller (2018)
13 Allen (2003a)
14 Robock (2008)
15 Wood (2009)
16 Wagner and Weitzman (2015), pp. 114–115; Caldeira and Bala (2017)

Chapter 19

1 John von Neumann, *Atomic Heritage Foundation*, www.atomicheritage.org/profile/john-von-neumann
2 Dyson (2012a), pp. 271–273
3 Whitman (2012), pp. 92
4 Macrae (1999), pp. 377–379
5 For example, Oreskes (2019), ch. 2 and also ch. 5 (O. Edenhofer and M. Kowarsch)
6 Nordhaus (2007a), p. 104
7 Kay and King (2020), p. 323
8 The Independent (2019)
9 Nordhaus (2007b)
10 Fleurbaey et al. (2019)
11 Nordhaus (2017)
12 Nordhaus (2007a), p. 105
13 Bastien-Olvera and Moore (2020)
14 Wagner and Weitzman (2015), p. 78
15 Yohe et al. (2004), Otto et al. (2015), Kaufman et al. (2020)
16 Kay and King (2020), p. 9
17 ibid., pp. 6–7

18 Interestingly, the probabilistic approach to dealing with risk traces its origins back to John von Neumann and his groundbreaking work on game theory and economics (in collaboration with Oskar Morgenstern).
19 The distinction between resolvable uncertainty (or probabilistic risk) and radical uncertainty made by Kay and King (2020) is somewhat analogous to the distinction we make in this book between well-known unknowns (or probabilistic prophecies) and poorly known unknowns.
20 Dessai and Hulme (2003), Curry (2018)
21 Beckage et al. (2020)
22 Shepherd et al. (2018)
23 Blair (1957)
24 Oreskes et al. (2010)
25 Wang et al. (2021)
26 For example, see Nielsen-Gammon et al. (2020), Nissan et al. (2020), and Fiedler et al. (2021)
27 Benestad (2016)
28 https://esgf.llnl.gov; see also Fiedler et al. (2021)
29 Cox and Stephenson (2007), Stainforth et al. (2007), Hawkins and Sutton (2009), Gettelman and Rood (2016), pp. 10–11
30 Hawkins and Sutton (2009)
31 Carton (2020)
32 Schmidt (2019)
33 Fiedler et al. (2021)
34 Stainforth et al. (2007)
35 Deser et al. (2012), Calel et al. (2020), Deser et al. (2020), Mankin et al. (2020)
36 Milinski et al. (2020)
37 Deser et al. (2020)
38 Maslin and Austin (2012)
39 Nissan et al. (2019), Palmer and Stevens (2019)
40 Hawkins and Sutton (2009), Doyle (2013), Knutti and Sedlaček (2013), Palmer and Stevens (2019), Voosen (2020b)
41 Nissan et al. (2019)
42 Roberts (2012), Colman (2021)
43 Roe and Bauman (2012)
44 See Calel et al. (2015) for a discussion of fat tails.
45 Smith and Stern (2011), King et al. (2015), Sutton (2019)
46 Deser (2020)
47 Dessai and Hulme (2003), Risbey (2004), Curry (2018), Nissan et al. (2019)
48 Nakićenović et al. (2000); Collins et al. (2013), p. 1036
49 Kunreuther et al. (2014), p. 177
50 Grubler and Nakicenovic (2001), Schneider (2002), Dessai and Hulme (2003), Risbey (2004)
51 Dessai and Hulme (2003), King et al. (2015)
52 Hausfather and Peters (2020)
53 Allen and Ingram (2002), Stainforth et al. (2007)
54 Knutti et al. (2010)
55 Weaver et al. (2017), Lawrence et al. (2020)
56 Roe (2013)
57 Schmidt (2020)
58 Tollefson (2020)

59 Schmidt (2018)
60 Norman et al. (2015)
61 Lombrana et al. (2020)

Chapter 20

1 Nuñez and Sweetser (2006)
2 Marvel (2018)
3 Pierrehumbert (2014), Lewandowsky et al. (2015)
4 Palmer and Stevens (2019)
5 Hausfather (2020b)
6 Knutti et al. (2016)
7 Smith (2016)
8 Matthews and Caldeira (2008)
9 Palm et al. (2020)
10 Revkin (2016), Goodell (2021)
11 Rogelj et al. (2014)
12 Otto et al. (2015)
13 Hayhoe (2015)
14 Kahn (2020)
15 Lawrence and Schäfer (2019), McLaren and Markusson (2020)
16 McGrath (2018)
17 NASEM (2019)
18 Prins and Rayner (2007)
19 Goodell (2021)
20 Nordhaus (2015)
21 Ip (2020), Puko (2020)
22 Andrijevic et al. (2020)

Epilogue

1 Andrijevic et al. (2020)
2 Ambrose (2020), Carrington (2020)
3 O'Neill and Nicholson-Cole (2009), Hayhoe (2018)
4 Somerville (1996)

Select Bibliography

This book is about climate prediction and climate modeling, but it does not get into the technical details of climate models. A good companion book that provides many of those details is:

Demystifying Climate Models
Andrew Gettelman and Richard Rood (SpringerOpen, 2016)
demystifyingclimate.org

A good textbook on the basics of climate change, aimed at the undergraduate level, is:

Introduction to Modern Climate Change, 2nd ed.
Andrew Dessler (Cambridge University Press, 2015)

For a more comprehensive history of climate modeling than what is presented in this book, see:

A Vast Machine: Computer Models, Climate Data, and the Politics of Global Warming
Paul Edwards (MIT Press, 2010)

For a more comprehensive history of climate change in general, see:

The Discovery of Global Warming
Spencer Weart (Harvard University Press, 2008)

A more technical description of the history of climate modeling, including details on modeling past climates, can be found in:

Beyond Global Warming: How Numerical Models Revealed the Secrets of Climate Change
Syukuro Manabe and Anthony Broccoli (Princeton University Press, 2020)

The early days of digital computing and John von Neumann's role in it are described in great detail by:

Turing's Cathedral: The Origins of the Digital Universe
George Dyson (Vintage, 2012)

Two recent books that discuss philosophical issues relevant to climate science are:

Philosophy and Climate Science.
Eric Winsberg (Cambridge University Press, 2018)
Why Trust Science?
Naomi Oreskes (Princeton University Press, 2019)

The website *CarbonBrief.org* is a very useful resource on many issues pertaining to climate change.
The following blogs contain interesting discussions of climate modeling and climate prediction, as well as related philosophical issues:

And Then There's Physics – andthentheresphysics.wordpress.com
Real Climate – realclimate.org
Serendipity – easterbrook.ca/steve
Isaac Held's Blog – gfdl.noaa.gov/blog_held

References

Abatzoglou, J. T., and A. P. Williams, 2016: Impact of anthropogenic climate change on wildfire across western US forests. *Proc. Natl. Acad. Sci.*, 113(42), 11770–11775, doi:10.1073/pnas.1607171113

Agrawala, S., 1998: Context and early origins of the Intergovernmental Panel on Climate Change. *Clim. Change*, 39, 605–620, doi:10.1023/A:1005315532386

AIP–Bryan, 1989: Interview of Kirk Bryan by Spencer Weart on December 20, 1989, College Park, MD, Niels Bohr Library & Archives, American Institute of Physics, www.aip.org/history-programs/niels-bohr-library/oral-histories/5068. College Park, MD, Niels Bohr Library & Archives, American Institute of Physics, www.aip.org/history-programs/niels-bohr-library/oral-histories/33642

AIP-Farman, 2009: Interview of Joseph Farman by Keynyn Brysse on March 16, 2009, Niels Bohr Library & Archives, American Institute of Physics, College Park, MD, www.aip.org/history-programs/niels-bohr-library/oral-histories/33642

AIP–Kasahara, 1998: Interview of Akira Kasahara by Paul Edwards on November 2, 1998, College Park, MD, Niels Bohr Library & Archives, American Institute of Physics, www.aip.org/history-programs/niels-bohr-library/oral-histories/32440-1

AIP–Manabe, 1989: Interview of Syukuro Manabe by Spencer Weart on December 20, 1989, College Park, MD, Niels Bohr Library & Archives, American Institute of Physics, www.aip.org/history-programs/niels-bohr-library/oral-histories/5040

1998a: Interview of Syukuro Manabe by Paul Edwards on March 14, 1998, College Park, MD, Niels Bohr Library & Archives, American Institute of Physics, www.aip.org/history-programs/niels-bohr-library/oral-histories/32158-1

1998b: Interview of Syukuro Manabe by Paul Edwards on March 15, 1998, College Park, MD, Niels Bohr Library & Archives, American Institute of Physics, www.aip.org/history-programs/niels-bohr-library/oral-histories/32158-2

AIP–Washington, 1998: Interview of Warren Washington by Paul Edwards on October. 28 and 29, 1998, College Park, MD, Niels Bohr Library & Archives, American Institute of Physics, www.aip.org/history-programs/niels-bohr-library/oral-histories/33098

Albritton, D. L., L. G. Meira Filho, U. Cubasch, et al., 2001: Technical summary. In: J. T. Houghton, Y. Ding, D. J. Griggs, et al. (eds.), *Climate change 2001: the scientific basis*. Cambridge University Press, 21–83.

Alcamo, J., J. M. Moreno, B. Nováky, et al., 2007: Europe. Climate change 2007: impacts, adaptation and vulnerability. *Contribution of working group II to the fourth assessment report of the Intergovernmental Panel on Climate Change.* Cambridge University Press

Allen, M., 2003a: Liability for climate change. *Nature*, 421, 891–892

2003b: Possible or probable? *Nature*, 425, 242

2019: Why protesters should be wary of "12 years to climate breakdown" rhetoric. *TheConversation.com*, https://theconversation.com/why-protesters-should-be-wary-of-12-years-to-climate-breakdown-rhetoric-115489

Allen, M., and D. Stainforth, 2002: Towards objective probabilistic climate forecasting. *Nature*, 419, 228, doi:10.1038/nature01092a

Allen, M. R., and W. J. Ingram, 2002: Constraints on future changes in climate and the hydrologic cycle. *Nature*, 419(6903), 224–232

Almanac.com, 2016: *The 2017 old farmer's almanac.* Yankee Publishing

Ambrose, J., 2020: "Hijacked by anxiety": how climate dread is hindering climate action. *The Guardian*, October 8, www.theguardian.com/environment/2020/oct/08/anxiety-climate-crisis-trauma-paralysing-effect-psychologists

Amos, J., 2019: UK's Halley Antarctic base in third winter shutdown. *BBC News*, February 28, www.bbc.com/news/science-environment-47408249

Anderson, P. W., 1972: More is different. *Science*, 177(4047), 393–396

Andrijevic, M., C. F. Schleussner, M. J. Gidden, D. L. McCollum, and J. Rogelj, 2020: COVID-19 recovery funds dwarf clean energy investment needs. *Science*, 370, 298–300

APS, 2015: American Physical Society, Statement on Earth's changing climate, aps.org/policy/statements/15_3.cfm

Archer, D., and R. Pierrehumbert, 2010: *The warming papers.* Wiley-Blackwell

Arrhenius, S., 1908: *Worlds in the making: the evolution of the universe.* Harper, https://archive.org/details/worldsinmakingev00arrhrich/mode/2up

Augustine, N. R., 1987: *Augustine's laws.* Penguin Books

Ayala, F. J., 2009: Darwin and the scientific method. *Proc. Natl. Acad. Sci.*, 106 (Supplement 1), 10033–10039

Baker, M., and G. Roe, 2009: The shape of things to come: why is climate change so predictable? *J. Clim.*, 22(17), 4574–4589

Balaji, V., 2021: Climbing down Charney's ladder: machine learning and the post-Dennard era of computational climate science. *Phil. Trans. R. Soc. A*, 379, 20200085, doi:10.1098/rsta.2020.0085

Ball, P., 2016: The tyranny of simple explanations. *The Atlantic*, August 11, www.theatlantic.com/science/archive/2016/08/occams-razor/495332/

Barnston, A. G., M. K. Tippett, M. Ranganathan, et al., 2019: Deterministic skill of ENSO predictions from the North American Multimodel Ensemble. *Clim. Dynam.*, 53, 7215–7234

Bastien-Olvera, B. A., and F. C. Moore, 2020: Use and non-use value of nature and the social cost of carbon. *Nat. Sustain.*, 4, 101–108, doi:10.1038/s41893-020-00615-0

Batterson, S., 2007: The vision, insight, and influence of Oswald Veblen. *Notices of the AMS*, 54, 606–618

Bauer, P., A. Thorpe, and G. Brunet, 2015: The quiet revolution of numerical weather prediction. *Nature*, 525, 47–57, doi:10.1038/nature14956

Baumberger, C., R. Knutti, and G. Hirsch Hadorn, 2017: Building confidence in climate model projections: an analysis of inferences from fit. *WIREs Clim. Change*, 8, e454

Beckage, B., K. Lacasse, J. M. Winter, et al., 2020: The earth has humans, so why don't our climate models? *Clim. Change*, 163, 181–188, doi:10.1007/s10584-020-02897-x

Benestad, R., 2016: *Downscaling climate information. Oxford research encyclopedia of climate science*, doi:10.1093/acrefore/9780190228620.013.27

Benjamin, S. G., J. M. Brown, G. Brunet, P. Lynch, K. Saito, and T. W. Schlatter, 2019: 100 years of progress in forecasting and NWP applications. *A Century of Progress in Atmospheric and Related Sciences: Celebrating the American Meteorological Society Centennial*, Meteor. Monogr., No. 59, Amer. Meteor. Soc., 13.1–13.67

Berardelli, J., 2020: Some new climate models are projecting extreme warming. Are they correct? *Yale Climate Connections*, July 1, https://yaleclimateconnections.org/2020/07/some-new-climate-models-are-projecting-extreme-warming-are-they-correct/

Betts, R., 2018: Is our planet headed toward a "Hothouse"? Here's what the science does – and doesn't – say. *Washington Post*, August 10, www.washingtonpost.com/news/capital-weather-gang/wp/2018/08/10/hothouse-earth-heres-what-the-science-actually-does-and-doesnt-say/

Betts, R. A., A. Lorenzo, B. Catherine, et al., 2018: Changes in climate extremes, fresh water availability and vulnerability to food insecurity projected at 1.5°C and 2°C global warming with a higher-resolution global climate model. *Phil. Trans. R. Soc. A.*, 37620160452, doi:10.1098/rsta.2016.0452

Betz, G., 2010: What's the worst case? The methodology of possibilistic prediction. *Analyse and Kritik*, 01, 87–106

Beucler, T., M. Pritchard, S. Rasp, J. Ott, P. Baldi, and P. Gentine, 2021: Enforcing analytic constraints in neural-networks emulating physical systems. *Phys. Rev. Lett.*, 126, 098302

Bindoff, N. L., P. A. Stott, K. M. AchutaRao, et al., 2013: Detection and attribution of climate change: from global to regional. In: *Climate change 2013: the physical science basis. Contribution of working group I to the fifth assessment report of the Intergovernmental Panel on Climate Change*. Cambridge University Press, 867–952.

Bintanja, R., R. S. W. v. d. Wal, and J. Oerlemans, 2005: A new method to estimate ice age temperatures. *Clim. Dynam.*, 24, 197–211, doi:10.1007/s00382-004-0486-x

Bjordal, J., T. Storelvmo, K. Alterskjær, et al., 2020: Equilibrium climate sensitivity above 5 °C plausible due to state-dependent cloud feedback. *Nat. Geosci.*, 13, 718–721, doi:10.1038/s41561-020-00649-1

Blackport, R., and J. A. Screen, 2020: Weakened evidence for mid-latitude impacts of Arctic warming. *Nat. Clim. Change*, 10, 1065–1066, doi:10.1038/s41558-020-00954-y

Blair, W. M., 1957: President draws planning moral. *The New York Times*, November 15, www.nytimes.com/1957/11/15/archives/president-draws-planning-moral-recalls-army-days-to-show-value-of.html

Blake, E. S., and D. A. Zelinsky, 2018: Hurricane Harvey. Tropical Cyclone Report, National Hurricane Center. May 8, www.nhc.noaa.gov/data/tcr/AL092017_Harvey.pdf

Bock, L., A. Lauer, M. Schlund, et al., 2020: Quantifying progress across different CMIP phases with the ESMValTool. *J. Geophys. Res. – Atmos.*, 125, e2019JD032321

Bojkov, R. D., and D. S. Balis, 2009: The history of total ozone measurements: the early search for signs of a trend and an update. In: C. Zerefos, G. Contopoulos, and G. Skalkeas (eds.), *Twenty years of ozone decline*. Springer, 73–110

Bony, S., B. Stevens, I. Held, et al., 2013: Carbon dioxide and climate: perspectives on a scientific assessment. In: J. W. Hurrell, and G. Asrar (eds.), *Climate science for serving society*. Springer Netherlands, 391–413, https://library.wmo.int/doc_num.php?explnum_id=7660

Boos, W. R., and T. Storelvmo, 2016: Near-linear monsoon response to range of forcings. *Proc. Nat. Acad. Sci.*, 113(6), 1510–1515

Borenstein, S., and C. K. Johnson, 2020: Modeling coronavirus: "uncertainty is the only certainty." Associated Press, April 7, https://apnews.com/article/public-health-health-us-news-ap-top-news-virus-outbreak-88866498ff5c908e5f28f7b5b5e5b695

Borges, J. L., 1975: On exactitude in science. In: *A universal history of infamy* (trans. Norman Thomas de Giovanni). London, Penguin Books, 325

Borland, J. B., M. Jamiolkowski, and C. Viggiani, 2003: The stabilisation of the Leaning Tower of Pisa. *Soils and Foundations – Tokyo*, 43(5), 63–80

Borowski, S., 2012: The origin and popular use of Occam's razor. American Association for the Advancement of Science, June 12, www.aaas.org/origin-and-popular-use-occams-razor

Bostrom, N., 2003: Are you living in a computer simulation? *Philosophical Quarterly*, 53(211), 243–255

Bova, S., Y. Rosenthal, Z. Liu, et al. 2021: Seasonal origin of the thermal maxima at the Holocene and the last interglacial. *Nature*, 589, 548–553

Box, G. E. P., 1976: Science and statistics. *Journal of the American Statistical Association*, 71(356), 791–799, doi:10.1080/01621459.1976.10480949

Brown, C., and R. Wilby, 2012: An alternate approach to assessing climate risks. *Eos*, 93(41), 401–402

Brown, P. T., M. B. Stolpe, and K. Caldeira, 2018: Assumptions for emergent constraints. *Nature*, 563, E1–E3

Bryan, K., 2006: Modeling ocean circulation. In: M. Jochum and R. Murtugudde (eds.), *Physical oceanography*. New York, Springer, 29–44. doi:10.1007/0-387-33152-2_3

Burgess, M., J. Ritchie, J. Shapland, and R. Pielke Jr, 2020: IPCC baseline scenarios have over-projected CO2 emissions and economic growth. *Environ. Res. Lett.*, 16, 014016

Burrakoff, M., 2019: NASA's study of astronaut twins creates a portrait of what a year in space does to the human body. *Smithsonian Magazine*, April 11, www.smithsonianmag.com/science-nature/nasas-twins-study-creates-portrait-human-body-after-year-space-180971945/

Caldeira, K., and G. Bala, 2017: Reflecting on 50 years of geoengineering research. *Earth's Future*, 5, 10–17, doi:10.1002/2016EF000454

Calel, R., S. C. Chapman, D. A. Stainforth, et al., 2020: Temperature variability implies greater economic damages from climate change. *Nat. Commun.*, 11, 5028

Calel, R., D. A. Stainforth, and S. Dietz, 2015: Tall tales and fat tails: The science and economics of extreme warming. *Clim. Change*, 132, 127–141

Canales, J., 2020: *Bedeviled: a shadow history of demons in science*. Princeton University Press

Carbon Brief, 2018: Q&A: How "integrated assessment models" are used to study climate change. *CarbonBrief.org*, October 2

Carrington, D., 2020: Climate "apocalypse" fears stopping people having children. *The Guardian*, November 27, www.theguardian.com/environment/2020/nov/27/climate-apocalypse-fears-stopping-people-having-children-study

Carroll, S., 2013: What is science? *Preposterous Universe* blog, July 3, www.preposterousuniverse.com/blog/2013/07/03/what-is-science/

2016: Maybe we do not live in a simulation: the resolution conundrum. *Preposterous Universe* blog, August 22, www.preposterousuniverse.com/blog/2016/08/22/maybe-we-do-not-live-in-a-simulation-the-resolution-conundrum/

Carton, W., 2020: Carbon unicorns and fossil futures. Whose emission reduction pathways is the IPCC performing? In: J. P. Sapinski, H. Buck, and A. Malm (eds.), *Has it come to this? The promises and perils of geoengineering on the brink*. Rutgers University Press

Ceruzzi, P. E., 1998: *A history of modern computing*. MIT Press

Charney, J. G., A. Arakawa, D. J. Baker, et al., 1979: *Carbon dioxide and climate: a scientific assessment*. National Academy of Sciences

Cho, A., 2012: Once again, physicists debunk faster-than-light neutrinos. *Science*, June 8, www.sciencemag.org/news/2012/06/once-again-physicists-debunk-faster-light-neutrinos

Christie, M., 2000: *The ozone layer: a philosophy of science perspective*. Cambridge University Press

Clark, D., 2020: Japanese supercomputer is crowned world's speediest. *The New York Times*, June 22, www.nytimes.com/2020/06/22/technology/japanese-supercomputer-fugaku-tops-american-chinese-machines.html

Clark, P., 2013: Scientist who beat NASA to the ozone hole. *Financial Times*, May 17, www.ft.com/content/4084f0b8-bd77-11e2-890a-00144feab7de

Clement, A., K. Bellomo, L. N. Murphy, et al., 2015: The Atlantic Multidecadal Oscillation without a role for ocean circulation. *Science*, 350(6258), 320, doi:10.1126/science.aab3980

Cohen, J., X. Zhang, J. Francis, et al., 2020: Divergent consensuses on Arctic amplification influence on midlatitude severe winter weather. *Nat. Clim. Change*, 10, 20–29, doi:10.1038/s41558-019-0662-y

Collins, M., R. Knutti, J. Arblaster, et al., 2013: Long-term climate change: projections, commitments and irreversibility. In: *Climate change 2013: the physical science basis. Contribution of working group I to the fifth assessment report of the Intergovernmental Panel on Climate Change*. Cambridge University Press, 1029–1136

Colman, Z., 2021: "Garbage" models and black boxes? The science of climate disaster planning. *Politico.com*, March 16, www.politico.com/news/2021/03/16/climate-change-murky-models-476316

Conway, E., 2008: What's in a name? Global warming vs. climate change. NASA, December 5, www.nasa.gov/topics/earth/features/climate_by_any_other_name.html

Copeland, B. J., and D. Proudfoot, 2011: Alan Turing: father of the modern computer. *The Rutherford Journal*, 4(1), https://espace.library.uq.edu.au/view/UQ:347665

Cox, P., C. Huntingford, and M. Williamson, 2018a: Emergent constraint on equilibrium climate sensitivity from global temperature variability. *Nature*, 553, 319–322

Cox, P., and D. Stephenson, 2007: Climate change – a changing climate for prediction. *Science*, 317, 207–208

Cox, P. M., M. S. Williamson, F. J. M. M. Nijsse, et al., 2018b: Cox et al. reply. *Nature*, 563, E10–E15

Crane, L., 2018: Terraforming Mars might be impossible due to a lack of carbon dioxide. *New Scientist*, July 30, www.newscientist.com/article/2175414-terraforming-mars-might-be-impossible-due-to-a-lack-of-carbon-dioxide/

Crowe, K., 2019: How "organized climate change denial" shapes public opinion on global warming. Canadian Broadcasting Corporation, Canada, September 27, www.cbc.ca/news/science/climate-change-denial-fossil-fuel-think-tank-sceptic-misinformation-1.5297236

Crowley, T. J., 2000: CLIMAP SSTs re-revisited. *Clim. Dynam.*, 16, 241–255

Crutzen, P., 2006: Albedo enhancement by stratospheric sulfur injections: a contribution to resolve a policy dilemma? *Clim. Change*, 77, 211–219

Cubasch, U., D. Wuebbles, D. Chen, et al., 2013: Introduction. In: *Climate change 2013: the physical science basis. Contribution of working group I to the fifth assessment report of the Intergovernmental Panel on Climate Change*. Cambridge University Press, 119–158

Curry, J., 2011: Reasoning about climate uncertainty. *Clim. Change*, 108, 723–732

2018: Climate uncertainty and risk. *US CLIVAR Variations*, 16, 1–5

Curry, J. A., and P. J. Webster, 2011: Climate science and the uncertainty monster. *Bull. Am. Meteorol. Soc.*, 92, 1667–1682

Dabberdt, W. F., R. Shellhorn, H. Cole, et al., 2003: Radiosondes. *Encyclopedia of atmospheric sciences*. Elsevier, 1900–1913

Das, S. R., 2008: The chip that changed the world. *The New York Times*, September 19, www.nytimes.com/2008/09/19/opinion/19iht-eddas.1.16308269.html

Dayton, L., 2016: Research chief cuts climate studies, sets new priorities. *Science*, 351 (6274), 649, doi:10.1126/science.351.6274.649

Deser, C., 2020: Certain uncertainty: the role of internal climate variability in projections of regional climate change and risk management. *Earth's Future*, doi:10.1029/2020EF001854

Deser, C., R. Knutti, S. Solomon, et al. 2012: Communication of the role of natural variability in future North American climate. *Nat. Clim. Change*, 2, 775–779, doi:10.1038/nclimate1562

Deser, C., F. Lehner, K. B. Rodgers, et al., 2020: Insights from Earth system model initial-condition large ensembles and future prospects. *Nat. Clim. Change*, 10, 277–286, doi:10.1038/s41558-020-0731-2

Dessai, S., and M. Hulme, 2003: Does climate policy need probabilities? *Tyndall Centre working paper no. 34*. Norwich, Tyndall Centre for Climate Change Research

Dessler, A. E., 2015: *Introduction to modern climate change, 2nd ed.*. Cambridge University Press

2018: The influence of internal variability on Earth's energy balance framework and implications for estimating climate sensitivity. *Atmos. Chem. Phys.*, 18, 5147–5155, doi:10.5194/acp-18-5147-2018

Dessler, A. E., and D. Cohan, 2018: We're scientists. We know the climate's changing. And we know why. *Houston Chronicle*, October 22, www.houstonchronicle.com/local/gray-matters/article/science-climate-change-combustion-fossil-fuels-13327165.php

Diamond, J., 1997: *Guns, germs and steel: the fate of human societies*. W. W. Norton

Dongarra, J., M. A. Heroux, and P. Luszczek, 2016: High-performance conjugate-gradient benchmark: a new metric for ranking high-performance computing systems. *Int. J. High Perform. Comput. Appl.*, 30(1), 3–10, doi:10.1177/1094342015593158

Donnelly, J. P., 2009: Paleotempestology, the sedimentary record of intense hurricanes. In: V. Gornitz (ed.), *Encyclopedia of paleoclimatology and ancient environments*, Encyclopedia of Earth Sciences Series. Dordrecht, Springer, 763–766. doi:10.1007/978-1-4020-4411-3_181

Doyle, A., 2013: Experts surer of manmade global warming but local predictions elusive. *Reuters*, August 16, www.reuters.com/article/us-climate-report/experts-surer-of-manmade-global-warming-but-local-predictions-elusive-idUSBRE97F0KM20130816

Duben, P. D., F. P. Russell, X. Niu, W. Luk, and T. N. Palmer, 2015: On the use of programmable hardware and reduced numerical precision in earth- system modeling. *J. Adv. Model. Earth Syst.*, 7, 1393–1408, doi:10.1002/2015MS000494.

Dudney, J., and K. N. Suding, 2020: The elusive search for tipping points. *Nat. Ecol. Evol.*, 4, 1449–1450, doi:10.1038/s41559-020-1273-8

Dunne, D., 2017: Hyperthermals: what can they tell us about modern global warming? *CarbonBrief.org*, October 9, www.carbonbrief.org/hyperthermals-what-can-they-tell-us-about-modern-global-warming

Dyson, F., 1988: *Infinite in all directions*. Harper Collins

Dyson, F. J., 2007: Heretical thoughts about science and society. *Edge*, August 7, edge.org/conversation/freeman_dyson-heretical-thoughts-about-science-and-society

Dyson, G., 2012a: *Turing's cathedral: the origins of the digital universe*. Vintage

2012b: The dawn of computing. *Nature*, 482, 459–460

Easterbrook, S., 2010: What's the pricetag on a Global Climate Model? *Serendipity* blog, September 3, www.easterbrook.ca/steve/2010/09/whats-the-pricetag-on-a-global-climate-model/

2011: One model to rule them all? *Serendipity* blog, November 6, www.easterbrook.ca/steve/2011/11/one-model-to-rule-them-all/

Edwards, J. R., 2012a: An early history of computing at Princeton. Princeton Alumni Weekly, April 4, https://paw.princeton.edu/article/early-history-computing-princeton

Edwards, P. N., 1999: Global climate science, uncertainty and politics: data-laden models, model-filtered data. *Science as Culture*, 8(4), 437–472

2010: *A vast machine: computer models, climate data, and the politics of global warming*. MIT Press

2012b: Entangled histories: climate science and nuclear weapons research. *Bull. At. Sci.*, 68(4), 28–40, doi:10.1177/0096340212451574

Emanuel, K. A., 1999: The power of a hurricane: an example of reckless driving on the information superhighway. *Weather*, 54, 107–108

2011: Edward Norton Lorenz 1917–2008: a biographical memoir. *National Academy of Sciences*, www.nasonline.org/publications/biographical-memoirs/memoir-pdfs/lorenz-edward.pdf

2020: The relevance of theory for contemporary research in atmospheres, oceans, and climate. *AGU Advances*, 1, e2019AV000129

Eyring, V., S. Bony, G. A. Meehl, et al., 2016: Overview of the Coupled Model Intercomparison Project Phase 6 (CMIP6) experimental design and organization. *Geosci. Model Dev.*, 9, 1937–1958, doi:10.5194/gmd-9-1937-2016

Farman, J. C., B. G. Gardiner, and J. D. Shanklin, 1985: Large losses of total ozone in Antarctica reveal seasonal ClOx/NOx interaction. *Nature*, 315, 207–210

Fasullo, J. T., 2020: Evaluating simulated climate patterns from the CMIP archives using satellite and reanalysis datasets using the Climate Model Assessment Tool (CMATv1). *Geosci. Model Dev.*, 13, 3627–3642

Feldman, M., 2019a: Dennard Scaling demise puts permanent dent in supercomputing. *The Next Platform*, June 18, www.nextplatform.com/2019/06/18/dennard-scaling-demise-puts-permanent-dent-in-supercomputing/

2019b: Exascale density pushes the boundaries of cooling. *The Next Platform*, November 26, www.nextplatform.com/2019/11/26/exascale-density-pushes-the-boundaries-of-cooling/

2020: HPC in 2020: AI is no longer an experiment. *The Next Platform*, January 9, www.nextplatform.com/2020/01/09/hpc-in-2020-ai-is-no-longer-an-experiment/

Feldstein, S. B., 2000: The timescale, power spectra, and climate noise properties of teleconnection patterns. *J. Clim.*, 13(24), 4430–4440

Ferreira, A. P., R. Nieto, and L. Gimeno, 2019: Completeness of radiosonde humidity observations based on the IGRA. *Earth Syst. Sci. Data*, 11, 603–627, doi:10.5194/essd-11-603-2019

Fiedler, S., T. Crueger, R. D'Agostino, et al., 2020: Simulated tropical precipitation assessed across three major phases of the Coupled Model Intercomparison Project (CMIP). *Mon. Wea. Rev.*, 148, 3653–3680, doi:10.1175/MWR-D-19-0404.1

Fiedler, T., A. J. Pitman, K. Mackenzie, et al. 2021: Business risk and the emergence of climate analytics. *Nat. Clim. Change*, doi:10.1038/s41558-020-00984-6

Fischer, E., and R. Knutti, 2015: Anthropogenic contribution to global occurrence of heavy-precipitation and high-temperature extremes. *Nat. Clim. Change*, 5, 560–564

Flato, G., J. Marotzke, B. Abiodun, et al., 2013: Evaluation of climate models. In: *Climate change 2013: the physical science basis. Contribution of working group I to the fifth assessment report of the Intergovernmental Panel on Climate Change*. Cambridge University Press, 741–866

Fleming, J. R., 2010: *Fixing the sky: the checkered history of weather and climate control*. Columbia University Press

Fleurbaey, M., M. Ferranna, M. Budolfson, et al., 2019: The social cost of carbon: valuing inequality, risk, and population for climate policy. *The Monist*, 102, 84–109, doi:10.1093/monist/ony023

Flynn, C., 2018: Forecasts in retrospect: a history of numerical weather prediction. *Metservice* blog, January 26, https://blog.metservice.com/HistoryNWP

Foote, E., 1856: Circumstances affecting the heat of the sun's rays: art. XXXI. *Am. J. Sci. Arts*, 2nd series, XXII(LXVI), November 1856, 382–383

Forster, P. M., A. C. Maycock, C. M. McKenna, et al., 2020: Latest climate models confirm need for urgent mitigation. *Nat. Clim. Change*, 10, 7–10

Fountain, H., 2017: Scientists link Hurricane Harvey's record rainfall to climate change. *The New York Times*, December 13, www.nytimes.com/2017/12/13/climate/hurricane-harvey-climate-change.html

Fourier, J., 1822: *Théorie analytique de la chaleur*. Didot

Fox, C. R., and G. Ulkümen, 2009: Distinguishing two dimensions of uncertainty. In: W. Brun, G. Keren, G. Kirkebøen, and H. Montgomery (eds.), *Perspectives on thinking, judging, and decision making*. Oslo, Universitetsforlaget.

Freedman, A., 2016: Nearly 3,000 climate scientists condemn Australia's dramatic research cuts. *Mashable*, February 10, https://mashable.com/2016/02/10/climate-scientists-australia-csiro-cuts/

Fried, B., 2014: What you missed in we the geeks: "weather is your mood and climate is your personality," *The White House*, January 10, https://obamawhitehouse.archives.gov/blog/2014/01/10/what-you-missed-we-geeks-weather-your-mood-and-climate-your-personality

Frigg, R., S. Bradley, H. Du, and L. A. Smith, 2014: Laplace's demon and the adventures of his apprentices. *Philos. Sci.*, 81, 31–59

Frisinger, H. H., 1973: Aristotle's legacy in meteorology. *Bull. Am. Meteorol. Soc.*, 54, 198–204

Fuhrer, O., T. Chadha, T. Hoefler, et al., 2018: Near-global climate simulation at 1 km resolution: establishing a performance baseline on 4888 GPUs with COSMO 5.0. *Geosci. Model Dev.,* 11, 1665–1681, doi:10.5194/gmd-11-1665-2018

Funtowicz, S., and J. Ravetz, 1990: *Uncertainty and quality in science for policy*. Kluwer Academic Publishers

Gardiner, L. S., 2019: What does climate sound and look like? *AGU Blogosphere*, March 11, https://blogs.agu.org/sciencecommunication/2019/03/11/what-does-climate-sound-and-look-like/

Gettelman, A., and R. B. Rood, 2016: *Demystifying climate models*. Berlin, Springer

Ghosh, P., 2017: Hawking says Trump's climate stance could damage earth. *BBC*, July 2, www.bbc.com/news/science-environment-40461726

Giorgi, F., and X. Bi, 2009: Time of emergence (TOE) of GHG-forced precipitation change hot-spots. *Geophys. Res. Lett.*, 36, L06709, doi:10.1029/2009GL037593.

GISTEMP Team, 2021: GISS surface temperature analysis (GISTEMP), version 4. *NASA Goddard Institute for Space Studies*, dataset accessed January 28, 2021, https://data.giss.nasa.gov/gistemp/

Glato, G. M., 2011: Earth system models: an overview. *WIREs Clim. Change*, 2, 783–800, doi:10.1002/wcc.148

Gleckler, P. J., K. E. Taylor, and C. E. Doutriaux, 2008: Performance metrics for climate models. *J. Geophys. Res. – Atmos.*, 113, D06104

Goddard, L., W. Baethgen, B. Kirtman, and G. Meehl, 2009: The urgent need for improved climate models and predictions. *Eos*, 90(39), 343–344

Godfrey-Smith, P., 2003: *Theory and reality: an introduction to the philosophy of science*. University of Chicago Press

Goenner, H., 1993: The reaction to relativity theory in Germany, III: "A hundred authors against Einstein." In J. Earman, M. Janssen, and J. D. Norton (eds.), *The attraction of gravitation: new studies in the history of general relativity*. Boston, Birkhäuser

Golaz, J. C., L. Horowitz, and H. Levy, 2013: Cloud tuning in a coupled climate model: impact on 20th century warming. *Geophys. Res. Lett.*, 40, 2246–2251, doi:10.1002/grl.50232

Goodell, J., 2018: Hothouse earth is merely the beginning of the end. *Rolling Stone*, August 9, www.rollingstone.com/politics/politics-features/hothouse-earth-climate-change-709470/

2021: Now is our last best chance to confront the climate crisis. *Rolling Stone*, April 14, www.rollingstone.com/politics/politics-features/climate-crisis-2050-goals-biden-administration-1154528/

Graham, E., 2018: Adventures in Fine Hall. *Princeton Alumni Weekly*, January 10

Grant, W. J., 2016: CSIRO needs to tackle the impact of climate change following its jobs shake-up. *TheConversation.com*, February 4, https://theconversation.com/csiro-needs-to-tackle-the-impact-of-climate-change-following-its-jobs-shake-up-54176

Griffies, S. M., R. J. Stouffer, A. J. Adcroft, et al., 2015: A historical introduction to MOM. NOAA/GFDL, www.gfdl.noaa.gov/wp-content/uploads/2019/04/mom_history_2017.09.19.pdf

Grossman, D., 2016: Why our intuition about sea-level rise is wrong. *Nautilus*, 33, February 18, http://nautil.us/issue/33/attraction/why-our-intuition-about-sea_level-rise-is-wrong

Grubler, A., and N. Nakicenovic, 2001: Identifying dangers in an uncertain climate. *Nature*, 412, 15

Guerrero, J. E., M. Sanguineti, and K. Wittkowski, 2020: Variable cant angle winglets for improvement of aircraft flight performance. *Meccanica*, 55, 1917–1947, doi:10.1007/s11012-020-01230-1

GWPF, 2020: Cost of "net zero" will be astronomical, new reports warn. *The Global Warming Policy Foundation*, February 24, www.thegwpf.org/cost-of-net-zero-will-be-ruinous-new-reports-warn/

Hagedorn, R., F. Doblas-Reyes, and T. N. Palmer, 2005: The rationale behind the success of multimodel ensembles in seasonal forecasting – I. Basic concept. *Tellus A*, 57, 219–233, doi:10.1111/j.1600-0870.2005.00103.x

Haigh, T., M. Priestley, and C. Rope, 2014: Los Alamos bets on ENIAC: nuclear Monte Carlo simulations, 1947–1948. *IEEE Ann. Hist. Comput.*, 36(3), 42–63. doi:10.1109/MAHC.2014.40

Hallam, R., 2019: *Common sense for the 21st century: only nonviolent rebellion can now stop climate breakdown and social collapse*. Chelsea Green

Halper, M., 2015: Supercomputing's super energy needs, and what to do about them. *Commun. ACM*, September 24, https://m-cacm.acm.org/news/192296-supercomputings-super-energy-needs-and-what-to-do-about-them/fulltext?mobile=true

Hansen, J., I. Fung, A. Lacis, et al., 1988: Global climate changes as forecast by Goddard Institute for Space Studies three-dimensional model. *J. Geophys. Res.*, 93, 9341–9364

Hansen, J., A. Lacis, R. Ruedy, and M. Sato, 1992: Potential climate impact of Mount Pinatubo eruption. *Geophys. Res. Lett.*, 19, 215–218

Hansen, J., R. Ruedy, M. Sato, and K. Lo, 2010: Global surface temperature change, *Rev. Geophys.*, 48, RG4004, doi:10.1029/2010RG000345

Hargreaves, J. C., and J. D. Annan, 2014: Can we trust climate models? *WIREs Clim. Change*, 5, 435–440, doi:10.1002/wcc.288

Harper, K., L. W. Uccellini, E. Kalnay, K. Carey, and L. Morone, 2007: 50th anniversary of operational numerical weather prediction. *Bull. Am. Meteorol. Soc.*, 88, 639–650, doi:10.1175/BAMS-88-5-639

Harris, D. C., 2010: Charles David Keeling and the story of atmospheric CO2 measurements. *Anal. Chem.*, 82, 7865–7870

Hartnett, K., 2018: Machine learning confronts the elephant in the room. *Quanta Magazine*, September 20, www.quantamagazine.org/machine-learning-confronts-the-elephant-in-the-room-20180920/

Harvey, F., 2013: Joe Farman obituary. *The Guardian*, May 16, www.theguardian.com/environment/2013/may/16/joe-farman

Hausfather, Z., 2017: Explainer: why the sun is not responsible for recent climate change. *CarbonBrief.org*, August 18, www.carbonbrief.org/why-the-sun-is-not-responsible-for-recent-climate-change

2019: CMIP6: the next generation of climate models explained. *CarbonBrief.org*, December 2, www.carbonbrief.org/cmip6-the-next-generation-of-climate-models-explained

2020a: CO2 emissions from fossil fuels may have peaked in 2019. *The Breakthrough Institute*, https://thebreakthrough.org/issues/energy/peak-co2-emissions-2019

2020b: Explainer: how the rise and fall of CO2 levels influenced the ice ages. *CarbonBrief.org*, July 2, www.carbonbrief.org/explainer-how-the-rise-and-fall-of-co2-levels-influenced-the-ice-ages

Hausfather, Z., and R. Betts, 2020: Analysis: how "carbon-cycle feedbacks" could make global warming worse. *CarbonBrief.org*, April 14, www.carbonbrief.org/analysis-how-carbon-cycle-feedbacks-could-make-global-warming-worse

Hausfather, Z., H. F. Drake, T. Abbott, and G. A. Schmidt, 2020: Evaluating the performance of past climate model projections. *Geophys. Res. Lett.*, 47, e2019GL085378, doi:10.1029/2019GL085378

Hausfather, Z., and G. P. Peters, 2020: Emissions – the "business as usual" story is misleading. *Nature*, 577, 618–620

Hawkins, E., R. S. Smith, J. M. Gregory, and D. A. Stainforth, 2015: Irreducible uncertainty in near-term climate projections. *Clim. Dynam.*, 46(11–12), 3807–3819, doi:10.1007/s00382-015-2806-8

Hawkins, E., and R. Sutton, 2009: The potential to narrow uncertainty in regional climate predictions. *Bull. Am. Meteorol. Soc.*, 90, 1095–1107

2012: Time of emergence of climate signals. *Geophys. Res. Lett.*, 39, L01702

Hayhoe, K., 2015: Who brings what to the global potluck? *Union of Concerned Scientists* blog, December 15, https://blog.ucsusa.org/science-blogger/at-cop21-who-brings-what-to-the-global-potluck

2018: Does messaging with fear really work? *Global Weirding Series*, January 31, www.youtube.com/watch?v=AeqAoozVyfQ

Heal, G., and A. Millner, 2014: Uncertainty and decision making in climate change economics. *Rev. Environ. Econ. Pol.*, 8, 120–37

Held, I. M., 2005: The gap between simulation and understanding in climate modeling. *Bull. Am. Meteorol. Soc.*, 86(11), 1609–1614

2011: Heat uptake and internal variability. Isaac Held's Blog, August 23, gfdl.noaa.gov/blog_held/16-heat-uptake-and-internal-variability

2014: Simplicity amid complexity. *Science*, 343(6176), 1206–1207

Held, I. M., and B. J. Soden, 2006: Robust responses of the hydrological cycle to global warming. *J. Clim.*, 19(21), 5686–5699

Held, I. M., and M. J. Suarez, 1994: A proposal for the intercomparison of the dynamical cores of atmospheric general circulation models. *Bull. Am. Meteorol. Soc.*, 75(10), 1825–1830

Henderson, L., 2018: The problem of induction. *Stanford encyclopedia of philosophy*, https://plato.stanford.edu/entries/induction-problem/

Hennessy, J. L., and D. A. Patterson, 2019: A new golden age for computer architecture. *Commun. ACM*, 62(2), 48–60, doi:10.1145/3282307

Hersh, S., 1972: Rainmaking is used as a weapon by U.S. *The New York Times*, July 3, www.nytimes.com/1972/07/03/archives/rainmaking-is-used-as-weapon-by-us-cloudseeding-in-indochina-is.html

Hilborn, R. C., 2004: Sea gulls, butterflies, and grasshoppers: a brief history of the butterfly effect in nonlinear dynamics. *Am. J. Phys.*, 72, 425, doi:10.1119/1.1636492

Hillebrand, H., I. Donohue, W. S. Harpole, et al., 2020: Thresholds for ecological responses to global change do not emerge from empirical data. *Nat. Ecol. Evol.*, 4, 1502–1509, doi:10.1038/s41559-020-1256-9

Holloway, J. L., A. W. Holt, J. W. Mauchly, and M. A. Woodbury, 1955: Topics in statistical meteorology. *Final report of the statistics project of the University of Pennsylvania*. Office of Naval Research, contract Nonr 551(07), 24 pp.

Hora, S., 1996: Aleatory and epistemic uncertainty in probability elicitation with an example from hazardous waste management. *Reliab. Eng. Syst. Safe.*, 54, 217–223

Hossenfelder, S., 2019: Has reductionism run its course? *Backreaction* blog, October 2, http://backreaction.blogspot.com/2019/10/has-reductionism-run-its-course.html

Hourdin, F., T. Mauritsen, A. Gettelman, et al., 2017: The art and science of climate model tuning. *Bull. Am. Meteorol. Soc.*, 98, 589–602

Huddleston, A., 2019: Happy 200th birthday to Eunice Foote, hidden climate science pioneer. *Climate.gov*, July 17, www.climate.gov/news-features/features/happy-200th-birthday-eunice-foote-hidden-climate-science-pioneer

Hulme, M., 2006: Chaotic world of climate truth. *BBC News*, November 4, http://news.bbc.co.uk/2/hi/science/nature/6115644.stm

2007: The appliance of science. *The Guardian*, March 13, www.theguardian.com/society/2007/mar/14/scienceofclimatechange.climatechange

2013: How climate models gain and exercise authority. In: K. Hastrup and M. Skrydstrup, (eds.), *The social life of climate change models: anticipating nature*. New York, Routledge, 30–44

2019: Is it too late (to stop dangerous climate change)? An editorial. *WIREs Clim. Change*, 11, e619

2020: Fetishising "the number": how not to govern pandemics, climate and biodiversity. *MikeHulme.org*, July 9, https://mikehulme.org/fetishising-the-number-how-not-to-govern-pandemics-climate-and-biodiversity/

Hunter, J., 2020: The "climate doomers" preparing for society to fall apart, *BBC News*, March 16, https://notalotofpeopleknowthat.wordpress.com/2020/03/17/the-climate-doomers-preparing-for-society-to-fall-apart/

Hutson, M., 2020: Why modeling the spread of COVID-19 is so damn hard, *IEEE Spectrum*, September 22, https://science.slashdot.org/story/20/09/25/2359231/why-modeling-the-spread-of-covid-19-is-so-damn-hard

Huybers, P., 2010: Compensation between model feedbacks and curtailment of climate sensitivity. *J. Clim.*, 23, 3009–3018

IAS, 2013: *Institute for Advanced Study: an introduction*. Princeton University Press, www.ias.edu/files/pdfs/publications/IASBook.pdf

Ingleby, B., M. Rodwell, and L. Isaksen, 2016: Global radiosonde network under pressure. *ECMWF Newsletter*, 149, 25–30, www.ecmwf.int/en/newsletter/149/meteorology/global-radiosonde-network-under-pressure

Ip, G., 2020: Business shifts from resistance to action on climate. *The Wall Street Journal*, September 16, www.wsj.com/articles/business-shifts-from-resistance-to-action-on-climate-11600233503

IPCC, 1990: *Climate change: the IPCC scientific assessment*, J. T. Houghton, G. J. Jenkins, and J. J. Ephraums (eds.). Cambridge University Press

1996: Summary for Policymakers. In: *Climate change 1995 – the science of climate change. Contribution of working group I to the second assessment report of the Intergovernmental Panel on Climate Change*. Cambridge University Press, 1–8

2001: Summary for Policymakers. In: *Climate change 2001: the scientific basis,* J. T. Houghton, Y. Ding, D. J. Griggs, et al. (eds.). Cambridge University Press, 1–20

2007: Summary for policymakers. In: *Climate change 2007: the physical science basis. Contribution of working group I to the fourth assessment report of the Intergovernmental Panel on Climate Change*. Cambridge University Press, 1–18

2013: Summary for policymakers. In: *Climate change 2013: the physical science basis. Contribution of working group I to the fifth assessment report of the Intergovernmental Panel on Climate Change*. Cambridge University Press, 3–29

2014a: Climate change 2014: synthesis report. *Contribution of working groups I, II and III to the fifth assessment report of the Intergovernmental Panel on Climate Change*. Geneva, IPCC

2014b: Summary for policymakers. In: *Climate change 2014: impacts, adaptation, and vulnerability. Part A: global and sectoral aspects. Contribution of working group II to the fifth assessment report of the Intergovernmental Panel on Climate Change*. Cambridge University Press, 1–32

2015: Special report global warming of 1.5° C, annex I: glossary

Jackson, C. S., M. K. Sen, G. Huerta, Y. Deng, and K. P. Bowman, 2008: Error reduction and convergence in climate prediction. *J. Clim.*, 21, 6698–6709

Jackson, R., 2020: Eunice Foote, John Tyndall and a question of priority. *Notes Rec.*, 74(1), 105–118

Jeevanjee, N., P. Hassanzadeh, S. Hill, and A. Sheshadri, 2017: A perspective on climate model hierarchies. *J. Adv. Model. Earth Syst.*, 9, 1760–1771, doi:10.1002/2017MS001038

Jehl, D., 2004: Judging intelligence: the report; senators assail C.I.A. judgments on Iraq's arms as deeply flawed. *The New York Times*, July 10, www.nytimes.com/2004/07/10/world/judging-intelligence-report-senators-assail-cia-judgments-iraq-s-arms-deeply.html

Kahn, D., 2020: California thought it could delay climate disaster. Now millions of acres are burning. *Politico.com*, October 8, www.politico.com/states/california/story/2020/10/08/california-thought-it-could-delay-climate-disaster-now-millions-of-acres-are-burning-1317641

Kandlikar, M., J. Risbey, and S. Dessai, 2005: Representing and communicating deep uncertainty in climate change assessments. *C. R. Geosci.*, 337(4), 443–455

Karmalkar, A. V., D. M. H. Sexton, J. M. Murphy, B. B. B. Booth, J. W. Rostron, and D. J. McNeall, 2019: Finding plausible and diverse variants of a climate model. Part II: development and validation of methodology. *Clim. Dynam.*, 53, 847–877, doi:10.1007/s00382-019-04617-3

Kaufman, N., A. R. Barron, W. Krawczyk, et al., 2020: A near-term to net zero alternative to the social cost of carbon for setting carbon prices. *Nat. Clim. Change*, 10, 1010–1014

Kay, J., and M. King, 2020: *Radical uncertainty: decision-making beyond the numbers*. Norton

Kay, J. E., C. Wall, V. Yettella, et al., 2016: Global climate impacts of fixing the Southern Ocean shortwave radiation bias in the Community Earth System Model (CESM). *J. Clim.*, 29, 4617–4636, doi:10.1175/JCLI-D-15-0358.1

Keeling, C. D., 1979: The Suess Effect: ^{13}carbon–^{14}carbon interrelations. *Environ. Int.*, 2, 229–300

Keeling, C. D., R. B. Bacastow, A. E. Bainbridge, et al., 1976: Atmospheric carbon dioxide variations at Mauna Loa Observatory, Hawaii. *Tellus*, 28, 6, 538–551

Keeling, C. D., S. C. Piper, R. B. Bacastow, et al., 2001: Exchanges of atmospheric CO_2 and $^{13}CO_2$ with the terrestrial biosphere and oceans from 1978 to 2000. I. Global aspects. *SIO reference series, No. 01–06*. San Diego, Scripps Institution of Oceanography

Keith, D. W., 2000: Geoengineering the climate: history and prospect. *Annu. Rev. Energy Environ.*, 25, 245–284, doi:10.1146/annurev.energy.25.1.245

Kerr, R. A., 1994: Climate modeling's fudge factor comes under fire. *Science*, 265 (5178), 1528

1997: Climate change: model gets it right – without fudge factors. *Science*, 276 (5315), 1041, doi:10.1126/science.276.5315.1041

Kiehl, J., 2006: Geoengineering climate change: treating the symptom over the cause? *Clim. Change*, 77, 227–228

King, D., D. Schrag, Z. Dadi, Y. Qui, and A. Ghosh, 2015: *Climate change: a risk assessment*. Cambridge, Cambridge University Centre for Science and Policy

Kirtman, B., S. B. Power, J. A. Adedoyin, et al., 2013: Near-term climate change: projections and predictability. In: *Climate change 2013: the physical science basis. Contribution of working group I to the fifth assessment report of the Intergovernmental Panel on Climate Change*. Cambridge University Press, 953–1028

Knight, F., 1922: *Risk, uncertainty and profit*. Houghton Mifflin

Knutson, T. R., and R. E. Tuleya, 2005: Reply to comments on "Impacts of CO2-induced warming on simulated hurricane intensity and precipitation: sensitivity to the choice of climate model and convective scheme." *J. Clim.*, 18, 5183–5187

Knutti, R., 2008a: Should we believe model predictions of future climate change? *Philos. T. R. Soc. A*, 366, 4647–4664

2008b: Why are climate models reproducing the observed global surface warming so well? *Geophys. Res. Lett.*, 35, L18704

2010: The end of model democracy? *Clim. Change*, 102, 395–404

2018: Climate model confirmation: from philosophy to predicting climate in the real world. In: E. A. Lloyd and E. Winsberg (eds.), *Climate modelling: philosophical and conceptual issues*, Springer, 325–359, doi:10.1007/978-3-319-65058-6_11

Knutti, R., C. Baumberger, and G. Hirsch Hadorn, 2019: Uncertainty quantification using multiple models – prospects and challenges. In: C. Beisbart and N. J. Saam (eds.), *Computer simulation validation: fundamental concepts, methodological frameworks, and philosophical perspectives*. Springer, 835–855

Knutti, R., R. Furrer, C. Tebaldi, J. Cermak, and G. Meehl, 2010: Challenges in combining projections from multiple climate models. *J. Clim.*, 23(10), 2739–2758

Knutti, R., J. Rogelj, J., Sedlaček, et al., 2016: A scientific critique of the two-degree climate change target. *Nat. Geosci.*, 9, 13–18

Knutti, R., M. Rugenstein, and G. Hegerl, 2017: Beyond equilibrium climate sensitivity. *Nat. Geosci.*, 10, 727–736

Knutti, R., and J. Sedlaček, 2013: Robustness and uncertainties in the new CMIP5 climate model projections. *Nat. Clim. Change*, 3, 369–373

Kolbert, E., 2005: The climate of man – II. *The New Yorker*, May 2, 64, www.newyorker.com/magazine/2005/05/02/the-climate-of-man-ii

2006: *Field notes from a catastrophe: man, nature, and climate change*. Bloomsbury

Koonin, S. E., 2014: Climate science is not settled. *The Wall Street Journal*, September 19, www.wsj.com/articles/climate-science-is-not-settled-1411143565

Korten, T., 2015: In Florida, officials ban term "climate change." *Miami Herald*, March 8, www.miamiherald.com/news/state/florida/article12983720.html

Kozlov, M., 2021: The arctic has a cloud problem. *The Atlantic*, February 27, www.theatlantic.com/science/archive/2021/02/arctic-has-cloud-problem/618159/

Kuhn, T., 1962: *The structure of scientific revolutions*. University of Chicago Press

Kunreuther H., S. Gupta, V. Bosetti, et al, 2014: Integrated risk and uncertainty assessment of climate change response policies. In: *Climate change 2014: mitigation of climate change. Contribution of working group III to the fifth assessment report of the Intergovernmental Panel on Climate Change*. Cambridge University Press, 151–205

Kusunoki, S., and O. Arakawa, 2015: Are CMIP5 models better than CMIP3 models in simulating precipitation over East Asia? *J. Clim.*, 28, 5601–5621, doi:10.1175/JCLI-D-14-00585.1

Lahsen, M., 2005: Seductive simulations? Uncertainty distribution around climate models. *Soc. Stud. Sci.*, 35(6), 895–922

2013: Anatomy of dissent: a cultural analysis of climate skepticism. *Am. Behav. Sci.*, 57(6), 732–753

Lane, A., 2004: Cold comfort – *The Day after Tomorrow*. *The New Yorker*, June 7, www.newyorker.com/magazine/2004/06/07/cold-comfort-4

Laplace, P. S., 1814: *A philosophical essay on probabilities*. Dover

Lawrence, J., M. Haasnoot, and R. Lempert, 2020: Climate change: making decisions in the face of deep uncertainty. *Nature*, 580, 456

Lawrence, M. G., and S. Schäfer, 2019: Promises and perils of the Paris Agreement. *Science*, 364, 829–830

Le Treut, H., R. Somerville, U. Cubasch, et al., 2007: Historical overview of climate change. In: *Climate change 2007: the physical science basis. Contribution of working group I to the fourth assessment report of the Intergovernmental Panel on Climate Change*. Cambridge University Press, 93–127

Legates, D., 2020: How the EPA's "endangerment finding" endangers you. *Townhall.com*, July 10, https://townhall.com/columnists/davidlegates/2020/07/10/how-the-epas-endangerment-finding-endangers-you-n2572275

Lenhard, J., and E. Winsberg, 2010: Holism, entrenchment, and the future of climate model pluralism. *Stud. Hist. Philos. M. P.*, 41, 253–262

Lenssen, N., G. Schmidt, J. Hansen, et al., 2019: Improvements in the GISTEMP uncertainty model. *J. Geophys. Res. – Atmos.*, 124, 12, 6307–6326, doi:10.1029/2018JD029522

Leonhardt, F., 1997: The committee to save the Tower of Pisa: a personal report. *Struct. Eng. Int.*, 7(3), 201 – 212, doi:10.2749/101686697780494734

Lewandowsky, S., T. Ballard, and R. D. Pancost, 2015: Uncertainty as knowledge. *Phil. Trans. R. Soc. A*, 373, 20140462

Lewis, J. M., 2008: Smagorinsky's GFDL: building the team. *Bull. Am. Meteorol. Soc.*, 89(9), 1339–1353

Lewis, N., and J. A. Curry, 2015: The implications for climate sensitivity of AR5 forcing and heat uptake estimates. *Clim. Dynam.*, 45, 1009–1023, doi:10.1007/s00382-014-2342-y

Lipman, D., 2014: D-Day at 70: the most important weather forecast in history. *Washington Post*, June 6, www.washingtonpost.com/news/capital-weather-gang/wp/2014/06/06/d-day-at-70-the-most-important-weather-forecast-in-the-history-of-the-world/

Live Science, 2020: How accurate are Punxsutawney Phil's Groundhog Day forecasts? *Live Science*, www.livescience.com/32974-punxsutawney-phil-weather-prediction-accuracy.html

Lloyd, E. A., 2010: Confirmation and robustness of climate models. *Philos. Sci.*, 77(5), 971–984

Loft, R., 2020: Earth system modeling must become more energy efficient. *Eos*, July 28, doi:10.1029/2020EO147051

Lohr, S., 2018: Move over, China: U.S. is again home to world's speediest supercomputer. *The New York Times*, June 8, www.nytimes.com/2018/06/08/technology/supercomputer-china-us.html

Lomborg, B., 2020a: The Lockdown's lessons for climate activism. *The Wall Street Journal*, July 11, www.wsj.com/articles/the-lockdowns-lessons-for-climate-activism-11594440061

2020b: Welfare in the 21st century: increasing development, reducing inequality, the impact of climate change, and the cost of climate policies. *Technol. Forecast. Soc.*, 156, 119981

Lombrana, L. M., A. Rathi, and H. Warren, 2020: It's a race against heat, and humanity is losing. *Bloomberg Green*, September 8, www.bloomberg.com/graphics/2020-race-against-heat/

Lorenz, E., 2006: Reflections on the conception, birth, and childhood of numerical weather prediction. *Annu. Rev. Earth Pl. Sc.*, 34, 37– 45

Lorenz, E. N. 1963a: Deterministic nonperiodic flow. *J. Atmos. Sci.*, 20, 130–141

1963b: The predictability of hydrodynamic flow. *Trans. N. Y. Acad. Sci.*, 25, 409–432

1969: The predictability of a flow which possesses many scales of motion. *Tellus*, 3, 290–307

1975: Climate predictability. In: B. Bolin, B. Döös, W. Godson, K. Hasselmann, J. Kutzbach, and J. Sawyer (eds.), *The physical basis of climate and climate modelling,* vol. 16, GARP Publication Series. Geneva, WMO, 132–136

1993: *The essence of chaos*. University of Washington Press

Luft, J., and H. Ingham, 1955: *The Johari window, a graphic model of interpersonal awareness. Proceedings of the Western Training Laboratory in Group Development*. Los Angeles, UCLA

Lynch, P., 2007: The origins of computer weather prediction and climate modeling. *J. Comput. Phys.*, doi:10.1016/j.jcp.2007.02.034

2008: The ENIAC forecasts: a re-creation. *Bull. Am. Meteorol. Soc.*, 89, 45–55

2011: From Richardson to early numerical weather prediction. In: L. Donner, W. Schubert, and R. Somerville (eds.), *The development of atmospheric general circulation models: complexity, synthesis and computation*. Cambridge University Press, 3–17

MacDougall, A. H., 2020: Is there warming in the pipeline? A multi-model analysis of the Zero Emissions Commitment from CO2. *Biogeosciences*, 17, 2987–3016

MacKenzie, D., 1998: The certainty trough. In: R. Williams and W. Faulkner (eds.), *Exploring expertise: issues and perspectives*. Palgrave Macmillan, 325–329

Mackey, R., 2009: Why Ahmadinejad voted against occupying the U.S. embassy in 1979. *The New York Times*, November 8, https://thelede.blogs.nytimes.com/2009/11/08/why-ahmadinejad-voted-against-occupying-the-us-embassy-in-1979/

Macrae, N., 1999: *John Von Neumann: the scientific genius who pioneered the modern computer, game theory, nuclear deterrence, and much more*. American Mathematical Society

Maher, P., E. P. Gerber, B. Medeiros, et al., 2019: Model hierarchies for understanding atmospheric circulation. *Rev. Geophys.*, 57, 250–280, doi:10.1029/2018RG000607

Maher, N., F. Lehner, and J. Marotzke, 2020: Quantifying the role of internal variability in the temperature we expect to observe in the coming decades. *Environ. Res. Lett.*, 15, 054014

Manabe, S., 2019: Role of greenhouse gas in climate change. *Tellus A: Dynamic Meteorology and Oceanography*, 71, 1, 1620078, doi:10.1080/16000870.2019.1620078

Manabe, S., and A. J. Broccoli, 2020: *Beyond global warming: how numerical models revealed the secrets of climate change*. Princeton University Press

Manabe, S., K. Bryan, and M. J. Spelman, 1975: A global ocean-atmosphere climate model. Part I. The atmospheric circulation. *J. Phys. Oceanogr.*, 5, 3–29

Mankin, J. S., F. Lehner, S. Coats, and K. A. McKinnon, 2020: The value of initial condition large ensembles to robust adaptation decision-making. *Earth's Future*, 8, e2012EF001610

Mann, M. E., 2020: The story about the "business as usual" story is misleading. *MichaelMann.net*, January 29, https://michaelmann.net/content/story-about-%E2%80%98business-usual%E2%80%99-story-misleading

Mann, M. E., S. J. Hassol, and T. Toles, 2017: Doomsday scenarios are as harmful as climate change denial. *Washington Post*, July 12, www.washingtonpost.com/opinions/doomsday-scenarios-are-as-harmful-as-climate-change-denial/2017/07/12/880ed002-6714-11e7-a1d7-9a32c91c6f40_story.html

Markow, T. A., 2015: The natural history of model organisms: the secret lives of Drosophila flies. *eLife*, 4, e06793, doi:10.7554/eLife.06793

Marvel, K., 2018: Thinking about climate on a dark, dismal morning. *Scientific American* blogs, December 25, https://blogs.scientificamerican.com/hot-planet/thinking-about-climate-on-a-dark-dismal-morning/

Marvel, K., R. Pincus, G. A. Schmidt, and R. L. Miller, 2018: Internal variability and disequilibrium confound estimates of climate sensitivity from observations. *Geophys. Res. Lett.*, 45, 1595–1601, doi:10.1002/2017GL076468

Maslin, M. A., and P. Austin, 2012: Uncertainty: climate models at their limit? *Nature*, 486, 183–184

Matthews, H. D., and K. Caldeira, 2008: Stabilizing climate requires near-zero emissions. *Geophys. Res. Lett.*, 35, L04705, doi:10.1029/2007GL032388

Mauchly, J. W., 1973: John W. Mauchly (1907–1980) interview: February 6, 1973, *Computer Oral History Collection*, Smithsonian Institution, https://amhistory.si.edu/archives/AC0196_mauc700622.pdf

Mauritsen, T., B. Stevens, E. Roeckner, et al., 2012: Tuning the climate of a global model. *J. Adv. Model. Earth Syst.*, 4, M00A01, doi:10.1029/2012MS000154.4

McAlpine, F., 2015: Never mind the groundhogs: happy Hedgehog Day! *BBC America*, www.bbcamerica.com/anglophenia/2015/02/never-mind-groundhogs-happy-hedgehog-day

McGrath, M., 2018: Caution urged over use of "carbon unicorns" to limit warming, *BBC*, October 5, www.bbc.com/news/science-environment-45742191

McKenna, S., A. Santoso, A. Sen Gupta, A. S. Taschetto, and W. Cai, 2020: Indian Ocean dipole in CMIP5 and CMIP6: characteristics, biases, and links to ENSO. *Sci. Rep.*, 10, 11500

McLaren, D., and N. Markusson, 2020: The co-evolution of technological promises, modelling, policies and climate change targets. *Nat. Clim. Change*, 10, 392–397

McSweeney, R., 2020: Explainer: nine "tipping points" that could be triggered by climate change. *CarbonBrief.org*, February 10, www.carbonbrief.org/explainer-nine-tipping-points-that-could-be-triggered-by-climate-change

McWilliams, J. C., 2007: Irreducible imprecision in atmospheric and oceanic simulations. *Proc. Nat. Acad. Sci.*, 104, 8709–8713, doi:10.1073/pnas.0702971104

Meehl, G. A., T. F. Stocker, W. D. Collins, et al., 2007: Global climate projections. In: *Climate change 2007: the physical science basis. Contribution of working group I to the fourth assessment report of the Intergovernmental Panel on Climate Change*. Cambridge University Press, 747–845

Meehl, G. A., C. A. Senior, V. Eyring, et al., 2020: Context for interpreting equilibrium climate sensitivity and transient climate response from the CMIP6 Earth system models. *Sci. Adv.*, 6(26), eaba1981, doi:10.1126/sciadv.aba1981

Mengel, M., A. Nauels, J. Rogelj, et al., 2018: Committed sea-level rise under the Paris Agreement and the legacy of delayed mitigation action. *Nat. Commun.*, 9, 601

Michaels, P., and R. Maue, 2018: Thirty years on, how well do global warming predictions stand up? *The Wall Street Journal*, June 21, www.wsj.com/articles/thirty-years-on-how-well-do-global-warming-predictions-stand-up-1529623442

Milinski, S., N. Maher, and D. Olonscheck, 2020: How large does a large ensemble need to be? *Earth Syst. Dynam.*, 11, 885–901, doi:10.5194/esd-11-885-2020

Mill, J. S., 1843: *A system of logic, ratiocinative and inductive*. London, J. W. Parker

Miller, R. L., G. A. Schmidt, L. S. Nazarenko, et al., 2014: CMIP5 historical simulations (1850–2012) with GISS ModelE2. *J. Adv. Model. Earth Syst.*, 6, 441–478

Milly, P. C. D., J. Betancourt, M. Falkenmark, et al., 2008: Stationarity is dead: whither water management? *Science*, 319(5863), 573–574, doi:10.1126/science.1151915

Miner, S., 2018: Newton didn't frame hypotheses. Why should we? *Physics Today*, April 24, doi:10.1063/PT.6.3.20180424a

Mirowski, P., 1992: Do economists suffer from physics envy? *Finn. Econ. Pap.,* 5(1), 61–68

Mitchell, J. F. B., S. Manabe, V. Meleshko, and T. Tokioka, 1990: Equilibrium climate change – and its implications for the future. In: J. T. Houghton, G. J. Jenkins, and J. J. Ephraums (eds.), *Climate change: the IPCC scientific assessment*. Cambridge University Press, 131–172

Molina, M., and F. Rowland, 1974: Stratospheric sink for chlorofluoromethanes: chlorine atom-catalysed destruction of ozone. *Nature*, 249, 810–812, doi:10.1038/249810a0

Moore, S., 2018: Follow the (climate change) money. *The Heritage Foundation, Commentary,* December 18, www.heritage.org/environment/commentary/follow-the-climate-change-money

Moss, R., J. Edmonds, K. Hibbard, et al., 2010: The next generation of scenarios for climate change research and assessment. *Nature*, 463, 747–756

Muller, J. Z., 2018: *The tyranny of metrics*. Princeton University Press

Nakićenović, N., J. Alcamo, G. Davis, et al., 2000: *Special report on emissions scenarios: a special report of the Intergovernmental Panel on Climate Change*. Cambridge, Cambridge University Press

NASA, 2004: How did navigators hit their precise landing target on mars? NASA Mars Exploration Rovers Spotlight, https://mars.nasa.gov/mer/spotlight/navTarget01.html

NASEM, 2016: *Attribution of extreme weather events in the context of climate change*. National Academies Press

2019: *Negative emissions technologies and reliable sequestration: a research agenda*. National Academies Press

NCEI, 2017: Groundhog Day forecasts and climate history. *National Centers for Environmental Information*, February 2, www.ncei.noaa.gov/news/groundhog-day-forecasts-and-climate-history

Nielsen-Gammon, J., J. Escobedo, C. Ott, J. Dedrick, and A. Van Fleet, 2020: Assessment of historic and future trends of extreme weather in Texas, 1900–2036. *Office of the Texas State Climatologist*, https://climatexas.tamu.edu/products/texas-extreme-weather-report/index.html

Nissan, H., L. Goddard, E. Coughlan de Perez, et al., 2019: On the use and misuse of climate change projections in international development. *WIREs Clim. Change*, 10, e579

Nissan, H., A. G. Muñoz, and S. J. Mason, 2020: Targeted model evaluations for climate services: a case study on heat waves in Bangladesh. *Clim. Risk Manag.*, 28, 100213

Noerdlinger, P. D., and K. R. Brower, 2007: The melting of floating ice raises the ocean level. *Geophys. J. Int.*, 170(1), 145–150, doi:10.1111/ j.1365-246X.2007.03472.x

Nordhaus, W., 2007a: *The challenge of global warming: economic models and environmental policy*. New Haven, CT, Yale University Press

2007b: *The Stern review on the economics of climate change*. New Haven, CT, Yale University Press

2015: Climate clubs: overcoming free-riding in international climate policy. *Am. Econ. Rev.*, 105(4), 1339–1370, doi:10.1257/aer.15000001

Nordhaus, W. D., 2017: Revisiting the social cost of carbon. *Proc. Natl. Acad. Sci.*, 114(7), 1518–1523, doi:10.1073/pnas.1609244114

Norman, J., R. Read, Y. Bar-Yam, and N. N. Taleb, 2015: Climate models and precautionary measures. *Issues Sci. Technol.*, 31(4), 1

NOVA, 1999: Fall of the leaning tower. *NOVA*, October 5, *PBS.org*, www.pbs.org/wgbh/nova/pisa/interventions2.html

Novak, M., 2011: Weather control as a cold war weapon. *Smithsonian Magazine*, December 5, www.smithsonianmag.com/history/weather-control-as-a-cold-war-weapon-1777409/

NRC, 1996: *The ozone depletion phenomenon*. National Academies Press

2015: *Climate intervention: reflecting sunlight to cool Earth*. National Academies Press

Nuñez, R. E., and E. Sweetser, 2006: With the future behind them: convergent evidence from Aymara language and gesture in the crosslinguistic comparison of spatial construals of time. *Cogn. Sci.*, 30, 401–450

O'Gorman, P. A., and J. G. Dwyer, 2018: Using machine learning to parameterize moist convection: potential for modeling of climate, climate change, and extreme events. *J. Adv. Model. Earth Syst.*, 10, 2548–2563

O'Neill, S., and S. Nicholson-Cole, 2009: "Fear won't do it": promoting positive engagement with climate change through visual and iconic representations. *Sci. Commun.*, 30, 355

O'Raifeartaigh, C., 2017: Einstein's greatest blunder? *Scientific American* Guest Blog, February 21, https://blogs.scientificamerican.com/guest-blog/einsteins-greatest-blunder/

Oestreicher, C., 2007: A history of chaos theory. *Dialogues Clin. Neurosci.*, 9(3), 279–289

Oppenheimer, M., 2007: How the IPCC got started. *Environmental Defense Fund* blogs, November 1, http://blogs.edf.org/climate411/2007/11/01/ipcc_beginnings/

Oppenheimer, M., B. C. O'Neill, and M. Webster, 2008: Negative learning. *Clim. Change*, 89, 155–117

Oreskes, N., 2007: The scientific consensus on climate change: how do we know we're not wrong? In: J. F. DiMento and P. Doughman (eds.), *Climate change: what it means for us, our children, and our grandchildren*. MIT Press, 90

2019: *Why trust science?* Princeton University Press

Oreskes, N., and K. Belitz, 2001: Philosophical issues in model assessment. In: M. Anderson and P. Bates (eds.), *Model validation: perspectives in hydrological science*. West Sussex, Wiley, 23–42

Oreskes, N., and E. M. Conway, 2011: *Merchants of doubt: how a handful of scientists obscured the truth on issues from tobacco smoke to climate change*. Bloomsbury Publishing

Oreskes, N., K. Shrader-Frechette, and K. Belitz, 1994: Verification, validation, and confirmation of numerical models in earth sciences. *Science*, 263, 641–646

Oreskes, N., L. Smith, and D. Stainforth, 2010: Adaptation to global warming: do climate models tell us what we need to know? *Philos. Sci.*, 77, 1012–1028

Ortiz, J. D., and R. Jackson, 2020: Understanding Eunice Foote's 1856 experiments: heat absorption by atmospheric gases. *Notes Rec.,* doi:10.1098/rsnr.2020.0031

Otto, F. E. L., D. Frame, A. Otto, and M. R. Allen, 2015: Embracing uncertainty in climate change policy. *Nat. Clim. Change*, 5, 917–920, doi:10.1038/nclimate2716

Overbye, D., 2017: The eclipse that revealed the universe. *The New York Times*, July 31, www.nytimes.com/2017/07/31/science/eclipse-einstein-general-relativity.html

Palm, R., T. Bolsen, and J. T. Kingsland, 2020: "Don't tell me what to do": resistance to climate change messages suggesting behavior changes. *Weather Clim. Soc.*, 12, 827–835, doi:10.1175/WCAS-D-19-0141.1

Palmer, B., 2012: Global warming would harm the earth, but some areas might find it beneficial. *Washington Post*, January 23, www.washingtonpost.com/national/health-science/global-warming-would-harm-the-earth-but-some-areas-might-find-it-beneficial/2012/01/17/gIQAbXwhLQ_print.html

Palmer, T., 2008: Obituary – Edward Norton Lorenz. *WMO Bulletin*, 57(3), https://public.wmo.int/en/bulletin/obituary-1

2016: A personal perspective on modelling the climate system. *Proc. R. Soc. A*, 472, 20150772

Palmer, T., A. Döring, and G. Seregin, 2014: The real butterfly effect. *Nonlinearity*, 27, R123

Palmer, T., and B. Stevens, 2019: The scientific challenge of understanding and estimating climate change. *Proc. Natl. Acad. Sci.*, 116 (49) 24390–24395

Palmer, T., B. Stevens, and P. Bauer, 2019: We need an international center for climate modeling. *Scientific American*, December 18, https://blogs.scientificamerican.com/observations/we-need-an-international-center-for-climate-modeling/

Palmer, T. N., 2009: Edward Norton Lorenz 23 May 1917 – 16 April 2008. *Biogr. Mems. Fell. R. Soc.*, 55, 139–155, doi:10.1098/rsbm.2009.0004

Panosetti, D., L. Schlemmer, and C. Schär, 2019: Bulk and structural convergence at convection-resolving scales in real-case simulations of summertime moist convection over land. *Q. J. R. Meteorol. Soc.*, 145, 1427–1443

Parker, W. S., 2006: Understanding pluralism in climate modeling. *Found. Sci*, 11, 349–368

2010: Predicting weather and climate: uncertainty, ensembles and probability. *Stud. Hist. Philos. M. P.*, 41, 263–272

Parker, W. S., and J. S. Risbey, 2015: False precision, surprise and improved uncertainty assessment. *Phil. Trans. R. Soc. A*, 373, 20140453

Pearce, F., 2010: Victory for openness as IPCC climate scientist opens up lab doors. *The Guardian*, February 9, www.theguardian.com/environment/2010/feb/09/ipcc-report-author-data-openness

Persson, A., 2005: Early operational numerical weather prediction outside the USA: an historical introduction. Part 1: Internationalism and engineering NWP in Sweden, 1952–69. *Meteorol. Appl.*, 12, 135–159, doi:10.1017/S1350482705001593

Peterson, T. C., W. M. Connolley, and J. Fleck, 2008: The myth of the 1970s global cooling consensus. *Bull. Am. Meteorol. Soc.*, 89, 1325–1338

Philip, S., S. Kew, G. J. van Oldenborgh, et al., 2020: A protocol for probabilistic extreme event attribution analyses. *Adv. Stat. Clim. Meterol. Oceanogr.*, 6, 177–203, doi:10.5194/ascmo-6-177-2020

Phillips, L., 2018: Turbulence, the oldest unsolved problem in physics. *Ars Technica*, October 10, https://arstechnica.com/science/2018/10/turbulence-the-oldest-unsolved-problem-in-physics/

Phillips, N. A., 1982: Jule Charney's influence on meteorology. *Bull. Am. Meteorol. Soc.*, 63(5), 492–498

2000: The start of numerical weather prediction in the United States. In: A. Spekat (ed), *50th anniversary of numerical weather prediction. Commemorative symposium*. Deutsche Meteorologische Gesellschaft, 13–28

Pielke Jr, R., 2001: Room for doubt. *Nature*, 410, 151

Pierrehumbert, R. T, 2014: Climate science is settled enough. *Slate*, October 1, https://slate.com/technology/2014/10/the-wall-street-journal-and-steve-koonin-the-new-face-of-climate-change-inaction.html

Pierrehumbert, R. T., H. Brogniez, and R. Roca, 2007: On the relative humidity of the atmosphere. In: T. Schneider and A. H. Sobel (eds.), *The global circulation of the atmosphere*. Princeton University Press, 143–185

Pinker, S., 1994: The game of the name, op-ed. *The New York Times*, April 5, www.nytimes.com/1994/04/05/opinion/the-game-of-the-name.html

Pipitone, J., and S. Easterbrook, 2012: Assessing climate model software quality: a defect density analysis of three models. *Geosci. Model Dev.*, 5, 1009–1022, doi:10.5194/gmd-5-1009-2012

Pirsig, R., 1974: *Zen and the art of motorcycle maintenance*. William Morrow

Pisa Committee, 2002: Safeguard and stabilisation of the Leaning Tower of Pisa 1990–2001. *The International Committee for the Safeguard of the Tower of Pisa*. In: Estrategias relativas al patrimonio cultural mundial. La salvaguarda en un mundo globalizado. Principios, practicas y perspectivas. 13th ICOMOS General Assembly and Scientific Symposium. Actas. Comité Nacional Español del ICOMOS, Madrid, pp. 199–207, http://openarchive.icomos.org/id/eprint/576/

Platzman, G., 1987: Conversations with Jule Charney, *NCAR Technical Note*, NCAR/TN-298+PROC, November, 127

Platzman, G. W., 1979: The ENIAC computation of 1950 – gateway to numerical weather prediction. *Bull. Am. Meteorol. Soc.*, 60(4), 302–312

Plumer, B., and J. Schwartz, 2020: These changes are needed amid worsening wildfires, experts say. *The New York Times*, September 10, www.nytimes.com/2020/09/10/climate/wildfires-climate-policy.html

Po-Chedley, S., C. Proistosescu, K. C. Armour, et al., 2018: Climate constraint reflects forced signal. *Nature*, 563, E6–E9

Poincaré, H., 1899: *Les Methodes Nouvelles de Ia Mécanique Celeste*, vols. 1–3. Paris, Gauthier Villars

Polvani, L. M., A. C. Clement, B. Medeiros, J. J. Benedict, and I. R. Simpson, 2017: When less is more: opening the door to simpler climate models. *Eos*, 98, doi:10.1029/2017EO079417

Popper, K., 1935: *Logik der Forschung*. Wien, J. Springer. Translated by Popper as *The Logic of Scientific Discovery*, London, Hutchinson, 1959

Prins, G., and S. Rayner, 2007: Time to ditch Kyoto. *Nature*, 449, 973–975

Prono, L., 2008: Smagorinsky, Joseph (1924–2005). In: S. G. Philander (ed.), *Encyclopedia of global warming and climate change*. SAGE, 903–903, doi:10.4135/9781412963893.n583

Puko, T., 2020: EPA proposes emissions limits for jet aircraft. *The Wall Street Journal*, July 22, www.wsj.com/articles/epa-proposes-emissions-limits-for-jet-aircraft-11595429280

Qiu, L., 2018: The baseless claim that climate scientists are "driven" by money. *The New York Times*, November 27, www.nytimes.com/2018/11/27/us/politics/climate-report-fact-check.html

Rahmstorf, S., 2013: Sea level in the 5th IPCC report. *RealClimate.org*, October 15, www.realclimate.org/index.php/archives/2013/10/sea-level-in-the-5th-ipcc-report/

Rainey, J., 2017: How forecasters nailed Harvey's massive rain dump. *NBC News*, August 31, www.nbcnews.com/news/us-news/how-forecasters-nailed-harvey-s-massive-rain-dump-n797506

Ramanathan, V., and A. W. Vogelmann, 1997: Greenhouse effect, atmospheric solar absorption and the earth's radiation budget: from the Arrhenius/Langley era to the 1990s. *Ambio*, 26(1), 38–46

Ramaswamy, V., O. Boucher, J. Haigh, et al., 2001: Radiative forcing of climate change. In: J. T. Houghton, Y. Ding, D. J. Griggs, et al. (eds.), *Climate change 2001: the scientific basis*. Cambridge University Press, 349–416

Randers, J., and U. Goluke, 2020: An earth system model shows self-sustained melting of permafrost even if all man-made GHG emissions stop in 2020. *Sci. Rep.*, 10, 18456, doi:10.1038/s41598-020-75481-z

Rasmussen, J. L., 1992: The world weather watch – a view of the future. *Bull. Am. Meteorol. Soc.*, 73(4), 477–481

Rasool, S. I., and S. H. Schneider, 1971: Atmospheric carbon dioxide and aerosols: effects of large increases on global climate. *Science*, 173(3992), 138–141

Ravilious, K., 2018: Thirty years of the IPCC. *Physics World*, October 8, https://physicsworld.com/a/thirty-years-of-the-ipcc/

Rayner, S., 2012: Uncomfortable knowledge: the social construction of ignorance in science and environmental policy discourses. *Econ. Soc.*, 41(1), 107–125

Readfearn, G., 2018: Earth's climate monsters could be unleashed as temperatures rise. *The Guardian*, October 5, www.theguardian.com/environment/planet-oz/2018/oct/06/earths-climate-monsters-could-be-unleashed-as-temperatures-rise

Reed, B. C., 2015: Kilowatts to kilotons: wartime electricity use at Oak Ridge. *History of Physics Newsletter*, XII(6), Spring, https://engage.aps.org/fhp/resources/newsletters/newsletter-archive/spring-2015

Reichstein, M., G. Camps-Valls, B. Stevens, et al., 2019: Deep learning and process understanding for data-driven Earth system science. *Nature*, 566, 195–204

Revkin, A., 2014: Certainties, uncertainties and choices with global warming. *Dot Earth, New York Times* blog, September 26, https://dotearth.blogs.nytimes.com/2014/09/26/certainties-uncertainties-and-choices-with-global-warming/

2016: My climate change. *Issues Sci. Technol.*, 32(2)

2017: There are lots of climate uncertainties. Let's acknowledge and plan for them with honesty. *Propublica.org*, May 2, www.propublica.org/article/climate-change-uncertainties-bret-stephens-column

Revkin, A. C., 2005: The daily planet: why the media stumble over the environment. In: D. Blum, M. Knudson, and R. M. Henig, (eds.), *A field guide for science writers*. Oxford University Press, 222–228

Riahi, K., D. P. van Vuuren, E. Kriegler, et al., 2017: The Shared Socioeconomic Pathways and their energy, land use, and greenhouse gas emissions implications: an overview. *Glob. Environ. Change*, 42, 153–168, doi:10.1016/j.gloenvcha.2016.05.009

Rich, N., 2018: Losing Earth: the decade we almost stopped climate change. *The New York Times Magazine*, August 1, www.nytimes.com/interactive/2018/08/01/magazine/climate-change-losing-earth.html?mtrref=www.google.com&gwh=F6DE46AFCA439A96F422E63B47140895&gwt=regi&assetType=REGIWALL

Risbey, J., 2008: The new climate discourse: alarmist or alarming? *Glob. Environ. Change*, 18(1), 26–37

Risbey, J. S., 2004: Agency and the assignment of probabilities to greenhouse emissions scenarios. *Clim. Change*, 67, 37–42

Risbey, J. S., and M. Kandlikar, 2007: Expressions of likelihood and confidence in the IPCC uncertainty assessment process, *Clim. Change*, 85, 19–31

Roberts, D., 2012: In a climate-crazed world, how can we plan for the future? *Grist*, September 28, https://grist.org/climate-energy/in-a-climate-crazed-world-how-can-we-plan-for-the-future/

Robock, A., 2008: 20 reasons why geoengineering may be a bad idea. *Bull. Atomic Sci.*, 64(2), 14–18, doi:10.2968/064002006

Rodhe, H., R. Charlson, and E. Crawford, 1997: Svante Arrhenius and the greenhouse effect. *Ambio*, 26(1), 2–5

Roe, G., 2013: Costing the Earth: a numbers game or a moral imperative? *Weather Clim. Soc.*, 5(4), 378–380, doi:10.1175/WCAS-D-12-00047.1

Roe, G. H., and Y. Bauman, 2012: Climate sensitivity: should the climate tail wag the policy dog? *Clim. Change*, 117, 647–662

Rogelj, J., M. Meinshausen, J. Sedlacek, and R. Knutti, 2014: Implications of potentially lower climate sensitivity on climate projections and policy. *Environ. Res. Lett.*, 9, 031003

Rogers, J. S., 1984: Planning models are for insight not numbers: a complementary modelling approach. In: B. Lev, (ed.), *Analytic techniques for energy planning*. Elsevier Science, 67–81

Rohr, M., 2016: Great storm of 1900 brought winds of change. *Houston Chronicle*, May 19, www.houstonchronicle.com/local/history/article/Great-Storm-of-1900-brought-winds-of-change-7724171.php

Roker, A., 2015: Blown away: Galveston hurricane, 1900. *HistoryNet*, October, www.historynet.com/blown-away.htm

Roston, E., and B. Migliozzi, 2015: What's really warming the world? *Bloomberg Businessweek*, June 24, www.bloomberg.com/graphics/2015-whats-warming-the-world/

Rowland, F. S., 2009: Stratospheric ozone depletion. In: C. Zerefos, G. Contopoulos, and G. Skalkeas (eds.), *Twenty years of ozone decline*. Springer, 23–66

Rypdal, M., H. Fredriksen, K. Rypdal, et al., 2018: Emergent constraints on climate sensitivity. *Nature*, 563, E4–E5

Salvia, S., 2017: From Archimedean hydrostatics to post-Aristotelian mechanics: Galileo's early manuscripts *De motu antiquiora* (ca. 1590). *Phys. Perspect.*, 19, 105–150, doi:10.1007/s00016-017-0202-y

Sanderson, B. M., and R. Knutti, 2012: On the interpretation of constrained climate model ensembles. *Geophys. Res. Lett.*, 39(16), L16708

Sarewitz, D., and R. Pielke Jr, 1999: Prediction in science and policy. *Technol. Soc.*, 21, 121–133

Schär, C., O. Fuhrer, A. Arteaga, et al., 2020: Kilometer-scale climate models: prospects and challenges. *Bull. Am. Meteorol. Soc.*, 101, E567–E587, doi:10.1175/BAMS-D-18-0167.1

Schmidt, G., 2007a: Hansen's 1988 projections. *RealClimate.org*, May 15, www.realclimate.org/index.php/archives/2007/05/hansens-1988-projections/

2007b: The physics of climate modeling. *Physics Today*, 60(1), 72, doi:10.1063/1.2709569

2014: The emergent patterns of climate change. *TED Talk*, March, www.ted.com/talks/gavin_schmidt_the_emergent_patterns_of_climate_change

2017: What did NASA know? and when did they know it? *RealClimate.org*, December 24, www.realclimate.org/index.php/archives/2017/12/what-did-nasa-know-and-when-did-they-know-it/

2018: 30 years after Hansen's testimony. *RealClimate.org*, June 21, www.realclimate.org/index.php/archives/2018/06/30-years-after-hansens-testimony/

2019: The best case for worst case scenarios. *RealClimate.org*, February 26, www.realclimate.org/index.php/archives/2019/02/the-best-case-for-worst-case-scenarios/

2020: Unknowability in climate science: chaos, structure, and society. *Social Research: An International Quarterly*, 87(1), 133–149, www.muse.jhu.edu/article/758637

Schmidt, G., D. Bader, L. J. Donner, et al., 2017: Practice and philosophy of climate model tuning across six US modeling centers. *Geosci. Model Dev.*, 10, 3207–3223, doi:10.5194/gmd-10-3207-2017

Schmidt, G., and S. Sherwood, 2015: A practical philosophy of complex climate modelling. *Eur. J. Phil. Sci.*, 5(2), 149–169

Schmitt, H. H., and W. Happer, 2013: In defense of carbon dioxide. *The Wall Street Journal*, May 8, www.wsj.com/articles/SB10001424127887323528404578452483656067190

Schneider, S. H., 2002: Can we estimate the likelihood of climatic changes at 2100? *Clim. Change*, 52, 441–451

Schneider, T., S. Lan, A. Stuart, and J. Teixeira, 2017b: Earth system modeling 2.0: a blueprint for models that learn from observations and targeted high-resolution simulations. *Geophys. Res. Lett.*, 44, 12396–12417

Schneider, T., J. Teixeira, C. Bretherton, et al., 2017a: Climate goals and computing the future of clouds. *Nat. Clim. Change*, 7, 3–5, doi:10.1038/nclimate3190

Schneider-Mayerson, M., and K. L. Leong, 2020: Eco-reproductive concerns in the age of climate change. *Clim. Change*, 163, 1007–1023

Schoeberl, M. R., and J. M. Rodriguez, 2009: The rise and fall of dynamical theories of the ozone hole. In: C. Zerefos, G. Contopoulos, and G. Skalkeas (eds.), *Twenty years of ozone decline*. Springer, 263–272. doi:10.1007/978-90-481-2469-5_19

Schrödinger, E., 1935: Die gegenwärtige Situation in der Quantenmechanik. *Naturwissenschaften*, 23, 807–812

Schulthess, T. C., P. Bauer, N. Wedi, O. Fuhrer, T. Hoefler, and C. Schär, 2019: Reflecting on the goal and baseline for exascale computing: a roadmap based on weather and climate simulations. *Comput. Sci. Eng.*, 21(1), 30–41, doi:10.1109/MCSE.2018.2888788

Schwalm, C. R., S. Glendon, and P. B. Duffy, 2020: RCP8.5 tracks cumulative CO2 emissions. *Proc. Natl. Acad. Sci.*, 117(33), 19656–19657

Scientific American, 1856: Scientific ladies. – Experiments with condensed gases. *Scientific American*, 12(1), (Sep. 13, 1856), 5

Scientific Reports, 2020: Climate change: ending greenhouse gas emissions may not stop global warming. *EurekaAlert! AAAS*, November 12, www.eurekalert.org/pub_releases/2020-11/sr-cce110520.php

Scientists, 2016: An open letter to the Australian Government and CSIRO, February 11, www.australasianscience.com.au/article/issue-janfeb-2016/open-letter-australian-government-and-csiro.html

Shackley, S., and B. Wynne, 1995: Integrating knowledges for climate change: pyramids, nets and uncertainties. *Glob. Environ. Change*, 5(2), 113–126

Shalett, B., 1946: Electronics to aid weather figuring. *The New York Times*, January 11, www.nytimes.com/1946/01/11/archives/electronics-to-aid-weather-figuring-scientists-tell-weather-bureau.html

Shanklin, J., 2010: Reflections on the ozone hole. *Nature*, 465, 34–35

Shaw, B., 2017: Weather forecasting: how does it work, and how reliable is it? *PrecisionAg.com*, November 14, www.precisionag.com/digital-farming/data-management/weather-forecasting-how-does-it-work-and-how-reliable-is-it/

2018: Betting on rain? The accuracy and reliability of precipitation forecasts. *PrecisionAg.com*, January 16, www.precisionag.com/digital-farming/data-management/betting-on-rain-the-accuracy-and-reliability-of-precipitation-forecasts/

Shaw, G. K., 2019: On ENIAC's anniversary, a nod to its female "computers." *Penn Today*, February 14, https://penntoday.upenn.edu/news/eniacs-anniversary-nod-its-female-computers

Shepherd, T. G., E. Boyd, R. A. Calel, et al., 2018: Storylines: an alternative approach to representing uncertainty in physical aspects of climate change. *Clim. Change*, 151, 555–557

Shepherd, T. G., and A. G. Sobel, 2020: Localness in climate change. *Comparative Studies of South Asia, Africa and the Middle East*, 40(1), 7–16, doi:10.1215/1089201X-8185983

Sherwood, S., and Q. Fu, 2014: A drier future. *Science*, 343, 737–739

Sherwood, S. C., M. J. Webb, J. D. Annan, et al., 2020: An assessment of Earth's climate sensitivity using multiple lines of evidence. *Reviews of Geophysics*, 58, e2019RG000678, doi:10.1029/2019RG000678

Shirani-Mehr, H., D. Rothschild, S. Goel, and A. Gelman, 2018: Disentangling bias and variance in election polls. *J. Am. Stat. Assoc.*, 113(522), 607–614, doi:10.1080/01621459.2018.1448823

Silver, N., 2012: *The signal and the noise: why most predictions fail – but some don't.* Penguin

2014: How FiveThirtyEight calculates pollster ratings. *FiveThirtyEight.com*, September 25, https://fivethirtyeight.com/features/how-fivethirtyeight-calculates-pollster-ratings/

Sima, R. J., 2020: Combining AI and analog forecasting to predict extreme weather. *Eos*, 101, March 4, doi:10.1029/2020EO140896

SIMIP Community, 2020: Arctic sea ice in CMIP6. *Geophys. Res. Lett.*, 47, e2019GL086749

Smagorinsky, J., 1963: General circulation experiments with the primitive equations. I. The basic experiment. *Mon. Weather Rev.*, 91, 99–164

1983: The beginnings of numerical weather prediction and general circulation modeling: early recollections. *Advances in geophysics*, vol. 25. Academic Press

Smith, D. M., A. A. Scaife, R. Eade, et al., 2020: North Atlantic climate far more predictable than models imply. *Nature*, 583, 796–800

Smith, L. A., 2002: What might we learn from climate forecasts? *Proc. Natl. Acad. Sci. USA*, 99, 2487–2492

Smith, L. A., and N. Stern, 2011: Uncertainty in science and its role in climate policy. *Phil. Trans. R. Soc. A*, 369, 4818–4841

Smith, S., 2016: Unfriendly climate. *Texas Monthly*, May, www.texasmonthly.com/articles/katharine-hayhoe-lubbock-climate-change-evangelist/

Sokal, A., 2015: Physics envy in psychology: a cautionary tale. *Seminar*, https://physics.nyu.edu/sokal/CCNY_lecture_Nov_19_15.pdf

Sokal, J., 2019: The hidden heroines of chaos. *Quanta Magazine*, May 20, www.quantamagazine.org/hidden-heroines-of-chaos-ellen-fetter-and-margaret-hamilton-20190520/

Solomon, S., 1997: Transcript of interview of Susan Solomon. American Meteorological Society Oral History Project. *NCAR Archives*, http://n2t.net/ark:/85065/d7959fz0

Solomon, S., D. Qin, M. Manning, et al., 2007: Technical Summary. In: *Climate change 2007: the physical science basis. Contribution of Working Group I to the Fourth Assessment Report of the Intergovernmental Panel on Climate Change.* Cambridge University Press, 19–91

Solomon, S., D. J. Ivy, D. Kinnison, M. J. Mills, R. R. Neely, and A. Schmidt, 2016: Emergence of healing in the Antarctic ozone layer. *Science*, 353, 269–274, doi:10.1126/science.aae0061

Somerville, R. C. J., 1996: *The forgiving air.* Berkeley, CA, University of California Press

Somerville, R. C. J., P. H. Stone, M. Halem, et al., 1974: The GISS model of the global atmosphere. *J. Atmos. Sci.*, 31, 84–117

Soon, W. W. H., 2005: Variable solar irradiance as a plausible agent for multidecadal variations in the Arctic-wide surface air temperature record of the past 130 years. *Geophys. Res. Lett.*, 32, L16712

Sorenson, R. P., 2011: Eunice Foote's pioneering research on CO2 and climate warming. *Search and Discovery*, Article # 70092, www.searchanddiscovery.com/pdfz/documents/2011/70092sorenson/ndx_sorenson.pdf.html

Spade, P. V., and C. Panaccio, 2019: William of Ockham. *Stanford encyclopedia of philosophy*, https://plato.stanford.edu/entries/ockham/

Spencer, R. W., and J. R. Christy, 1990: Precise monitoring of global temperature trends from satellites. *Science*, 247, 1558

Spiegelhalter, D. J., and H. Riesch, 2011: "Don't know, can't know": embracing deeper uncertainties when analysing risks. *Phil. Trans. R. Soc. A*, 369, 4730–4750

Stainforth, D., T. Aina, C. Christensen, et al., 2005: Uncertainty in predictions of the climate response to rising levels of greenhouse gases. *Nature*, 433, 403–406

Stainforth, D. A., M. R. Allen, E. R. Tredger, and L. A. Smith, 2007: Confidence, uncertainty and decision-support relevance in climate predictions. *Phil. Trans. R. Soc. A*, 3652145–3652161

Steffen, W., J. Rockström, K. Richardson, et al., 2018: Trajectories of the earth system in the anthropocene. *Proc. Natl. Acad. Sci.*, 115(33), 8252–8259

Stephens, B., 2017: Climate of complete certainty. *The New York Times*, April 28, www.nytimes.com/2017/04/28/opinion/climate-of-complete-certainty.html

Stirling, A., 1998: Risk at a turning point? *Journal of Risk Research*, 1, 97–109

Stocker, T. F., D. Qin, G. K. Plattner, et al., 2013: Technical summary. In: *Climate change 2013: the physical science basis. Contribution of working group I to the fifth assessment report of the Intergovernmental Panel on Climate Change.* Cambridge University Press, 33–115

Stormfax, 2020: Groundhog Day. *Stormfax Weather Almanac,* www.stormfax.com/ghogday.htm

Subramanian, A., S. Juricke, P. Dueben, and T. Palmer, 2019: A stochastic representation of subgrid uncertainty for dynamical core development. *Bull. Am. Meteorol. Soc.*, 100, 1091–1101, doi:10.1175/BAMS-D-17-0040.1

Supran, G., and N. Oreskes, 2017: Assessing ExxonMobil's climate change communications (1977–2014). *Environ. Res. Lett.*, 12, 084019

Sutton, R. T., 2019: Climate science needs to take risk assessment much more seriously. *Bull. Am. Meteorol. Soc.*, 100 (9), 1637–1642

Sutton, R. T., and E. Hawkins, 2020: ESD ideas: global climate response scenarios for IPCC assessments. *Earth Syst. Dynam.*, 11, 751–754, doi:10.5194/esd-11-751-2020

Taylor, K. E., R. J. Stouffer, and G. A. Meehl, 2012: An overview of CMIP5 and the experiment design. *Bull. Am. Meteorol. Soc.*, 93, 485–498

The Economist, 2017: Language: finding a voice. *The Economist*, *Technology* Quarterly, May 1, www.economist.com/technology-quarterly/2017-05-01/language

The Independent, 2019: Lloyds of London suffers £1bn loss after year of devastating hurricanes and wildfires. *The Independent*, March 27, www.independent.co.uk/news/business/news/lloyds-london-losses-hurricanes-wildfires-natural-disasters-a8841576.html

Thompson, C., 2019: The secret history of women in coding. *The New York Times*, February 13, www.nytimes.com/2019/02/13/magazine/women-coding-computer-programming.html

Thompson, E., 2013: *Modelling North Atlantic storms in a changing climate*, Ph.D. thesis. London, Imperial College

Thompson, N. C., and S. Spanuth, 2018: The decline of computers as a general purpose technology: why deep learning and the end of Moore's Law are fragmenting computing. *Commun. ACM*, 64(3), 64–72, doi:10.1145/3430936

Thompson, P. D., 1957: Uncertainty of initial state as a factor in the predictability of large scale atmospheric flow patterns. *Tellus*, 9(3), 275–295

1983: A history of numerical weather prediction in the United States. *Bull. Am. Meteorol. Soc.*, 64, 755–769

Thorne, P. W., J. R. Lanzante, T. C. Peterson, D. J. Seidel, and K. P. Shine, 2011: Tropospheric temperature trends: history of an ongoing controversy. *WIREs Clim. Change*, 2, 66–88, doi:10.1002/wcc.80

Tierney, J. E., J. Zhu, J. King, et al., 2020: Glacial cooling and climate sensitivity revisited. *Nature*, 584, 569–573

Tollefson, J., 2020: How hot will Earth get by 2100? *Nature*, 580, 443–445

Trenberth, K., 2010: More knowledge, less certainty. *Nat. Clim. Change*, 1, 20–21, doi:10.1038/climate.2010.06

Trenberth, K., and R. Knutti, 2017: Yes, we can do "sound" climate science even though it's projecting the future. *TheConversation.com*, April 5, https://theconversation.com/yes-we-can-do-sound-climate-science-even-though-its-projecting-the-future-75763

Trenberth, K. E., 2007: Predictions of climate. Climate Feedback Blog, *Nat. Clim. Change*, June 4, http://blogs.nature.com/climatefeedback/2007/06/predictions_of_climate.html

Trenberth, K. E., J. T. Fasullo, and J. Kiehl, 2009: Earth's global energy budget. *Bull. Am. Meteorol. Soc.*, 90(3), 311–324

Trimmer, J. D., 1980: The present situation in quantum mechanics: a translation of Schrödinger's "Cat Paradox" paper. *Proc. Am. Philos. Soc.*, 124(5), 323–338

Tropp, H. S., 2003: Mauchly, John W. *ACM Digital Library*, https://dl.acm.org/doi/pdf/10.5555/1074100.1074582

Trumbla, R., 2007: *The Great Galveston Hurricane of 1900. NOAA*, https://celebrating200years.noaa.gov/magazine/galv_hurricane/welcome.html

Turing, A. M., 1950: I. – Computing machinery and intelligence. *Mind*, LIX(236), 433–460, doi:10.1093/mind/LIX.236.433

United Nations, 1992: Rio declaration on environment and development. *The United Nations Conference on Environment and Development*, www.un.org/en/development/desa/population/migration/generalassembly/docs/globalcompact/A_CONF.151_26_Vol.I_Declaration.pdf

Valdes, P., 2011: Built for stability. *Nat. Geosci.*, 4, 414–416

Vallis, G. K., 2016: Geophysical fluid dynamics: whence, whither and why? *Proc. R. Soc. A,* 47220160140, doi:10.1098/rspa.2016.0140

van der Sluijs, J., J. van Eijndhoven, S. Shackley, and B. Wynne, 1998: Anchoring devices in science for policy: the case of consensus around climate sensitivity. *Soc. Stud. Sci.*, 28, 291–323

van Oldenborgh, G. J., E. Mitchell-Larson, G. A. Vecchi, et al., 2019: Cold waves are getting milder in the northern midlatitudes. *Environ. Res. Lett.*, 14, 114004 van Oldenborgh, G. J., K. van der Wiel, A. Sebastian, et al., 2017: Attribution of extreme rainfall from Hurricane Harvey, August 2017. *Environ. Res. Lett.*, 12, 124009

Veisdal, J., 2020: John von Neumann's 1935 letter to Oswald Veblen. *Medium.com/cantors-paradise*, April 18, www.cantorsparadise.com/john-von-neumanns-1935-letter-to-oswald-veblen-3acbe1b69098

Vitello, P., 2013: Joseph Farman, 82, is dead; discovered ozone hole. *The New York Times*, May 18, www.nytimes.com/2013/05/19/science/earth/joseph-farman-82-is-dead-discovered-ozone-hole.html

Vlamis, K., 2021: After extreme cold events in 1989 and 2011, Texas was warned to winterize power plants – but many still froze in the latest storms. *Yahoo! News*, February 19, https://news.yahoo.com/extreme-cold-events-1989-2011-051906264.html

Voiland, A., 2010: Aerosols: tiny particles, big impact. NASA Earth Observatory, November 2, https://earthobservatory.nasa.gov/features/Aerosols

Voosen, P., 2016: Climate scientists open up their black boxes to scrutiny. *Science*, 354 (6311), 401–402

2019: A world without clouds? Hardly clear, climate scientists say. *Sciencemag.org*, February 26, www.sciencemag.org/news/2019/02/world-without-clouds-hardly-clear-climate-scientists-say

2020a: Europe is building a "digital twin" of Earth to revolutionize climate forecasts. *Science*, October 1, doi:10.1126/science.abf0687

2020b: Hidden predictability in winds could improve climate forecasts. *Science*, 369, 490–491

Wagner, G., and M. L. Weitzman, 2015: *Climate shock*. Princeton University Press

2018: Potentially large equilibrium climate sensitivity tail uncertainty. *Econ. Lett.*, 168, 144–146, doi:10.1016/j.econlet.2018.04.036

Wagner, G., and R. J. Zeckhauser, 2016: Confronting deep and persistent climate uncertainty. Discussion paper, 2016–84. *Harvard project on climate agreements*. Belfer Center, July

Wagner, T. J. W., and I. Eisenman, 2015: How climate model complexity influences sea ice stability. *J. Clim.*, 28(10), 3998–4014

Waldman, S., 2019: Why a high-profile climate science opponent quit Trump's White House. *Science*mag.org, September 12, doi:10.1126/science.aaz4845

Walker, J. C. G., P. B. Hays, and J. F. Kasting, 1981: A negative feedback mechanism for the long-term stabilization of the Earth's surface temperature. *J. Geophys. Res.*, 86(C10), 9776–9782

Wallace-Wells, D., 2017: The uninhabitable Earth. *New York Magazine*, July 9, https://nymag.com/intelligencer/2017/07/climate-change-earth-too-hot-for-humans.html

2019: We're getting a clearer picture of the climate future – and it's not as bad as it once looked. *New York Magazine*, December 20, https://nymag.com/intelligencer/2019/12/climate-change-worst-case-scenario-now-looks-unrealistic.html

2020: What climate alarm has already achieved. *New York Magazine*, August 14, https://nymag.com/intelligencer/2020/08/what-climate-alarm-has-already-achieved.html

Wang, C., B. J. Soden, W. Yang, and G. A. Vecchi, 2021: Compensation between cloud feedback and aerosol-cloud interaction in CMIP6 models. *Geophys. Res. Lett.*, 48, e2020GL091024, doi: 10.1029/2020GL091024

Washington, W. M., 1970: On the simulation of the Indian monsoon and tropical easterly jet stream with the NCAR general circulation model. *Proc. Symp. Tropical Meteor.,* Honolulu, HI, Amer. Meteor. Soc., J VI 1–5

Washington, W. M., and A. Kasahara, 2011: The evolution and future research goals for general circulation models. In L. Donner, W. Schubert, and R. Somerville (eds.), *The development of atmospheric general circulation models.* Cambridge University Press, 18–36

Weart, S. R., 2008: *The discovery of global warming*. Harvard University Press, https://history.aip.org/climate/index.htm

Weaver, C. P., R. H. Moss, K. L. Ebi, et al., 2017: Reframing climate change assessments around risk: recommendations for the US National Climate Assessment. *Environ. Res. Lett.*, 12(8), 080201

Webster, B., 2020: Unhaltable global warming press release withdrawn by Scientific Reports journal. *The Times*, November 13, www.thetimes.co.uk/article/unhaltable-global-warming-claim-withdrawn-by-scientific-reports-journal-rtxjz9m6f

Weik, M. H., 1961: The ENIAC Story. *Ordnance*. Washington, DC, American Ordnance Association, January–February 1961

Weiss, E. B., 2009: The Vienna Convention for the protection of the ozone layer and the Montreal Protocol on substances that deplete the ozone layer, https://legal.un.org/avl/ha/vcpol/vcpol.html

Wentz, F. J., and M. Schabel, 1996: Effects of satellite orbital decay on MSU lower tropospheric temperature trends. *Nature*, 394, 661–664

Wheeling, K., 2020: The debate over the United Nations' energy emissions projections. *Eos.org*, 101, December 18, https://eos.org/articles/the-debate-over-the-united-nations-energy-emissions-projections

Whitman, M., 2012: *The Martian's daughter: a memoir*. University of Michigan Press

Wild, M., 2020: The global energy balance as represented in CMIP6 climate models. *Clim. Dynam.*, 55, 553–577

Winsberg, E., 2018a: *Philosophy and climate science*. Cambridge University Press

2018b: Communicating uncertainty to policymakers: the ineliminable role of values. In: E. A. Lloyd and E. Winsberg (eds.), *Climate modelling: philosophical and conceptual issues*. Springer Verlag, 381–412

Winsberg, E., and W. M. Goodwin, 2016: The adventures of climate science in the sweet land of idle arguments. *Stud. Hist. Philos. Sci. B: Stud. Hist. Philos. M. P.*, 54, 9–17, doi:10.1016/j.shpsb.2016.02.001

Witman, S., 2017: Meet the computer scientist you should thank for your smartphone's weather app. *Smithsonian Magazine*, June 16, www.smithsonianmag.com/science-nature/meet-computer-scientist-you-should-thank-your-phone-weather-app-180963716/

WMO Bulletin, 1996: The Bulletin interviews Professor Edward N. Lorenz. *WMO Bulletin*, 45(2), April

Woetzel, J., D. Pinner, H. Samandari, et al., 2020: Climate risk and response: Physical hazards and socioeconomic impacts. *McKinsey Global Institute*, www.mckinsey.com/business-functions/sustainability/our-insights/climate-risk-and-response-physical-hazards-and-socioeconomic-impacts

Wood, G., 2009: Re-engineering the earth. *The Atlantic, July/August*, www.theatlantic.com/magazine/archive/2009/07/re-engineering-the-earth/307552/

Worland, J., 2021: The Texas power grid failure is a climate change cautionary tale. *Time*, February 18, https://time.com/5940491/texas-power-outage-climate/

Yan, X. H., T. Boyer, K. Trenberth, et al., 2016: The global warming hiatus: slowdown or redistribution? *Earth's Future*, 4, 472–482, doi:10.1002/2016EF000417

Yoder, K., 2019: Is it time to retire "climate change" for "climate crisis"? *Grist.org*, June 17, https://grist.org/article/is-it-time-to-retire-climate-change-for-climate-crisis/

Yohe, G., N. Andronova, and M. Schlesinger, 2004: To hedge or not against an uncertain climate future? *Science*, 306(5695), 416–417

Yong, E., 2015: How reliable are psychology studies? *The Atlantic*, August 27, www.theatlantic.com/science/archive/2015/08/psychology-studies-reliability-reproducability-nosek/402466/

Žižek, S., 2004: Iraq's false promises. *Foreign Policy*, 140 (January/February), 42–49

Zworykin, V. K., 1945: Outline of weather proposal (with historical introduction by James Fleming). In: J. R. Fleming (ed.), *History of meteorology*, vol. 4, 57–78, 2008

Index

Note: Page numbers in **bold** indicate main entries

Printed in the United States
by Baker & Taylor Publisher Services